Probabilistic Foundations of Statistical Network Analysis

MONOGRAPHS ON STATISTICS AND APPLIED PROBABILITY

Editors: F. Bunea, P. Fryzlewicz, R. Henderson, N. Keiding, T. Louis, R. Smith, and W. Wong

Stochastic Analysis for Gaussian Random Processes and Fields
With Applications
Vidyadhar S. Mandrekar and Leszek Gawarecki (2015) 145

Semialgebraic Statistics and Latent Tree Models
Piotr Zwiernik (2015) 146

Inferential Models
Reasoning with Uncertainty
Ryan Martin and Chuanhai Liu (2016) 147

Perfect Simulation
Mark L. Huber (2016) 148

State-Space Methods for Time Series Analysis
Theory, Applications and Software
Jose Casals, Alfredo Garcia-Hiernaux, Miguel Jerez, Sonia Sotoca, and A. Alexandre Trindade (2016) 149

Hidden Markov Models for Time Series
An Introduction Using R, Second Edition
Walter Zucchini, Iain L. MacDonald, and Roland Langrock (2016) 150

Joint Modeling of Longitudinal and Time-to-Event Data
Robert M. Elashoff, Gang Li, and Ning Li (2016) 151

Multi-State Survival Models for Interval-Censored Data
Ardo van den Hout (2016) 152

Generalized Linear Models with Random Effects
Unified Analysis via H-likelihood, Second Edition
Youngjo Lee, John A. Nelder, and Yudi Pawitan (2017) 153

Absolute Risk
Methods and Applications in Clinical Management and Public Health
Ruth M. Pfeiffer and Mitchell H. Gail (2017) 154

Asymptotic Analysis of Mixed Effects Models
Theory, Applications, and Open Problems
Jiming Jiang (2017) 155

Missing and Modified Data in Nonparametric Estimation
With R Examples
Sam Efromovich (2017) 156

Probabilistic Foundations of Statistical Network Analysis
Harry Crane (2018) 157

For more information about this series please visit:
https://www.crcpress.com/Chapman--HallCRC-Monographs-on-Statistics--Applied-Probability/book-series/CHMONSTAAPP

Monographs on Statistics and Applied Probability 157

Probabilistic Foundations of Statistical Network Analysis

Harry Crane

Rutgers University

New Jersey, USA

CRC Press
Taylor & Francis Group
Boca Raton London New York

CRC Press is an imprint of the
Taylor & Francis Group, an **informa** business
A CHAPMAN & HALL BOOK

CRC Press
Taylor & Francis Group
6000 Broken Sound Parkway NW, Suite 300
Boca Raton, FL 33487-2742

Printed on acid-free paper
Version Date: 20180313

International Standard Book Number-13: 978-1-1385-8599-7 (Hardback)
International Standard Book Number-13: 978-1-1386-3015-4 (Paperback)

Visit the Taylor & Francis Web site at
http://www.taylorandfrancis.com

and the CRC Press Web site at
http://www.crcpress.com

To Howard and John

Contents

Preface

As I complete this book on the last day of 2017, the scope of 'network science' continues to expand throughout the social, biological, and physical sciences. Meanwhile, the focus of statistical network analysis remains tethered to a few standard methods and a limited class of models. Community detection, asymptotic analysis, and minimax estimation for networks assumed to follow the stochastic blockmodel, exponential random graph model, and graphon model litter the statistical literature. By now it should be clear that these results are mostly of an incremental nature, lacking both the insight and the motivation necessary to address substantive questions in network science. In *Probabilistic Foundations of Statistical Network Analysis*, I seek to reverse these current trends by proposing to view 'network analysis' as the base case of a new statistical theory for complex data analysis. As I emphasize in Chapter 1, the perspective needed to address this new class of complex data problems cannot be achieved by recasting classical techniques and regurgitating old ideas. In this text, I hope to first convince the reader of the need for such a new perspective, and then to enlist the reader in fulfilling this vision.

Many of the ideas presented here are adapted from my prior work on network analysis, exchangeability theory, and graph-valued stochastic processes [43, 44, 45, 46, 47, 48, 49, 50, 51, 52, 53, 54, 57, 58, 59]. Still others, e.g., the general formulation of sampling consistency (Chapter 3), the principle of coherence (Chapter 5), and the specification of random sampling contexts for network data (Sections 3.9, 9.8, 10.7, and 11.2.2), are new additions which have not yet appeared in the published literature. Without question, the most important ideas in this book (Chapters 5, 9, and 10) are either attributed to or have been greatly inspired by my ongoing conversations with Walter Dempsey [52, 53, 54]. Walter first proposed the idea of edge exchangeability to me in December 2014 at Rutgers. Though at first I resisted—the idea was too new—Walter's 'E2 mindset' has profoundly influenced my thinking ever since.

Chapter synopses and reading guide

One of my primary goals in writing this book is to expose readers from a wide range of disciplines to the core statistical ideas underlying network analysis. The book is therefore intended for social scientists and psychologists just as much as it is for professional statisticians and mathematical probabilists. Although understanding the accompanying technical work in [43, 44, 45, 46, 47, 48, 49, 50, 51, 52, 53, 54, 57, 58, 59] requires a high level of mathematical sophistication, appreciating the conceptual ideas underlying this technical work does not. Because simple concepts too often get lost in mathematical formalism, I have tried to every extent possible to strike the delicate balance between conceptual and technical, in hope that those with the conceptual understanding will grow more accustomed to the technicalities and those with the mathematical training will come to better appreciate the practical motivations. Whether I have succeeded in this mission is not for me to judge, and I welcome the reader's feedback on how I could have done better.

Realizing that readers might come to these pages with different motivations, I give a brief synopsis of each chapter before beginning the main text. Readers interested in specific network model classes are referred to Chapters 2 and 6–11. Readers interested in invariance principles and exchangeability are referred to Chapters 6–11. Readers interested in time-varying networks and graph-valued stochastic processes are referred to Chapter 11. Readers interested in the philosophical underpinnings of this work are referred to Chapters 1 and 3–5. Together, the framework of Chapter 5 and the discussion in Chapters 1, 9, and 10 capture the essence of the book.

Chapter 1: Orientation

Though 'network data' presents a number of new challenges for statistical theory and methods, 'statistical network analysis' remains focused on a limited number of inference problems which have been adapted from classical statistics. Community detection, asymptotic analysis, and minimax estimation for networks generated from stochastic blockmodels, exponential random graph models, and graphon models have been studied by extending well-established ideas in clustering, large sample theory, and nonparametric regression. But those analyses, for all their rigor, tend to overlook the big picture and broader implications motivating the study of networks in the first place. In this opening chapter, I discuss the limitations of the conventional 'networks-as-graphs' perspective of network analysis, and I emphasize network analysis not merely as a new discipline within statistics, on par with high-dimensional statistics or machine learning, but rather as a new statistical paradigm for complex data analysis which ought to be viewed and developed in parallel to classical statistical theory. As modern data grows more complex, not in terms of size but rather in terms of dependence and heterogeneity, the usefulness of classical statistical thinking diminishes. Rather than strain the limits of classical theory to fit these problems, I argue in favor of developing new theory and methods that can better handle the complexity of modern data structures. Such a theory does not yet exist. In writing this book, I hope to convince the reader of the need for such a theory, and to offer some ideas for how to make progress in this new direction.

Chapter 2: Binary relational data

Though network science is seen as a modern field, the study of network-type data dates at least to the early 1900s, with the initial sociometric studies of Moreno and other quantitative social scientists. The classical network models, namely exponential random graph models and stochastic blockmodels, originated in the early theoretical developments of social network analysis. These models are still used today, but their adequacy for modern network data problems is limited by their inability to account for observed heterogeneity (e.g., sparsity, power law) in network data or the sampling scheme by which these networks are obtained. In this chapter, I review some early social network models, namely the p_1 model and exponential random graph model, discuss their initial motivations, and explore their limitations for modern applications. A discussion of sampling issues in the p_1 and exponential random graph models sets the stage for the next chapter on network sampling.

Chapter 3: Network sampling

The large size of modern networks and the varied circumstances under which network data arise necessitate a theory for network analysis in the presence of sampling. Although a sampling theory for social network analysis had been developed in the social networks literature by Frank and his coauthors, its relevance to modern network analysis is unclear. Modern network data is often sampled, but rarely according to any well-articulated or well-understood design mechanism. In such cases, classical techniques, e.g., the Horvitz–Thompson estimator, which often assume a careful and relatively straightforward sampling mechanism such as simple random vertex sampling, may not be applicable. In many applications, it is known that the network has been sampled, but it is not known how, thus adding additional uncertainty over and above ordinary statistical variation. More realistic network sampling schemes, e.g., vertex, edge, path, snowball sampling, etc., introduce potential biases that should be accounted for in the model. This chapter also discusses how other standard statistical considerations, such as the sample size, study design, and the basic observational units, feature in network analysis.

Chapter 4: Generative models

Whereas sampled networks are obtained as part of a larger population structure, evolving networks (e.g., the World Wide Web and some social networks) grow according to a (possibly unknown) generating mechanism. For evolving network data, the objective of analysis is to understand the mechanism by which the network is evolving in order to better predict or anticipate future updates. The preferential attachment model [14] is a well-known generative model for describing the emergence of power law structure in real-world networks. In this brief chapter, I discuss some basic properties of the preferential attachment and other generative models.

Chapter 5: Statistical modeling paradigm

Previous chapters highlight the two main considerations of statistical network modeling: (i) describe variability and uncertainty in the observed network and (ii) articulate the context in which inferences are to be interpreted. Although conventional approaches to statistical modeling only address consideration (i), contextual factors, such as sampling design and other circumstances surrounding data collection, are essential to sound network analysis. The modeling paradigm introduced in this chapter brings components (i) and (ii) together in the concept of *model coherence*. Along with the coherence condition, the paradigm presented in this chapter provides a general modeling framework which can serve as a foundation for future theoretical and methodological developments in complex data analysis.

Chapter 6: Vertex exchangeable models

Invariance principles often play a central role in expressing how the observed data is assumed to 'represent' the system from which it has been observed. This 'representativeness' assumption establishes the logical link by which inferences about the (observed) data can be extended to the (unobserved) population. But in many circumstances, the theoretical justification for extending inferences based on the assumed invariance principle does not align with the context in which the inferences are being made. Within the conventional networks-as-graphs paradigm of network analysis, vertex exchangeability is a prominent invariance principle on which a great deal of theoretical work has been based. In a vertex exchangeable model any two graphs that are isomorphic up to relabeling of their vertices are assigned equal probability. Implicit in this model property is the assumption that the observed network is that of a representative sample of vertices from the population. In this chapter I discuss some fundamental aspects of vertex exchangeability, including graphon models, the Aldous–Hoover theorem, the theory of dense graph limits, and how the homogeneity properties implied by vertex exchangeability raise doubts about the use of graphons in modern network analysis.

Chapter 7: Getting beyond graphons

In Chapter 6 I highlighted several limitations of vertex exchangeability and graphon models, especially when it comes to modeling networks that exhibit sparsity and power law degree distributions. Although vertex exchangeability makes certain theoretical and computational aspects of network analysis tractable, the theory and computations made possible by this assumption are of little practical value because they fail to consider the data in a realistic context. Some initial attempts to move beyond graphons include so-called 'sparse graphon' approaches and the Caron–Fox model based on completely random measures. This chapter covers both approaches, with additional discussion of the graphex representation and sampling interpretations of the Caron–Fox model.

Chapter 8: Relatively exchangeable models

Relative exchangeability refines vertex exchangeability (Chapter 6) by defining the invariance of a network in terms of the symmetries of some other (fixed, known) structure on its vertices. Stochastic blockmodels (SBMs) are a canonical example of a relatively exchangeable network model, with the underlying structure given by a classification of vertices into distinct groups (or 'blocks'). More generally, relatively exchangeable models can account for heterogeneity caused by an underlying social network, covariate information (as in the latent space model), or generic combinatorial structure in the population. The theory for relatively exchangeable random graphs mirrors that of Chapter 6, with the Ackerman–Crane–Towsner theorem refining the Aldous–Hoover theorem. Further extensions and applications of the theory presented here are left as topics for future research.

Chapter 9: Edge exchangeable models

The Crane–Dempsey edge exchangeable framework posits a new model class for networks constructed by repeated sampling of binary interactions. Edge exchangeable models are defined analogously to vertex exchangeable models (Chapter 6) by assigning equal probability to any two edge-labeled graphs that are isomorphic with respect to relabeling of their edges. The major breakthrough of edge exchangeability is its new perspective, which more faithfully captures the way in which many interaction networks are observed (i.e., by sampling edges instead of vertices). The canonical edge exchangeable model, called the Hollywood model, easily accounts for a range of empirical behaviors commonly found in modern networks, including sparse, power law structure. Edge exchangeability in general, and the Hollywood model in particular, provides fertile ground for future methodological developments in network analysis.

Chapter 10: Relationally exchangeable models

Relational exchangeability refines edge exchangeability (Chapter 9) by modeling networks constructed by repeated sampling of generic interactions (e.g., email communications, scientific coauthorships, and paths in the Internet). So whereas edge exchangeable models are limited to networks constructed from binary interactions, such as caller-receiver pairs in a phone call database, relationally exchangeable models allow for multiway interactions between email recipients, paths between routers in the Internet, etc. With its inclusion of hyperedge exchangeable and path exchangeable models, relational exchangeability is thus a general variant on the theme of edge exchangeability, and leads to an analogous theory.

Chapter 11: Dynamic network models

Previous chapters focus mostly on data for a single network, such as networks formed by friendships among high school students, social media interactions, professional collaborations, and connectivity in the Internet. But many networks, in addition to representing complex dependencies and interactions, change with respect to time. Such *dynamic networks* introduce a temporal dimension into network analysis which, on top of the considerations of exchangeability and other invariance principles from previous chapters, must be incorporated into the model in a coherent way. At the time of publication, statistical methods for dynamic network analysis are relatively underdeveloped. To give a sense of some basic challenges and considerations in this emerging area, I focus this chapter on one specific temporal invariance principle, called Markovian projectivity [44, 48, 57], which acts as a dynamic version of the consistency under subsampling property discussed in earlier chapters (e.g., Section 3.9). I present Markovian projectivity in the context of two classes of dynamic networks, called rewiring processes and graph-valued Lévy processes, both of which are ripe for future developments in theory, methodology, and applications.

Final comment

As the field of network science is quickly expanding, with thousands of articles published in the short time since I started writing this book, the references cited are far from complete. Although I have acknowledged any work relevant to the specific topics discussed here, I could not possibly give comprehensive coverage to the entire body of literature. For example, I do not discuss recent work at the interface of causal inference and network analysis, see, e.g., [15, 142]. I apologize in advance for any topics or references I may have overlooked.

Throughout the text, I highlight a number of open-ended questions as 'Research Problems'. These are suggestions for future inquiry by the curious reader. To the best of my knowledge, none of these problems has a satisfactory answer as of yet, and many of them are being posed here for the first time. Several of these problems would make a good Ph.D. thesis topic. Researchers who make progress on any of these or related open problems are encouraged to contact me so that I can document their status on the book's website http://www.harrycrane.com/networks.html. Another 30 or so 'Exercises' supplement the main text, with the solution to each exercise given at the end of its respective chapter.

I welcome readers with questions, comments, criticisms, or suggestions, including typos and overlooked citations, to contact me on Twitter (@HarryDCrane) or by email (hcrane@stat.rutgers.edu).

Harry Crane
December 31, 2017
West New York, NJ

Acknowledgments

My sincere thanks to John Kimmel, who has been involved in this process since the beginning. Before meeting John for the first time during the 2016 IMS meeting in Toronto, I had never considered writing a book. His expertise has made this first publishing experience as smooth as I could have imagined.

Over the past few years, I have benefited greatly from conversations with Elie Ayache, Lamiae Azizi, Shankar Bhamidi, Peter Bickel, Benjamin Bloem-Reddy, Joshua Cape, Eddy Keming Chen, Tirthankar Dasgupta, Stephen DeSalvo, Nick Fazzari, Branden Fitelson, Cameron Freer, Chao Gao, Andrew Gelman, Edinah Gnang, Joe Guinness, Dick Gundy, Nils Hjort, Peter Hoff, Alan Huang, Alan Izenman, Lancelot James, Eric Kolaczyk, Eric Laber, Steve Lalley, Antonio Lijoi, Chuanhai Liu, Regina Liu, Barry Loewer, Ryan Martin, Peter McCullagh, George Michailidis, Subhadeep Mukhopadhyay, Andrew Nobel, Sofia Olhede, Peter Orbanz, Carey Priebe, Matt Reimherr, Don Richards, Daniel Rock, Karl Rohe, Dan Roy, Matteo Ruggiero, Siddhartha Sahi, Teddy Seidenfeld, Glenn Shafer, Stephen Sherman, Nozer Singpurwalla, Neil Sloane, Tom Snijders, Rebecca Steorts, Dan Sussman, Nassim Nicholas Taleb, Henry Towsner, Veronica Vinciotti, Volodya Vovk, Isaac Wilhelm, Ernst Wit, Sandy Zabell, and Doron Zeilberger. I also thank the numerous people with whom I've corresponded over the Internet, mostly on Twitter, for exposing me to ideas far more interesting and conversations far more stimulating than I could ever experience in the academy. Aside from that, I thank anyone whose questions or criticisms have forced me to think harder, and hopefully better. To those whose names I have accidentally omitted, I apologize deeply.

The students in my Fall 2017 special topics class at Rutgers asked many insightful questions and made a number of suggestions which have greatly improved the exposition. For this I owe special thanks to Yiwei Shen, Ellie Small, and Don Walpola.

Warren Ewens has been encouraging since the very beginning of my career. I'll never forget his first email to me in March 2011. Just a few words, but they meant a lot, and marked the beginning of a great friendship.

Dimitris Tsementzis commented on numerous early drafts of this work. But most importantly, our many lengthy and wide-ranging conversations, about Alain Badiou, Bitcoin, homotopy type theory, category theory, probabilities, shapes, probabilities as shapes, and (yes) networks, have played no small part in what this book has ultimately become.

I am eternally indebted to Walter Dempsey, whose involvement in the core ideas presented here goes without saying. Our work on edge exchangeability remains among the most exciting discoveries of my early career. Without our successful collaboration and many fruitful conversations over the past five years, the main content in Chapters 3–5, 9, and 10 would be missing.

I thank my closest friends Jud Crandal, Adrian Di Antonio, Gus Joo, Adam Kusowski, and Mark Tyson for nothing in particular.

Above all, I thank my family for reminding me of what's important, and what isn't. My parents Harry and Regina, my sister Kayla, brother-in-law Jay, and nephew Little Jim have supported me through the years when nobody else did. And my parents-in-law Huizhi Ren and Shouming Zeng, whose unconditional support I can feel from halfway around the world. They all deserve far more thanks than I could ever express.

Finally, my wife Jie, who inspires me to be the best version of myself, who keeps me going when I want to stop, and whose steady support has made so many things possible over the past seven years. Under the adverse circumstances I faced when writing this book, she didn't flinch, so neither did I. We've come such a long way, and still have a long way to go. For all of this and more, I can never thank her enough.

Chapter 1

Orientation

In recent years there has been an explosion of network data *— that is, measurements that are either of or from a system conceptualized as a network — from seemingly all corners of science.* (Kolaczyk [106])

Empirical studies and theoretical modeling of networks have been the subject of a large body of recent research in statistical physics and applied mathematics. (Newman and Girvan [83])

Networks have in recent years emerged as an invaluable tool for describing and quantifying complex systems in many branches of science. (Clauset, Moore and Newman [38])

Prompted by the increasing interest in networks in many fields [...]. (Bickel and Chen [19])

Networks are fast becoming part of the modern statistical landscape. (Wolfe and Olhede [155])

The rapid increase in the availability and importance of network data [...]. (Caron and Fox [32])

Network analysis is becoming one of the most active research areas in statistics. (Gao, Lu and Zhou [79])

Networks are ubiquitous in science. (Fienberg [74])

Networks are ubiquitous in science and have become a focal point for discussion in everyday life. (Goldenberg, Zheng, Fienberg, and Airoldi [84])

"Networks are everywhere"

There is currently no shortage of interest in 'network science', 'network data', 'complex networks', or just about anything else that invokes the term 'network'; see, e.g., recent popular books on the topic [13, 151]. In writing this book, I have done my part in furthering this trend; and in reading it, so have you. But as it was never my intention to become part of the networks hype—a hype reflected in the quotes at the top of this page—I do not set out here to celebrate the importance of network science or its great 'successes' in better understanding the complexities of our world. To the contrary, while I acknowledge the potential of network science for gaining better insights about complex data structures and the systems that produce them, I also recognize that this potential has not yet been realized. Especially within statistics, the study of 'networks' has been greatly limited by a lack of appreciation for the complexity of 'network data' and a lack of creativity in developing new ways to think about those complexities. By now these limitations are woven so deeply into the fabric of statistical thinking that overcoming them is easier done by starting a new fabric, rather than modifying the existing one. So, in addition to clarifying the current limitations of statistical network analysis (in Chapters 1–4 and 6–7), I set out here along a new path with the hope of catching a glimpse of what lies ahead. And while certain parts of this book (e.g., Chapters 5, 8–11) do represent substantial progress in this direction, I make no claim to overcome all of these limitations here.

With these objectives in mind, this book is not intended as a survey of existing models or a catalog of currently available techniques for analyzing network data. The book is instead a *perspective* on how to better represent, model, and think about complex, heterogeneous data structures that arise in modern applications. The current ways of doing things, and their various extensions, are insufficient for this purpose. I discuss some early attempts at gaining such a new perspective throughout Chapters 7–11, but surely the future of statistical network analysis lies almost entirely beyond these pages, in a yet-to-be-celebrated breakthrough.

In venturing beyond the conventional graph-theoretic representation of networks and its associated random graph models, I am confident that the later chapters are a step in the right direction. But just as it is wrongheaded to believe that the current graph-theoretic convention is the 'correct', 'best', or 'only' way to think about network data, it would be foolish to suggest that any of these new approaches is absolutely superior to more conventional methods. To be sure, there are ways in which these new approaches provide a better perspective on network data of a certain kind. For example, the perspective of edge exchangeability (Chapter 9) allows us to express and extract properties from interaction data that standard vertex-centric approaches cannot. Such an expansion of the prevailing mindset, regardless of whether it proves 'useful' in any practical domain, is necessary to broaden the scope of statistical thinking beyond the traditional paradigm. Continued sharpening of perspective and enrichment of mindset, far beyond what came before and what lies within these pages, motivates everything that follows.

1.1 *Analogy*: Bernoulli trials

Network analysis is no more about studying Facebook, or Twitter, or the loyalties of karate club members [161] than classical statistics is about tossing coins. And yet, the theory of coin tossing, as formalized by infinite sequences of independent, identically distributed (i.i.d.) Bernoulli trials, lays the groundwork for much of classical statistical theory; see, e.g., [71]. For an analogy, coin tossing is to the statistical analysis of simple, unstructured data as networks are to the statistical analysis of complex, dependent data:

$$\text{coin tossing} : \text{unstructured data} :: \text{network analysis} : \text{complex, structured data}.$$

From this analogy, I make a few initial observations.

First, just as i.i.d. Bernoulli trials are an entry point into classical statistics, through the law of large numbers, central limit theorem, etc., so too is network modeling an entry point into modern complex data analysis. Much like the classical theory of statistical inference is erected on the scaffolding of the i.i.d. sequence, the modern theory of inference from complex data will be built on the probabilistic foundations of statistical network analysis.

Second, instead of heralding the ubiquity of 'networks', as in the opening quotations, we would be better off recognizing the emergence of *complexity* in modern data science, where 'complexity' is used here to mean dependence, structure, heterogeneity, and the like. At present, networks are the primary vehicle for representing complex data structures and network analysis is the predominant method for understanding complexity, dependence, and heterogeneity.

Third, given the ubiquity of complexity and its many forms, statisticians can no longer rely on a limited toolbox of classical techniques and old ideas. New foundations for the statistical analysis of complex data must be forged; and these foundations cannot be derivative on the classical theory of linear models, i.i.d. sequences, etc. The newness of modern networks problems is paradigm-shifting, and thus warrants a shift in the paradigm within which we think about, discuss, and analyze such data. I clarify this point of view in the coming several pages, with special focus on the statistical foundations of network analysis, where they currently stand and where they are headed.

Probabilistic Foundations of Statistical Network Analysis emphasizes *modeling* (as a verb, the *act* of specifying a model), not *models* (the noun, those models which already exist). The reasons are manifold:

- One, the *act* of modeling should be thought of as an act of *imposing* structure on the data (and thus on the world). One does not simply *choose* a model from an existing class of acceptable choices. One instead *posits* a model, and in doing so declares how the data behaves and how that behavior fits into a bigger picture. Classical statistics, which deals primarily with data having little or no internal structure (i.e., sequences and sets), has conditioned the statistician to behave rather lazily when choosing a model. Since there is little structure in many classical datasets, the act of modeling involves little more than identifying a family of probability distributions to describe a (nearly) structureless collection of measurements. (To be clear, I am not claiming that classical data sets lack structure; rather, I am observing that their conventional representation, most often as sets of points in $\mathbb{R}^d$, and the models chosen to describe them, e.g., often i.i.d. or exchangeable models, tend to minimize the impact of this structure on data analysis.) When dealing with structured data—and in the case of network data, the structure *is* the data—the act of imposing structure (via modeling) should be taken much more seriously.
- Two, most of the network models that already exist are inadequate for modern network data structures. They do not live up to their name as 'models' in the vast majority of situations. We encounter several examples throughout Chapters 2 and 6–8.

- Three, even though existing network models (i.e., stochastic blockmodels, exponential random graph models, graphons) are known to suffer serious drawbacks for modern applications, their appearance throughout the theoretical and applied literature remains pronounced. I have no desire to continue this trend.
- Four, a major reason for the continued use of these limited models seems to be a general lack of interest in positing new ones. The canonical statistics curriculum focuses primarily on the analysis and application of standard models (Binomial, Poisson, Gaussian, Exponential) but without emphasizing the principles that make these models 'good' in any given situation. Rather than fret over the technicalities and nuances of constructing better models, students and researchers are instead indoctrinated with the Boxian trope, "All models are wrong, but some are useful" [26], without any clarity as to why models are 'wrong' or what makes them 'useful'. With Box's proverb comes the demotion of models and modeling, and the elevation of estimation, prediction, approximation, and computation.

Perhaps the Boxian proverb does little harm in the classical paradigm, where laws of large numbers, the central limit theorem, and asymptotic approximations abound. But it is untenable within the emerging paradigm of network analysis, in which there are few reliable asymptotic results; and those asymptotic results that do exist are hard to make sense of, e.g., minimax rates for graphon models, consistency properties for stochastic blockmodels and exponential random graph models, and asymptotic sparsity properties of so-called 'sparse graphon' models (Section 7.2). Bear in mind: the model is what the researcher puts in. Everything else is either given (i.e., data) or derived (i.e., inferred). The choices made while modeling—how one chooses to 'look at' and 'think about' the data—are most critical to determining whether the resulting inferences are 'useful', in Box's parlance. As I emphasize with the statistical modeling paradigm of Chapter 5, whether the result of an analysis 'is useful' or 'makes sense' or 'is valid' cannot be assessed solely on whether the estimators are unbiased, consistent, efficient, etc., as these diagnostics are meaningless unless grounded by an internally coherent model. No matter how much statistical inference is presented as an 'objective' approach to data analysis, modeling is undoubtedly a subjective and personal activity. And so it ought to be taken personally.

With the discussion below, I hope more than anything else to restore modeling to its role at the center of the statistical paradigm, bridging the divide between data collection and inference. Along the way I will carefully consider Box's admonition—to employ models that are 'useful'—along with other foundational topics (i.e., symmetry and exchangeability) at the heart of statistical inference. For the most part, I have chosen to deemphasize technical aspects of network analysis in favor of high-level concepts, both in the remainder of this chapter and throughout the book. For the rest of this opening chapter, I discuss the guiding principles of statistical network analysis at a high level. Although the technical aspects of this chapter are light, the concepts are subtle, and are essential in order to appreciate the core ideas motivating this book.

1.2 *What it is*: Graphs vs. Networks

In these pages, the term 'network' refers to a specific instantiation of what can be vaguely understood as 'complex data'. But even in the specific case of 'network data', it is important to distinguish between the fundamental objects of study (i.e., networks) and the conventional mathematical representation of those objects (i.e., graphs). The distinction between networks and graphs marks the initial divergence between the perspective put forward here and the prevailing 'networks-as-graphs' perspective found throughout the literature.

To be clear: *networks are not graphs*. A *graph* is a mathematical object consisting of a set of vertices V and a set of edges $E \subseteq V \times V$. This mathematical concept can be extended in several ways to allow for multiple edges, hyperedges, and multiple layers, but all of these objects, i.e., graphs, multigraphs, hypergraphs, multilayer graphs, etc., are mathematical entities. They are also distilled entities, in that they can be discussed independently of any presumed statistical or scientific context. From this point of view, graphs can be regarded as a 'syntax' for communicating about network data. But this graph-theoretic syntax is just one language with which to communicate about networks. And like any language, it is limited in what it can express. In becoming too attached to this one language for talking about networks, we limit the nature of insights that can be gleaned from network data. A sure sign of progress in the foundations of network analysis is the development of new ways to express and understand network data. See Chapters 9 and 10 for one such new approach.

A *network*, on the other hand, is an abstract concept referring to a system of interrelated entities. For us, the concept of 'network' is neither concrete nor well-defined, but rather is vague and amorphous, emerging from an intuitive judgment about perceived structure in an observed system. For example, import-export partnerships between countries, social relationships among high school students, patterns in phone call activity, connectivity among Internet servers, and interactions among genes all invoke the concept of a *network* of relationships or interactions in a particular context. Although it is sometimes reasonable to represent these networks mathematically as graphs, the systems are not graphs in themselves. For example, the Internet is a physical structure consisting of wires, servers, and routers. A graph is a set V together with another set $E \subseteq V \times V$. The physical Internet invokes the concept of a 'network', and some aspects of it can be represented or modeled as a graph, but the Internet is not a graph.

Moving beyond graphs

The reader who has read the word 'network' and every time envisioned a 'graph' faces a steep *unlearning* curve to appreciate the richness of structure encoded in the concept of 'network'. If there is to be progress in understanding complex, structured data, then the conventional ways of thinking about ordinary, unstructured data—the data sequences and arrays that fill statistics textbooks—must be purged from memory, or at least demoted from their default status in data analysis. To think about networks properly, one must strongly resist any temptation to embed networks in Euclidean space, or use the terms 'network' and 'graph' interchangeably, or any similar such urge to impose the flat view of data taken by classical statistics on the voluminous and rich structure which the concept of network calls into being.

Though I strongly advocate this point, it is with great regret that almost all of the 'networks' discussed in this book are treated as ordinary 'graphs', an exception being the important class of edge and relationally exchangeable network models in Chapters 9 and 10. This antithetical presentation can be explained by the extraordinary primitiveness in the current state of affairs. The concept of 'network data' is itself a very special case—the base case—of what can be understood as 'complex data'. The mathematical language of graph theory studies the even more restricted class of 'networks' which can be represented as pairs (V,E) consisting of a set V of vertices and a set $E \subseteq V \times V$ of edges. The recent proposal of edge-labeled networks (Chapter 9) breaks free of this traditional view and inspires hope for expanding the scope of 'network analysis' beyond what is currently conceivable, but there is still a long way to go.

1.3 *How to look at it*: Labeling and representation

Think of statistical analysis as the act of discerning the nature of some large, complex object in a dark room. You only have a flashlight, which can illuminate just a small piece of that object. In this analogy, the illuminated piece is the data on which your inference about the large, unobservable object is to be based. Different angles of shining the flashlight can be understood as different ways of looking at, or representing, the data. For example, the representation of a network as a vertex-labeled graph (Figure 1.1(b)) corresponds to the shadow cast by shining the light from one angle; the edge-labeled graph (Figure 1.1(c)) is the shadow cast from a different angle. Both are shadows of the same object, namely Figure 1.1(a), but the angle from which the light is shone (i.e., the perspective from which the data is viewed) affects which attributes are visible and which are obscured, and thus which inferences the data supports and which it does not.

Because in many classical applications there is just one canonical angle from which to look at the data, it is easy to overlook the role played by 'perspective' in complex data analysis. In a sequence, for example, the measurements $X_1, X_2, \ldots$ contain the primary information. Changing the 'angle' from which we shine our proverbial flashlight on this data (e.g., by converting pounds to kilograms, or feet to inches) does not change the nature of the measurements $X_1, X_2, \ldots$. But the significance of this 'angle' cannot be overstated when handling networks and other complex data structures. In these latter instances, the structure *is* the data, and different aspects of this structure may be visible depending on the angle from which the light is shone.

In practice, this 'angle' is manifested first and foremost in how the network is represented, for which the choice of labeling is a basic consideration. In Figure 1.1, for example, the 'unlabeled' structure in Figure 1.1(a) is the object of interest. Ideally, we would treat this 'unlabeled' structure as the data and analyze it directly, but this is not possible. Unlabeled structures cannot be treated as data because unlabeled objects cannot be *represented*. To analyze data one must be able to talk about it; and to be able to talk about something, one must assign names to whatever parts of that thing

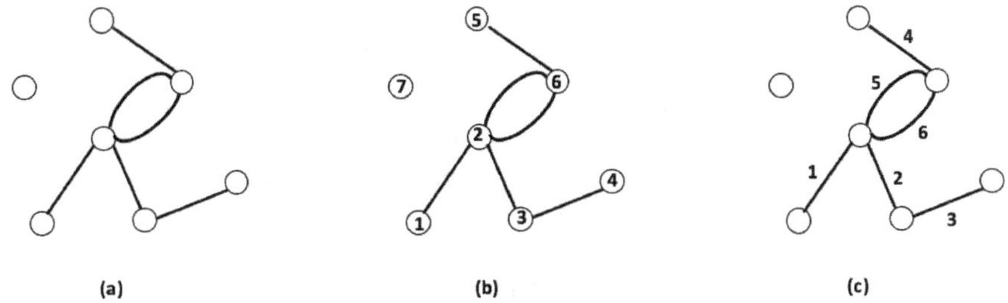

Figure 1.1 *Two perspectives on network data. (a) Represents the essential structure of 'unlabeled' network data. (b) Represents the structure in (a) by assigning labels to its vertices (i.e., vertex-centric perspective). (c) Represents the structure in (a) by assigning labels to its edges (i.e., edge-centric perspective).*

are being discussed. For networks, this 'naming' comes in the form of labeling the constituent parts of the data. Without such a labeling, we cannot even utter a word.

To make this point clear, realize that the object in Figure 1.1(a) merely *represents* the abstract notion of an unlabeled network. But the object itself is labeled by the spatial orientation of its edges and vertices on the page. This spatial orientation allows one to speak (i.e., 'utter') about this network by referring to the relative positions of vertices/edges, e.g., by pointing or describing the positions in words. Mathematically, such 'unlabeled' structures are typically represented by 'removing the labels' and working with a class of structures that are equivalent up to relabeling. But the appropriate notion of equivalence itself depends on the perspective from which two networks are to be treated as 'equivalent'. For example, the equivalence class of vertex-labeled networks (as in Figure 1.1(b)) differs from the equivalence class of edge-labeled networks (as in Figure 1.1(c)), because the corresponding notions of equivalence differ based on the chosen perspective. Which perspective is appropriate for a given application depends on the context.

1.4 *Where it comes from*: Context

Given the diverse scenarios in which networks arise, there can be no single 'correct' approach to network analysis. Instead, what makes 'network analysis' relevant to a given problem depends on the *context*. And this context should be accounted for at every stage of the analysis, beginning with the way in which the data is represented, continuing through model specification, and culminating in inference. As emphasized in the previous section: the representation of network data reflects the perspective from which it is being analyzed, which in turn determines what inferences can be drawn from the analysis. To elicit the best available insights from the data, we want to shine our flashlight (i.e., represent and model the data) from the optimal angle, and the optimal angle in any given application depends crucially on the context.

Consider the structures in Figure 1.1. Do they represent the same network? Perhaps. Assuming they do represent the same network, do they provide the same *representation* of that network? Of course not. Figure 1.1(a) represents the 'shape' of the network, without explicitly identifying any of its other components, e.g., vertices or edges. Figure 1.1(b) identifies each vertex with a distinct label. Figure 1.1(c) identifies each edge with a distinct label, leaving vertices unlabeled. But what's the difference? The difference, we will see throughout the coming chapters, is a matter of perspective. In labeling the vertices, Figure 1.1(b) asserts a 'vertex-centric' view of the shape in Figure 1.1(a), and this vertex-centric view differs from the 'edge-centric' point of view taken in Figure 1.1(c). Even though these may be different representations of the same network, the choice of representation reflects the perspective of the data analyst and the context of the application, both of which affect inference.

1.5 *Making sense of it all*: Coherence

There are primarily two aspects to network modeling. The model first *describes* the observed data from the perspective of the statistician. And then, to draw inferences beyond the observed data, the model specifies a *context* in which to interpret the data. With this, the model has two components:

- a *descriptive* component consisting of the family of candidate probability distributions for describing variability in the observed data, and
- an *inferential* component explaining how the observed data fits into a larger context.

Both components are essential to proper model specification and sound statistical inference.

Returning to the Boxian proverb, "All models are wrong, but some are useful," I regard 'making sense' as the first step towards 'being useful'. To make sense, the inferences based on the model should be interpretable within a single (coherent) context. This observation culminates in the formal concept of *coherence*, by which the *description* of the model 'fits coherently' into its *context* in a sense made precise in Definitions 5.2–5.3. (See Chapter 5 for a more formal discussion of coherence and its significance for statistical inference.)

Beyond coherence, there are often practical considerations regarding whether or not the presumed context is suitable, or whether the specified model can actually be used (i.e., computed) in a given application. But such practical matters should be considered only after minimal logical conditions, such as coherence, are met. Without coherence, any computational or practical techniques which enhance the analysis are of little use, precisely because the model which they will have enhanced does not make sense.

1.6 *What we're talking about*: Examples of network data

Throughout these pages, we will encounter a number of scenarios under which different modeling considerations are appropriate. Whenever possible, I try to motivate these scenarios by real (or realistic) applications for which canonical examples already exist. I survey some of these common scenarios below. For the most part, these examples are not interesting on their own, and are offered here only to illustrate how basic principles of network analysis arise in practice.

1.6.1 Internet

Several early developments in network science grew out of empirical observations taken on the Internet, defined as the network of servers and systems connected by physical wires. Of all the datasets discussed here, the Internet is one of the only 'real' networks in the sense that it corresponds to an actual physical object. Explaining the observed power law structure in sampled data from the Internet and World Wide Web was one of the primary motivations for Barabási and Albert's preferential attachment model (Section 4.2). The widespread empirical observations of power law degree distribution, both in the Internet and other real-world networks, remains one of the most evocative illustrations of the effects of sampling on network analysis, which have been mostly overlooked until recently [52, 54, 112, 127, 154]. I discuss the role of sampling further in Chapter 3.

Because of its physical nature, the Internet network invokes a notion of 'ground truth' that is absent from other familiar applications in network science. For example, community detection in social networks seeks an optimal clustering of vertices into (disjoint) communities based on their network connectivity. As the concept of 'social network' is itself a nebulous one, in many cases there is no 'true' division of vertices against which to assess the inferred clustering. (A notable exception is the karate club network of Zachary [161], see Section 1.6.3. But in modern network analysis, the karate club network is treated more as a meme than as a serious dataset.[1])

1.6.2 Social networks

In social network analysis, vertices represent individuals and edges represent social ties between their adjacent vertices. The network does not correspond to anything physical, as in the Internet, but rather represents invisible social forces driving interactions within a population, e.g., shared recreational interests, common political views, or professional relationships. I discuss some scenarios of social network modeling in Chapters 2, 3, 6, 7, and 8.

1.6.3 Karate club

The karate club dataset [161] records social interactions among 34 members of a university karate club for the three-year period spanning 1970–1972. Represented as a network with multiple edges, each vertex corresponds to a different member of the club and each edge corresponds to a different social interaction between the corresponding club members. Since all club members have been observed, the dataset

[1] Since 2013, the 'Zachary Karate Club Club' (ZFCC) trophy has been presented, as a joke, at various conferences to the speaker who first mentions the karate club network in his or her presentation. See http://networkkarate.tumblr.com/ for more information.

exhibits no vertex sampling or growth. Zachary's initial analysis highlighted the division of members into two factions, caused by a rift between the club's two leaders. This known separation of its vertices makes the karate club network a canonical testbed for community detection methods. The standard analyses of the karate club also demonstrate a common pitfall of network analysis, which I discuss further in Section 9.10.

1.6.4 Enron email corpus

The Enron email corpus [104] consists of email activity for 150 employees at the Enron corporation. The dataset contains not only information about the senders and recipients of emails but also textual content, timestamps, etc.[2] Most relevant for our purposes is the network structure induced by the exchange of emails between employees, which we construct by letting each edge correspond to a different email in the corpus. An important difference from the karate club network (Section 1.6.3) is that a single edge (i.e., email) can involve more than two vertices (i.e., sender/recipient). For example, an email sent from employee A to employees B, C, and D corresponds to a single (hyper)edge in the network representation. Interaction networks such as this and the collaboration networks discussed next are the subject of Chapters 9 and 10.

1.6.5 Collaboration networks

Collaboration networks between actors [104, 134], scientists, authors, and other communities of professionals have much in common with the above Enron dataset. In an actors network, for example, each actor corresponds to a different vertex and each movie corresponds to a different edge consisting of the set of all vertices whose associated actors play a role in that movie. A common feature of the karate club, Enron, and collaboration networks is their growth by sequential addition of new edges, in the form of interactions, as opposed to sequential addition of new vertices, as in the preferential attachment model (Section 4.2). This feature of interaction networks figures prominently in Crane and Dempsey's framework of edge exchangeability (see [54] and Chapters 9–10).

1.6.6 Blockchain and cryptocurrency networks

Cryptocurrencies, such as Bitcoin [10, 122], Ethereum [29], and RChain [40], combine several innovative ideas in an effort to revolutionize economic activity through the use of peer-to-peer networks, blockchain technology, and smart contracts. These 'digital currencies', e.g., Bitcoin, operate on a blockchain, which records all transactions in a 'ledger' that stores the complete history of all Bitcoin transactions. This ledger is maintained by a distributed peer-to-peer network, which updates the blockchain by adding blocks according to a majority voting consensus protocol. Peer-to-peer networks also play an important role in decentralizing control of the network

[2]See http://www.cs.cmu.edu/~enron/ and [131] for some applications involving this dataset.

away from a centralized authority toward a distributed collection of nodes in the network. All of these components come together to create a complex network of transactions between addresses on the blockchain. As this revolutionary new technology matures, blockchain data should serve as a fertile testbed for model development at the frontiers of complex data analysis.

1.6.7 Other networks

In addition to the above examples, there are networks from social media platforms such as Facebook and Twitter [18, 117], brain networks [82], gene regulatory networks, telecommunications networks, wireless sensor networks [100, 101], etc. All of these are just a small selection of the many structures that are now referred to as 'network data'. Because I focus in this book on establishing the foundations of network analysis, I do not undertake any detailed application of a particular network dataset. These examples do, however, provide concrete modeling 'scenarios' within which to discuss different modeling approaches. The 'scenarios' accompanying each new class of models are meant to provide additional context for the more technical aspects of network analysis covered throughout the text.

1.6.8 Some common scenarios

As the scope of networks expands to encompass problems in new disciplines, so too must the mathematical and statistical techniques available to address these problems. I conclude this section with a brief review of some of the basic contexts for network modeling in social science, epidemiology, and national security. In the near future, it seems inevitable that the relevance of networks will continue to expand to include a wider range of disciplines as human behaviors and complex systems become ever more entangled through the growth of the Internet, social media, and other emergent technologies, such as blockchain.

Social science. Social network analysis was the primary domain of statistical network analysis until the mid-1990s. By all known accounts, the study of social networks began with Moreno's invention of the sociogram in 1930 [121]. Still today, many common network models (e.g., stochastic blockmodels (SBMs) [89] and exponential random graph models (ERGMs) [78, 90]) were originally motivated by sociological applications. With the growth of online communities and social media as a way to consume and disseminate information, traditional social networks have given way to networks with much more complex structure than traditional social network models, namely SBMs and ERGMs, are equipped to handle.

Epidemiology. Stochastic process models for disease spread on networks garner substantial interest in applied probability and statistical physics. The now classical SI (susceptible-infected), SIS (susceptible-infected-susceptible), and SIR (susceptible-infected-recovered) models describe how infections spread in a population whose interactions are represented by a graph. In the SIR model, for example, each node fluctuates among three states: susceptible to infection (S), infected (I), or recovered

(R). As time evolves, infected individuals randomly transmit the disease to their susceptible neighbors. Infected individuals recover, and are henceforth immune from infection, according to another random process. Basic questions center around how the different combinations of network structure and disease dynamics affect disease spread. For example, given certain initial conditions, what is the probability of an epidemic, i.e., the disease spreads to a non-negligible fraction of the population? One can also imagine how such models could be useful for designing effective advertising strategies or for modeling how information percolates through social networks.

National security. Networks arise in national security in at least two different ways. There are physical networks, such as the Internet, the U.S. Power Grid, and the transportation network of roads, bridges, and highways, all of which must be protected against failure or targeted attack. In fact, many experts [37] now regard cyberspace as the primary battlefield of modern warfare and national security, making resilience to targeted network attacks critical to national security interests. These concerns over cybersecurity, and the role of network science in resolving them, will continue to grow as more economic and social activity transitions to cyberspace.

Non-physical networks also play a role in national security, as terrorist organizations rely on complex webs of social, financial, and political interactions in order to evade detection [135]. As long as critical national infrastructure is controlled by a centralized, bureaucratic government, the governed society is vulnerable to both external attacks (e.g., hacking) and internal attacks (e.g., leaks), both of which have become increasingly prevalent and widely publicized in recent times. As a countermeasure to the vulnerability and antiquity of centralized authority, blockchain technology and cryptocurrencies (Section 1.6.6) distribute control of currency and other critical information to a "trustless" peer-to-peer network [10, 122]. The use of networks for this purpose is likely to have potential national security implications moving forward.

1.7 Major open questions

The probabilistic foundations of statistical network analysis currently face a few major open questions that are worth keeping in mind over the coming chapters.

1.7.1 Sparsity

Early interest in network science grew out of several concurrent empirical observations of sparsity and so-called 'scale-free' structure in real-world networks [1, 5, 14, 70, 111, 113]. (Refer to Chapters 4, 7, and 9 for a more detailed discussion of sparsity and power law, i.e., scale-free, properties.) For present purposes, it is sufficient to interpret 'sparsity' to mean that the network has few connections relative to its size. For example, when represented as a graph with n vertices, a network is sparse if it has a negligible number of edges compared to the number n^2 of all possible edges. In statistics, sparsity draws interest for two competing reasons, which together capture the tension between empirical properties of network data and logical principles of statistical modeling. First, many sparse networks are observed to be well-connected as a result of heterogeneous patterns of connectivity (e.g., 'scale-free' structure). So while the network is in one sense poorly connected (because it is sparse), it is at the same time well-connected (because of its complex patterns). Second, the prevailing approaches to network modeling (e.g., stochastic blockmodels, graphons, and exponential random graph models) are unable to account for these observed empirical behaviors. These competing elements of network modeling have stalled progress in statistical network analysis for nearly a decade, primarily due to unrecognized limitations of conventional approaches. Chapters 9–10 present one attempt to address this challenge, which interested readers are encouraged to build upon.

1.7.2 Modeling network complexity

In addition to sparsity, other heterogeneous features of real-world networks, such as power law degree distributions, clustering, and the 'small-world' property [152], confound attempts to analyze network data with standard models. In this opening chapter, I have emphasized the need for new tools to conceptualize the complexity of modern data structures. Above all, we seek to work with the complexity of network data, rather than fight against it by reducing complexity to something with less structure. This latter attitude of 'flattening' network structure is common throughout statistical analysis, and especially in network community detection, where non-overlapping subsets (i.e., communities) are sought to provide a 'low resolution' summary of much richer network structure. Community detection has become a cottage industry among statisticians interested in network analysis, but it is mostly counterproductive for understanding data complexity. I discuss models for community detection in the context of relative exchangeability (Chapter 8).

1.7.3 Sampling issues

Understanding the impact of sampling is one of the longest standing challenges in modern network science. Empirical observations of power law degree distribution in the Internet and other real-world networks [1, 5, 14, 70, 111] raise the question of whether these observed properties reflect the actual network structure or are merely an artifact of sampling bias [27, 112, 154]. This question is of central importance to statistical network analysis, for which the mode of sampling establishes the essential link between observed and unobserved parts of the network needed for inference. But even as interest in network analysis has grown among statisticians, there has not been much effort to incorporate sampling into the theoretical foundations of the subject. Much of the work on network analysis promoted by flagship statistics journals consists of asymptotic results and standard analyses under models that are known to be inadequate for most serious applications (e.g., graphons, stochastic blockmodels, and exponential random graph models). Remarkably few of these analyses acknowledge the importance of sampling to network analysis; and those that do, e.g., [138], assume a stylized form of sampling by vertex selection (Section 3.2) which does not even remotely resemble the way in which real-world networks are sampled. I discuss these issues at length throughout Chapters 3–5, and again in Chapters 6 and 9.

1.7.4 Modeling network dynamics

While much of this book is dedicated to modeling single instances of a network, there is emerging interest in analyzing dynamic network data, such as temporal observations of brain activity and social media interactions. But so far statistical work on dynamic networks is mostly confined to theory and applications for the temporal exponential random graph model or other *ad hoc* approaches. Because network dynamics add another dimension to the already challenging problem of network modeling, the foundations of dynamic network analysis are even more technically and conceptually challenging than their non-dynamic counterpart. Chapter 11, in which I give a brief non-technical overview of some otherwise technical work from the stochastic processes literature [44, 48, 57], offers a potential starting point for a more general theory of dynamic network modeling. More in depth coverage of dynamic network analysis is beyond the scope of this book and is left as a topic worthy of its own book length treatment.

1.8 Toward a Probabilistic Foundation for Statistical Network Analysis

In this opening chapter I have laid out a vision for network analysis as the foundation for what I am calling *complex data analysis*. As of yet, this vision has not been realized, but it is my hope in this book to clarify the major tenets underlying this vision and, if possible, to light the path toward its ultimate fulfillment. If nothing else, I hope to convince readers that real progress in the analysis of complex data will be limited as long as the field continues to seek incremental advances within the networks-as-graphs orthodoxy. The ideas in Chapters 5 and 9–11 offer some first attempts to get beyond these limitations, but many challenges still lie ahead.

Chapter 2

Binary relational data

We begin with network data that takes the form of a relation among individuals in a well-defined population. For this we write R to denote a relation on the set of $n \geq 1$ elements $[n] = \{1, \ldots, n\}$. Here each element $1, \ldots, n$ is assumed to uniquely label one of the n individuals in a population. (We will have ample opportunity to discuss the significance of labeling throughout Chapters 6–10. For now it is assumed that the labels $1, \ldots, n$ carry no additional significance other than to distinguish between individuals.)

In concrete applications, R is understood as the nature of relationship under consideration, with $(i, j) \in R$ indicating that (i, j) is an instance of relation R, i.e., i exhibits relation R to j. For example, if $1, \ldots, n$ label students in a high school, then $(i, j) \in R$ might indicate that 'i considers j as a friend', or that 'i voted in favor of j for election to student government', or that 'i and j are friends on social media'. Or if $1, \ldots, n$ label countries, then $(i, j) \in R$ might indicate that 'country i imports goods from country j', or that 'country i is a military ally of country j', or the like. Data with this structure includes networks built from interactions on social media such as Facebook and Twitter, the karate club dataset, and other sociometric datasets.

The instances of the relation R mentioned above are all binary, i.e., defined for pairs of individuals, but networks can also be built from higher-order relations, as when interactions involve more than two individuals. In a coauthorship network, for example, each published article represents an interaction of 'coauthorship' among its authors, in which case the labels $1, \ldots, n$ correspond to authors and $(i_1, \ldots, i_k) \in R$ indicates an article coauthored by $i_1, \ldots, i_k$. Additional examples include collaborations among movie actors, as obtained from the Internet Movie Database[1] (IMDb), and email communications among employees in a company, as recorded in the Enron email corpus [104, 134]. While there is little conceptual difference between binary relations, i.e., each relationship involves exactly two elements, and higher-order relations, indicating coauthorship, professional collaboration, or email communication, the two different cases raise important practical considerations in light of how they are typically treated in the network science literature. In Chapter 10, we discuss some subtleties of analyzing higher-order relations. But for now we specialize to binary relations.

[1] http://www.imdb.com/

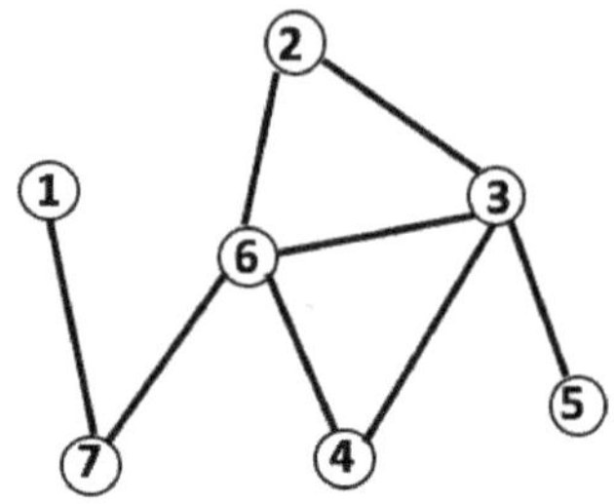

Figure 2.1 *Undirected graph corresponding to adjacency matrix in* (2.2).

From any binary relation R, one can construct an *adjacency matrix* $\mathbf{Y} = (Y_{ij})_{1 \leq i,j \leq n}$ with

$$Y_{ij} = \begin{cases} 1, & (i,j) \in R, \\ 0, & \text{otherwise}. \end{cases} \tag{2.1}$$

The matrix $\mathbf{Y}$ defined in (2.1) can be envisioned as a *graph* with vertex set $[n]$ and a (directed) edge from i to j if $Y_{ij} = 1$. (The graph is undirected if $Y_{ij} = Y_{ji}$ for all $1 \leq i, j \leq n$. We discuss both directed and undirected graphs throughout the text, often without distinguishing between the two cases.) Figure 2.1 depicts the undirected graph with corresponding adjacency matrix

$$\mathbf{y} = \begin{pmatrix} 0 & 0 & 0 & 0 & 0 & 0 & 1 \\ 0 & 0 & 1 & 0 & 0 & 1 & 0 \\ 0 & 1 & 0 & 1 & 1 & 1 & 0 \\ 0 & 0 & 1 & 0 & 0 & 1 & 0 \\ 0 & 0 & 1 & 0 & 0 & 0 & 0 \\ 0 & 1 & 1 & 1 & 0 & 0 & 1 \\ 1 & 0 & 0 & 0 & 0 & 1 & 0 \end{pmatrix}. \tag{2.2}$$

The representation of relational data as a binary array, in (2.1), and its visualization as a graph, in Figure 2.1, facilitates the commonplace 'networks-as-graphs' perspective of network data which I mentioned in Section 1.2 and will discuss again throughout the coming chapters.

Throughout the text I use $\mathbf{Y}$ and $\mathbf{y}$ to denote networks, with $\mathbf{Y}$ being random and $\mathbf{y}$ being generic or fixed. As emphasized in Chapter 1, these networks need not always take the same form, e.g., as a $\{0,1\}$-valued array in (2.2) or as a graph in Figure 2.1, but when working with arrays, as in (2.1), $\mathbf{Y}$ and $\mathbf{y}$ are understood to represent $\{0,1\}$-valued arrays of arbitrary size. When wishing to emphasize that $\mathbf{Y}$ or $\mathbf{y}$ is $n \times n$, I write $\mathbf{Y}_n$ or $\mathbf{y}_n$, as appropriate.

Notice that we allow the random array $\mathbf{Y}$ in (2.1) to be asymmetric ($Y_{ij} \neq Y_{ji}$) and to have loops ($Y_{ii} = 1$). In practice, the decision of whether to allow for directed edges or loops depends on the nature of the relation under study. For example, directed edges arise in the international trade scenario of Section 2.1, in which $(i,j) \in R$ indicates that country i imports goods from country j. Since this relation need not

Table 2.1 *Array* **Y** *describing binary import/export relationships between countries.*

	USA	Russia	China	England	⋯
USA	–	0	1	1	⋯
Russia	1	–	1	0	⋯
China	1	1	–	0	⋯
England	1	0	0	–	⋯
⋮	⋮	⋮	⋮	⋮	⋱

be reciprocated, i.e., j need not import from i even though i imports from j, the relation R is asymmetric. On the other hand, if $(i,j) \in R$ indicates that i is a 'friend' or 'follower' of j on a social media platform, then the relation is sometimes symmetric, e.g., on Facebook i can only be a friend of j (i.e., $(i,j) \in R$) if j is a friend of i (i.e., $(j,i) \in R$), and sometimes asymmetric, e.g., on Twitter it is possible for i to 'follow' j without j following i. Loops occur when vertices bear the relation R relative to themselves. For example, if $(i,j) \in R$ indicates that 'i voted for j' in an election, then $(i,i) \in R$ indicates that vertex i voted for himself/herself. For the most part these distinctions pose no technical difficulty, and so we allow both asymmetry and loops in R as a default, unless stated otherwise.

Though not exclusive to social network analysis, relational datasets frequently arise in classical sociometric studies, of which both the high school social network and international relations network discussed above are specific examples. In the sociometric context, statistical models serve primarily as a way to summarize the structure in the observed data **Y**, rather than to infer patterns in a larger population of individuals. Although it is possible in some cases to draw inferences beyond the observed data, sociometric analysis is more often an exercise in descriptive statistics and exploratory data analysis. And while exploratory data analysis has proven useful in the study of social networks, it is not where the future of complex data analysis lies, and thus is not the appropriate setting for this discussion on the foundations of network analysis. Nevertheless, this context motivates the more in-depth treatment of network sampling which begins in Chapter 3 and continues throughout the text.

2.1 Scenario: Patterns in international trade

Let $[n]$ index a set of countries (e.g., USA, England, China, Russia, etc.) and let $\mathbf{Y} = (Y_{ij})_{1 \leq i,j \leq n}$ represent binary relational data with '$Y_{ij} = 1$' indicating that 'country i imports from country j' and $Y_{ij} = 0$ otherwise. For the time being we assume that **Y** is observed without any further information about the countries, such as GDP, geographical location, etc. In this way the data consists only of the import-export relation **Y** among the n observed countries. The goal of the analysis is to detect patterns in the trade relationships among these countries.

2.1.1 Summarizing network structure

In many sociometric studies, the number of vertices n is often small or moderate. In the above international trade scenario, for example, the number of countries can be no larger than a few hundred (i.e., the number of countries in the world), but in general will be much less, such as the 20 countries with the largest economies. Even though these datasets are small by modern standards, it may still be difficult to represent the structure induced by Table 2.1 visually (e.g., as a graph with vertex set $[n]$ and edge $i \to j$ whenever $Y_{ij} = 1$) in a way that makes interesting patterns apparent. So while we do not seek to draw inferences beyond the sample in this scenario, a network model may be useful for summarizing the observed network structure in terms of a few sufficient statistics of interest. In this case a model with easily interpretable parameters which capture the essence of the desired network properties is most valuable. More sophisticated models that take into account aspects of network sampling or network generation are important to other settings of network analysis but add little or no value in the present scenario.

One common and straightforward approach to describing patterns in $\mathbf{Y}$ is to compute certain summary statistics of interest, such as *reciprocity* between countries—both i and j import from one another—and *differential attractiveness*—the attractiveness of j is measured as the number of countries that import goods from j. (The term 'differential' here indicates that attractiveness is measured *relative* to the other countries in the population. There is no absolute measure of attractiveness.) Other measures include *transitivity*—if i imports from (or exports to) j and j imports from (or exports to) k, then i also imports from (or exports to) k. The dyad independence model (Section 2.2) was designed for the purpose of summarizing these, and similar, aspects of network structure.

2.2 Dyad independence model

Holland and Leinhardt [90] introduced the so-called p_1 *model* for the purpose of describing patterns in sociometric data. The p_1 model arises as a special case of the *dyad independence model* for binary relational arrays in $\{0,1\}^{n\times n}$ as follows.

For any $\mathbf{Y} = (Y_{ij})_{1\le i,j\le n}$, the *dyad* D_{ij} is the pair (Y_{ij}, Y_{ji}) for each $1 \le i \ne j \le n$. Thus, if $\mathbf{Y}$ is represented graphically, then each dyad D_{ij} describes how vertices i and j are related to one another, with

$$\begin{aligned} D_{ij} = &\ (0,0) \text{ indicating no relationship between } i \text{ and } j,\\ &\ (1,0) \text{ indicating a relationship in the direction from } i \text{ to } j \text{ but not } j \text{ to } i,\\ &\ (0,1) \text{ indicating a relationship in the direction from } j \text{ to } i \text{ but not } i \text{ to } j, \text{ and}\\ &\ (1,1) \text{ indicating a relationship both from } i \text{ to } j \text{ and from } j \text{ to } i. \end{aligned}$$

In full generality, the dyad independence model assigns a probability distribution p_{ij} to each dyad, i.e.,

$$\Pr(D_{ij} = (z,z')) = p_{ij}(z,z'), \quad z,z' \in \{0,1\}, \tag{2.3}$$

and assumes that these dyads behave independently. With $\mathbf{p} = (p_{ij})_{1\leq i\neq j\leq n}$ and the assumption that all pairs $(D_{ij})_{1\leq i<j\leq n}$ behave independently according to (2.3), $\mathbf{Y}$ has distribution

$$\Pr(\mathbf{Y}=\mathbf{y};\mathbf{p}) = \prod_{1\leq i<j\leq n} p_{ij}(y_{ij},y_{ji}), \quad \mathbf{y}\in\{0,1\}^{n\times n}. \tag{2.4}$$

Throughout the text, Pr denotes a generic probability operator and $\Pr(\cdot;\theta)$ denotes a probability distribution parameterized by θ.

The *dyad independence model* in (2.4) can be alternatively expressed in terms of parameters $(\rho_{ij})_{1\leq i<j\leq n}$ and $(\theta_{ij})_{1\leq i\neq j\leq n}$ by

$$\begin{aligned}&\Pr(\mathbf{Y}=\mathbf{y};(\rho_{ij})_{1\leq i<j\leq n},(\theta_{ij})_{1\leq i\neq j\leq n})\propto \\ &\qquad \propto \exp\left\{\sum_{1\leq i<j\leq n}\rho_{ij}y_{ij}y_{ji}+\sum_{1\leq i\neq j\leq n}\theta_{ij}y_{ij}\right\}\end{aligned} \tag{2.5}$$

for each $\mathbf{y}=(y_{ij})_{1\leq i,j\leq n}\in\{0,1\}^{n\times n}$, where

$$\rho_{ij}=\log\left(\frac{p_{ij}(0,0)p_{ij}(1,1)}{p_{ij}(0,1)p_{ij}(1,0)}\right) \quad \text{and}$$
$$\theta_{ij}=\log(p_{ij}(1,0)/p_{ij}(0,0)).$$

Note that the distributions in (2.4) and (2.5) are equivalent, with (2.5) expressing the dyad independence model as an exponential family model with natural parameters $(\rho_{ij})_{1\leq i<j\leq n}$ and $(\theta_{ij})_{1\leq i\neq j\leq n}$.

In their initial development of this model, Holland and Leinhardt focused specifically on the properties of reciprocity and differential attractiveness, prompting them to streamline the expression in (2.4) by specifying parameters θ,ρ, $\alpha=(\alpha_i)_{1\leq i\leq n}$, and $\beta=(\beta_i)_{1\leq i\leq n}$ and setting

$$\rho_{ij}=\rho, \quad 1\leq i<j\leq n, \quad \text{and}$$
$$\theta_{ij}=\theta+\alpha_i+\beta_j, \quad 1\leq i\neq j\leq n.$$

With these parameters, the so-called p_1 *model* with parameter $(\rho,\theta,\alpha,\beta)$ on $\{0,1\}^{n\times n}$ assigns probability

$$\Pr(\mathbf{Y}=\mathbf{y};\rho,\theta,\alpha,\beta)=\frac{\exp\left\{\rho\sum_{1\leq i<j\leq n}y_{ij}y_{ji}+\theta y_{\bullet\bullet}+\sum_{i=1}^n\alpha_i y_{i\bullet}+\sum_{j=1}^n\beta_j y_{\bullet j}\right\}}{\prod_{1\leq i<j\leq n}\eta_{ij}} \tag{2.6}$$

to each $\mathbf{y}\in\{0,1\}^{n\times n}$, where $y_{i\bullet}=\sum_{j=1}^n y_{ij}$ is the *out-degree* of vertex i, $y_{\bullet j}=\sum_{i=1}^n y_{ij}$ is the *in-degree* of vertex j, $y_{\bullet\bullet}=\sum_{i,j=1}^n y_{ij}$ is the *total degree* of $\mathbf{y}$, and the product over

$$\eta_{ij}=1+e^{\rho+\alpha_i+\beta_j}+e^{\rho+\alpha_j+\beta_i}+e^{\rho+2\theta+\alpha_i+\alpha_j+\beta_i+\beta_j}, \quad 1\leq i<j\leq n, \tag{2.7}$$

determines the normalizing constant. The *reciprocity* parameter ρ in (2.6) captures the relative probability that two generic vertices 'reciprocate' their relation to one another in the observed network; and the *differential attractiveness* parameters α_i and β_i of each vertex i capture how likely (relative to other vertices) i is to have outgoing links (α_i) and incoming links (β_i).

For our purposes, the p_1 model plays a limited but important role: first in motivating the more general class of ERGMs (Section 2.3); and second in giving a non-trivial example of a model that is consistent under selection (Section 2.5). Some recommended further reading about the p_1 model and ERGMs can be found in Section 2.6.

2.3 Exponential random graph models (ERGMs)

For real-valued parameters $\theta_1, \ldots, \theta_k \in \mathbb{R}$ and statistics $T_1, \ldots, T_k : \{0,1\}^{n\times n} \to \mathbb{R}$, the *exponential random graph model* (ERGM) with (natural) parameter $\theta = (\theta_1, \ldots, \theta_k)$ and (canonical) sufficient statistic $T = (T_1, \ldots, T_k)$ assigns probability

$$\Pr(\mathbf{Y} = \mathbf{y}; \theta, T) \propto \exp\{\sum_{i=1}^{k} \theta_i T_i(\mathbf{y})\} \tag{2.8}$$

to each $\mathbf{y} \in \{0,1\}^{n\times n}$. (The proportionality constant

$$\sum_{\mathbf{y}^* \in \{0,1\}^{n\times n}} \exp\left\{\sum_{i=1}^{k} \theta_i T_i(\mathbf{y}^*)\right\} \tag{2.9}$$

of the distribution in (2.8) cannot, in general, be expressed in closed form, and is therefore omitted.) Under (2.8), T is a sufficient statistic for the distribution of $\mathbf{Y}$ in the sense that (2.8) assigns equal probability to any two realizations $\mathbf{y}, \mathbf{y}' \in \{0,1\}^{n\times n}$ with $T(\mathbf{y}) = T(\mathbf{y}')$, for any choice of θ.

The p_1 model in (2.6) has the form of (2.8), as does the famed Erdős–Rényi–Gilbert model [68, 81], under which each Y_{ij} is an i.i.d. draw from the Bernoulli distribution with success probability $p \in (0,1)$. More specifically, the Erdős–Rényi–Gilbert distribution with parameter p is expressed as

$$\Pr(\mathbf{Y} = \mathbf{y}; p) = \prod_{1\leq i\neq j\leq n} p^{y_{ij}}(1-p)^{1-y_{ij}}, \quad \mathbf{y} \in \{0,1\}^{n\times n}. \tag{2.10}$$

In the form of (2.8), the sufficient statistic of the Erdős–Rényi–Gilbert distribution counts the number of edges in $\mathbf{y}$,

$$T(\mathbf{y}) = \sum_{1\leq i\neq j\leq n} y_{ij}, \quad \mathbf{y} \in \{0,1\}^{n\times n},$$

the natural parameter is the log-odds ratio $\theta = \log(p/(1-p))$, and the probability in (2.10) is given by

$$\begin{aligned}
\Pr(\mathbf{Y} = \mathbf{y}; \theta, T) &\propto \exp\{\theta T(\mathbf{y})\} \\
&= \exp\{T(\mathbf{y})\log(p) - T(\mathbf{y})\log(1-p)\} \\
&= p^{T(\mathbf{y})}(1-p)^{-T(\mathbf{y})},
\end{aligned}$$

which is equivalent to (2.10) after computing the normalizing constant in (2.9).

For summarizing the structure in a relational data matrix $\mathbf{Y}$, the ERGM in (2.8) is quite a bit more general than the p_1 model in (2.6). The sufficient statistics in (2.8) allow for a wider range of network properties to be incorporated into the model description, e.g., transitive closure as measured by the number of triangles $\sum_{i,j,k=1}^{n} y_{ij}y_{jk}y_{ki}$. The viability of ERGMs for statistical inference, however, is limited by computational as well as logical constraints. The potential computational issues should be apparent by the presence of the unspecified normalizing constant in (2.8). If the normalizing constant in (2.9) can be computed in closed form or the computational burden can be otherwise resolved, then the class of ERGMs seems to be reasonably flexible for summarizing patterns in the international trade data of Section 2.1 and similar sociometric datasets. But in the grand scheme of modern network science, sociometric studies are a niche topic, and are out of step with the vast majority of situations for which statistical techniques are used to draw inferences based on a partial observation or sample from a population. The modeling challenges brought about by sampling considerations will occupy our attention for most of the chapters that follow. As a precursor to Chapters 3–5, I now briefly discuss some limitations of ERGMs for modeling sampled networks.

2.4 Scenario: Friendships in a high school

Consider a social network of friendships among N high school students, of which we take a sample of $n < N$ students and observe the friendships among them. Unlike in Section 2.1, the observed relationships in this scenario comprise a partial sample of the friendships in which we are interested. We wish to use the binary relational data $\mathbf{Y}_n$ for the n sampled students to draw inferences about the friendship patterns among all N students in the high school. But doing so requires an assumption about how the sampled students are related to the population of all students, raising the question:

> *In what way is the observation $\mathbf{Y}_n$ representative of the population network $\mathbf{Y}_N$?*

This question is at the center of much confusion and debate in the network science literature, as well as Big Data and complex data analysis more broadly. Most of this book is dedicated to clarifying the role of sampling in network analysis, with special focus on how traditional modeling assumptions impose unspoken, and often undesirable, sampling constraints on network analysis.

2.5 Network inference under sampling

The distinction between population and sample did not arise in the international trade scenario of Section 2.1. In that scenario we were only interested in the patterns among observed countries, regardless of whether or not those countries comprised a sample of all countries in the world. But the issue of sampling does arise in Section 2.4.

To be clear when discussing situations like the one in Section 2.4, we index $\mathbf{Y}$ by the number of individuals to which it pertains, so that $\mathbf{Y}_N$ is the adjacency matrix for all N students and $\mathbf{Y}_n$ is the matrix for the sample of $n \leq N$ students. Since we seek to draw inferences about $\mathbf{Y}_N$ based on $\mathbf{Y}_n$, whatever model we invoke must both describe the behavior of $\mathbf{Y}_n$ and articulate the logical link between $\mathbf{Y}_n$ and $\mathbf{Y}_N$ (via sampling).

To appreciate the non-trivial interplay between sampling and modeling, consider how the observed friendships and the inferences they support would change under the following scenarios:

1. n students are sampled uniformly among all freshmen, i.e., first-year students, in the school;
2. n students are sampled uniformly among all seniors, i.e., final-year students, in the school;
3. n students are sampled uniformly among all students in the school; or
4. all students who write for the school newspaper, of which there are n in total, are sampled.

Under scenarios 1–3, the sampling mechanism is the same, i.e., students are chosen uniformly at random, but the populations are different, i.e., freshmen, seniors, and all students, respectively. Under scenario 4, the population is the same as in scenario 3 (i.e., all students in the school), but the sampling mechanism is such that the sampled students are known to already have similar interests, i.e., writing for the newspaper, and therefore would seem more likely to be friends than a random selection of students from the school (as in scenario 3). We should not expect the friendship patterns observed under scenario 4 to resemble the friendship patterns among students in the entire school, and this distinction ought to be incorporated into our model. Notice also that in scenario 4 the number of sampled individuals is determined by the number of students who write for the school newspaper; it is not specified *a priori* by the data analyst, as in scenarios 1–3. This last point hints that the number of students 'n' observed in scenario 4 has a different significance than the number n in scenarios 1-3, and this difference may well affect how we specify the model and interpret any subsequent inferences. (See Sections 3.7–3.8 for further discussion of this last point.)

Over the next several chapters, we will have ample opportunity to discuss the nuances presented by these and other sampling scenarios. To conclude the present chapter, I remark that although the observation/sampling mechanism must be incorporated into any well-specified statistical model (not just network models), it is commonly overlooked when specifying models whose observation mechanism is obvious. For example, the observation mechanism for i.i.d. sequence data establishes an implicit relationship between observed data and the rest of the population—in particular, the observations are independent and follow a common distribution. But because the concept of i.i.d. models is well-trodden and standard, the meaning of this specification is understood *by convention.*

A key takeaway from the next few chapters, especially Chapter 5, is that in network analysis there is no prevailing convention, and therefore network models must be specified more explicitly than is customary in more classical applications. While there has been some progress recently toward accounting for sampling issues in network modeling, most work has focused on modeling under selection sampling. As we discuss beginning in the next chapter, ordinary selection sampling for graphs is just one of many ways in which network datasets can be sampled, and in most situations selection sampling is far from realistic. Just as time series and Markov models have been developed to address the deficiency of i.i.d. models for many applications, new network models are needed to address the deficiencies of stochastic blockmodels, ERGMs, and graphon models. I cover a few such new network modeling frameworks throughout Chapters 7–11.

2.6 Further reading

Readers familiar with the statistical networks literature may wonder why ERGMs garner so much attention if they are of such limited use in modern applications. To understand the disconnect, realize that exponential family distributions have long been a fixture of statistical inference, and the apparent flexibility and familiarity of exponential family distributions made them a natural choice in the early work on social network analysis by Holland and Leinhardt, Frank and Strauss, and others. So while I focus here on the usefulness, or lack thereof, of ERGMs in modern applications, it is important to bear in mind that they were not originally intended for such problems. Since the circumstances under which ERGMs are now being applied (e.g., social media, the Internet, World Wide Web) did not exist when ERGMs were first conceived, any critique of ERGMs given here is directed toward their inappropriate use in modern applications, not their initial conception or their, perhaps appropriate, use in the applications for which they were originally designed.

When viewed from the perspective of modern network science applications, these observations raise doubts about claims touting the importance of ERGMs in present-day network analysis. For example, in [138, p. 509], ERGMs are described as "undoubtedly one of the most important and popular classes of statistical models of network structure." Popular, yes, but the *importance* of ERGMs is dubious given their many practical and conceptual drawbacks highlighted above and in the coming chapters. If one interprets the 'importance' of an idea based solely on how many articles are published on it or how many researchers study it, then such claims quickly become self-fulfilling: ERGMs are 'important' because a lot of articles have been written about them, which in turn inspires even more articles on ERGMs, *ad infinitum.* (This phenomenon is a real-life manifestation of the 'rich get richer' phenomenon discussed in Section 4.2 below. For another instance of 'rich get richer', refer to the widespread declarations of the 'ubiquity' and 'importance' of networks which have been offered as justification for studying networks, as mentioned in the opening section of Chapter 1.) This herd mentality is the antithesis of the viewpoint espoused here, in which I emphasize repeatedly the importance of thinking independently of, so as not to become blinded by, conventional perspectives of network analysis found throughout the literature.

The orthodox mindset is epitomized in a recently proposed 'foundational' approach to network analysis [136] which, in contrast to the conceptual perspective emphasized here, confines itself to exponential family models, seemingly unaware and unconcerned by the incredibly narrow perspective offered by ERGMs in the context of modern network analysis. In light of the numerous concerns about ERGMs voiced elsewhere and echoed here, the proposal in [136] gives a skewed treatment of how ERGMs fit within the broader scope of network science.

Because of the excessive attention paid to ERGMs elsewhere, I shall discuss ERGMs sparingly in the rest of this text. The reader interested in more about the p_1 model and ERGMs can consult the original papers of Holland and Leinhardt [90], Frank and Strauss [78], and Wasserman and Pattison [150], and the abundance of followup work which has appeared in the decades since. Particularly, the works of [17, 34, 138], and references therein, should give a well-rounded presentation of ERGMs.

Chapter 3

Network sampling

The content of Section 2.5 and other observations about the impact of sampling on inferred network properties [112, 154, 162] makes a sampling theory for statistical network analysis one of the main priorities for developing the probabilistic foundations of the field; see Section 1.7.3 for further discussion. Over the next three chapters, I discuss network sampling and a number of related issues in some depth, culminating in a two-stage formulation of a statistical model as a family of candidate distributions describing the *uncertainty* and *variability* in the data together with the *context* in which to interpret inferences made under the assumed model. These two aspects of modeling (uncertainty and context) come together in the concept of *coherence*. Incorporating the *context* and articulating the condition of *coherence* are two novelties of the framework presented below, cf. [52]. (See Chapter 5 for further discussion of uncertainty, context, and coherence. Refer back to Sections 1.4–1.5 for a high-level overview of these essential modeling components.) To streamline the presentation, I tailor this framework to network models set either in a sampling context—called *sampling models*—or in a generative context—called *generative models*. I begin in this chapter with sampling models, and defer generative models to Chapter 4.

3.1 Opening example

Before delving into fundamental issues of network sampling it is instructive to consider how sampling figures into more traditional data analysis. Let $X_1,\dots,X_N$ represent the sizes of N households in a population, so that each X_i counts the number of residents of household i. In this population suppose that household sizes behave as i.i.d. random variables from the 1-shifted Poisson distribution with parameter $\lambda > 0$:

$$\Pr(X_i = k+1; \lambda) = \lambda^k e^{-\lambda}/k!, \quad k = 0, 1, \dots. \tag{3.1}$$

(The 'shift' by 1 reflects the assumption that each household must be inhabited by at least 1 person.) We would like to estimate the parameter λ on the basis of a sample $X_1^*,\dots,X_n^*$ of household sizes from $X_1,\dots,X_N$. How should we model $X_1^*,\dots,X_n^*$?

Consider how the sampling mechanism affects the choice of model for $X_1^*,\dots,X_n^*$ in the following two cases.

1. Suppose $X_1^*,\dots,X_n^*$, $n \le N$, have been sampled uniformly without replacement from $X_1,\dots,X_N$. Then since $X_1,\dots,X_N$ are i.i.d. and $X_1^*,\dots,X_n^*$ have been chosen

independently of the realized values of $X_1, \ldots, X_N$, it follows that $X_1^*, \ldots, X_n^*$ are also i.i.d. from the 1-shifted Poisson distribution (3.1) with the same parameter $\lambda > 0$.

2. Suppose $X_1^*, \ldots, X_n^*$, $n \leq N$, have been observed by sampling individuals in the population uniformly at random and recording the size of the household in which each sampled individual lives. Since a household of size k can be chosen in k different ways, one for each of the k individuals in the household, the probability of choosing a household of size k is proportional to $k\Pr(X_i = k; \lambda)$. Each of the sampled household sizes $X_1^*, \ldots, X_n^*$ is marginally distributed according to the size-biased distribution associated to (3.1), as given by

$$\Pr(X_i^* = k+1; \lambda) = \frac{(k+1)\lambda^k e^{-\lambda}}{(\lambda+1)k!}, \quad k = 0, 1, \ldots. \tag{3.2}$$

Comparing these two scenarios makes clear that the distribution of the data $X_1^*, \ldots, X_n^*$ is affected by not only the distribution of household sizes in the population $X_1, \ldots, X_N$ but also the sampling mechanism used in obtaining $X_1^*, \ldots, X_n^*$ from $X_1, \ldots, X_N$. Under either scenario, the parameter 'λ' governing the population translates into a parameter, also denoted 'λ', governing the sample. But the use of the same Greek letter 'λ' in the parameterization of both models does not imply that the parameters in these models can be regarded as 'the same' in the sense of being literally identical or having the same meaning.

To illustrate this last point, let us denote by $\mathscr{M}_{\text{pop}}$ the set of candidate distributions governing the population of household sizes, as given by the family of distributions parameterized by $\lambda > 0$ in (3.1). Further, let us denote by $\mathscr{M}_{\text{sb}}$ the set of candidate distributions induced by $\mathscr{M}_{\text{pop}}$ under size-biased sampling in scenario 2. With this notation, $\mathscr{M}_{\text{pop}}$ is the set of all distributions in (3.1) and $\mathscr{M}_{\text{sb}}$ is the set of all distributions in (3.2). But even though $\mathscr{M}_{\text{pop}} \neq \mathscr{M}_{\text{sb}}$, the relationship between population and sample, as described by the sampling operation in scenario 2, still permits inference about $\mathscr{M}_{\text{pop}}$ based on inferences for $\mathscr{M}_{\text{sb}}$.

For a simple demonstration, suppose the sample consists of a single data point $X_1^* = x$, for some $x = 1, 2, \ldots$. Then under the model in (3.2), the maximum likelihood estimate for λ can be computed by

$$\hat{\lambda}_{\text{MLE}} = \frac{(x-3) + \sqrt{(x-1)^2 + 4}}{2}, \tag{3.3}$$

giving an estimate for λ in both the population and sample models.[1] The key point here is that the estimate for λ in (3.3) is obtained by fitting the data to $\mathscr{M}_{\text{sb}}$. Fitting $X_1^* = x$ instead to the model $\mathscr{M}_{\text{pop}}$ would lead to the erroneous estimate $\hat{\lambda}_{\text{MLE}} = x$, which fails to account for the size-biased sampling in scenario 2.

The plot thickens when one considers possible reparameterizations of the model

[1] This estimate is obtained by maximizing the log-likelihood for the observation $X_1^* = x$:

$$\log L(\lambda; x) = k\log(\lambda) - \lambda - \log(\lambda+1) + \log(k+1) - \log(k!).$$

for $X_1^*,\ldots,X_n^*$. For example, by taking the transformation $\theta \leftrightarrow \lambda+1$, one could equivalently express the set of distributions in (3.2) by

$$\Pr(X_i^* = k;\theta) = \frac{k(\theta-1)^{k-1}e^{1-\theta}}{\theta(k-1)!}, \quad k = 1,2,\ldots, \tag{3.4}$$

for $\theta > 1$. And since the parameter θ in (3.4) is merely a scalar ranging over $(1,\infty)$, we can express the same family of distributions in terms of a parameter λ by

$$\Pr(X_i^* = k;\lambda) = \frac{k(\lambda-1)^{k-1}e^{1-\lambda}}{\lambda(k-1)!}, \quad k = 1,2,\ldots, \tag{3.5}$$

for $\lambda > 1$. With $\mathscr{M}_{\text{sb-2}}(\theta)$ and $\mathscr{M}_{\text{sb-2}}(\lambda)$ denoting the sets of distributions parameterized by θ and λ in (3.4) and (3.5), respectively, we see that

$$\mathscr{M}_{\text{sb-2}}(\theta) = \mathscr{M}_{\text{sb-2}}(\lambda) = \mathscr{M}_{\text{sb}}$$

as *sets* of probability distributions. As we have already argued that inferences about $\mathscr{M}_{\text{sb}}$ can be extended to inferences about $\mathscr{M}_{\text{pop}}$, it should follow that inferences based on either of $\mathscr{M}_{\text{sb-2}}(\theta)$ or $\mathscr{M}_{\text{sb-2}}(\lambda)$ can be extended to inferences about $\mathscr{M}_{\text{pop}}$. But the relationship between the models for $X_1^*,\ldots,X_n^*$ and $X_1,\ldots,X_N$ must be carefully articulated by taking account of the context within which the two are related, and in particular the way in which the respective models are parameterized.

Following up on the discussion surrounding (3.3), we can estimate λ from the model in (3.5). But the relationship between the parameter 'λ' governing this model and the 'λ' in the population model must be kept straight. In particular, 'λ' in the population model (3.1) is not the "same λ" as in (3.4): the MLE $\hat{\lambda}_{\text{MLE}}$ obtained under (3.5) translates to an estimate $\hat{\lambda}_{\text{MLE}}+1$ for the model in (3.1).

Though the data and context of a typical network application are far more complex than the example given here, the same principles apply. One difference between the above case and a typical scenario of network analysis is that in network analysis the relationship between population and sample can rarely be expressed as precisely as through the size-biased sampling relationship that links (3.1) and (3.2).

Though network sampling has been previously studied in the quantitative social science literature [75, 76, 77, 85, 143, 144], many more far-reaching issues have not yet been recognized or even formulated. The reader is especially urged to study Lee, Kim, and Jeong's empirical analysis of the effects of sampling on network inference [112]. From the example in this section and the empirical analysis in [112], it should be clear that sampling plays a pivotal role in modeling and inference. But it is one thing to recognize that sampling affects observed network properties, and something else entirely to incorporate these effects into a workable network modeling framework. It is toward this latter objective that this and the coming two chapters are directed.

3.2 Consistency under selection

Consistency under selection is a specific kind of consistency under subsampling

(Section 3.9) which is relevant for modeling network data obtained by *selection sampling*.[2] For an example, consider the scenario of Section 2.4 with N high school students labeled uniquely $1,\ldots,N$ in a way that does not depend on their social relationships given in $\mathbf{Y}_N$. (One way is to imagine that each student is randomly assigned a unique identifier $1,\ldots,N$, with all possible assignments being equally likely.) Given such a labeling, the sampled network of $n < N$ students can be obtained by observing the social relationships among those students assigned labels $1,\ldots,n$. This way of sampling is called *selection (of* $[n]$ *from* $[N]$*)*, indicating that the sampled elements were chosen merely by *selecting* a predefined set of elements.

With $\mathbf{Y}_N = (Y_{ij})_{1\le i,j\le N}$ denoting the relational array for all N students, the observation $\mathbf{Y}_n$ obtained by selection sampling corresponds to the restriction of $\mathbf{Y}_N$ to its first n rows and columns, written $\mathbf{Y}_N|_{[n]} = (Y_{ij})_{1\le i,j\le n}$. In general, for an array $\mathbf{A} = (A_{ij})_{i,j\ge 1}$ and any subset $S \subseteq \mathbb{N}$, we write $\mathbf{A}|_S = (A_{ij})_{i,j\in S}$ to denote the restriction of $\mathbf{A}$ to the rows and columns indexed by S.[3] For example, for $\mathbf{A} = (A_{ij})_{1\le i,j\le N}$ given by

$$\begin{pmatrix} A_{11} & A_{12} & \cdots & A_{1n} & \cdots & A_{1N} \\ A_{21} & A_{22} & \cdots & A_{2n} & \cdots & A_{2N} \\ \vdots & \vdots & \ddots & \vdots & \ddots & \vdots \\ A_{n1} & A_{n2} & \cdots & A_{nn} & \cdots & A_{nN} \\ \vdots & \vdots & \ddots & \vdots & \ddots & \vdots \\ A_{N1} & A_{N2} & \cdots & A_{Nn} & \cdots & A_{NN} \end{pmatrix},$$

the restriction $\mathbf{A}|_{[n]}$ is the upper $n \times n$ submatrix

$$\begin{pmatrix} A_{11} & A_{12} & \cdots & A_{1n} \\ A_{21} & A_{22} & \cdots & A_{2n} \\ \vdots & \vdots & \ddots & \vdots \\ A_{n1} & A_{n2} & \cdots & A_{nn} \end{pmatrix}.$$

We denote the operation of *selecting* $[n]$ *from* $[N]$ as a function $\mathbf{S}_{n,N}$ between the population and sample spaces, i.e.,

$$\begin{aligned} \mathbf{S}_{n,N} : & \{0,1\}^{N\times N} \to \{0,1\}^{n\times n} \\ & \mathbf{y} \mapsto \mathbf{S}_{n,N}(\mathbf{y}) = \mathbf{y}|_{[n]}. \end{aligned} \tag{3.6}$$

To emphasize that (3.6) is intended as a sampling operation, we refer to the action $\mathbf{y} \mapsto \mathbf{S}_{n,N}(\mathbf{y})$ as *selection sampling*. The action in (3.6) is sometimes called *projection* or *restriction* by other authors. (It is sometimes convenient, e.g., in (3.7) below, to

[2] Statistically-minded readers should not confuse 'sampling consistency' with the statistical term 'consistency', i.e., asymptotic convergence of a statistical estimator toward its true parameter value. This latter notion of consistency does not appear anywhere in this book.

[3] When network data takes a form other than an array, as it does in later chapters, the restriction operation $\cdot|_S$ is defined analogously by restricting the network to the structure induced on units labeled by S. This operation will be defined precisely in whatever context it appears.

write $\mathbf{S}_{n,N}\mathbf{y}$ instead of $\mathbf{S}_{n,N}(\mathbf{y})$, and I will often do so whenever it causes no confusion.)

As we saw in Section 3.1, the relationship between sample and population plays an essential role in population-level inference from sampled data. In the context of binary arrays, the population network $\mathbf{Y}_N$ is assumed to follow some distribution on $\{0,1\}^{N\times N}$ and the observed data $\mathbf{Y}_n$ is assumed to have been obtained from $\mathbf{Y}_N$ via a sampling scheme. Under selection sampling, the observation $\mathbf{Y}_n$ corresponds to $\mathbf{S}_{n,N}\mathbf{Y}_N$ and the distribution of $\mathbf{Y}_N$ induces a distribution on $\{0,1\}^{n\times n}$ by

$$\Pr(\mathbf{S}_{n,N}\mathbf{Y}_N=\mathbf{y})=\Pr(\mathbf{Y}_N\in\mathbf{S}_{n,N}^{-1}(\mathbf{y})),\quad \mathbf{y}\in\{0,1\}^{n\times n}, \tag{3.7}$$

where

$$\mathbf{S}_{n,N}^{-1}(\mathbf{y})=\{\mathbf{y}^*\in\{0,1\}^{N\times N}:\mathbf{S}_{n,N}(\mathbf{y}^*)=\mathbf{y}\}$$

is the set of all arrays in $\{0,1\}^{N\times N}$ which could have given rise to the observation $\mathbf{y}$ under selection sampling. (In words, $\mathbf{S}_{n,N}^{-1}(\mathbf{y})$ is the set of all possible realizations of $\mathbf{Y}_N$ which would have resulted in the observation $\mathbf{Y}_n=\mathbf{y}$ under the assumed selection sampling scheme.) On the one hand, the distribution of the observed data $\mathbf{S}_{n,N}\mathbf{Y}_N$ is implicitly specified by the assumed population model and sampling scheme through (3.7). On the other hand, it is also common to specify a distribution $\Pr(\mathbf{Y}_n=\cdot)$ explicitly for every possible observation $\mathbf{Y}_n$, $1\le n\le N$. This explicitly specified distribution $\Pr(\mathbf{Y}_n=\cdot)$ is *consistent* with the induced distribution (3.7) only if the two coincide.

Definition 3.1 (Consistency under selection) *Let $\mathbf{Y}_n$ and $\mathbf{Y}_N$, $n\le N$, be random $\{0,1\}$-valued arrays, and let $\mathbf{S}_{n,N}$ be the selection sampling operation defined in (3.6). Then $\mathbf{Y}_n$ and $\mathbf{Y}_N$ are* consistent under selection *if they satisfy the distributional identity*

$$\mathbf{S}_{n,N}\mathbf{Y}_N=_{\mathscr{D}}\mathbf{Y}_n, \tag{3.8}$$

where $=_{\mathscr{D}}$ denotes equality in distribution. *More explicitly, Y_n and Y_N are consistent under selection if*

$$Pr(\mathbf{S}_{n,N}\mathbf{Y}_N=\mathbf{y})=Pr(\mathbf{Y}_n=\mathbf{y})\quad \textit{for all } \mathbf{y}\in\{0,1\}^{n\times n}. \tag{3.9}$$

3.2.1 *Consistency of the p_1 model*

Suppose that both $\mathbf{Y}_N$ and $\mathbf{Y}_n$ are assumed to follow the p_1 model (2.6) on $\{0,1\}^{N\times N}$ and $\{0,1\}^{n\times n}$, respectively, with the distribution of $\mathbf{Y}_N$ parameterized by $\rho,\theta,\alpha=(\alpha_1,\ldots,\alpha_N)$, and $\beta=(\beta_1,\ldots,\beta_N)$ and the distribution of $\mathbf{Y}_n$ parameterized by $\rho,\theta,\alpha|_{[n]}=(\alpha_1,\ldots,\alpha_n)$, and $\beta|_{[n]}=(\beta_1,\ldots,\beta_n)$. (The notation $\alpha|_{[n]}$ indicates the projection of $\alpha=(\alpha_1,\ldots,\alpha_N)$ to the components labeled by $[n]$, and similarly for $\beta|_{[n]}$.) With the same notation as in Section 2.2, this explicit description gives

$$\Pr(\mathbf{Y}_N=\mathbf{y};\rho,\theta,\alpha,\beta)=\frac{\exp\left\{\rho\sum_{1\le i<j\le N}y_{ij}y_{ji}+\theta y_{\bullet\bullet}+\sum_{i=1}^N\alpha_i y_{i\bullet}+\sum_{j=1}^N\beta_j y_{\bullet j}\right\}}{\prod_{1\le i<j\le N}\eta_{ij}},$$

for each $\mathbf{y} \in \{0,1\}^{N\times N}$, and

$$\Pr(\mathbf{Y}_n = \mathbf{y}; \rho, \theta, \alpha|_{[n]}, \beta|_{[n]}) = \\ = \frac{\exp\left\{\rho \sum_{1\le i<j\le n} y_{ij}y_{ji} + \theta y_{\bullet\bullet} + \sum_{i=1}^n \alpha_i y_{i\bullet} + \sum_{j=1}^n \beta_j y_{\bullet j}\right\}}{\prod_{1\le i<j\le n} \eta_{ij}},$$

for each $\mathbf{y} \in \{0,1\}^{n\times n}$, where η_{ij} is defined in (2.7). Under the further assumption that the data is sampled from $\mathbf{Y}_N$ by selection $\mathbf{S}_{n,N}$, the implicit model for $\mathbf{Y}_n$ is given by the induced distribution of $\mathbf{S}_{n,N}\mathbf{Y}_N$ as in (3.7),

$$\begin{aligned} \Pr(\mathbf{S}_{n,N}\mathbf{Y}_N = \mathbf{y}) &= \Pr(\mathbf{Y}_N \in \mathbf{S}_{n,N}^{-1}(\mathbf{y})) \\ &= \sum_{\mathbf{y}^* \in \{0,1\}^{N\times N}: \mathbf{y}^*|_{[n]} = \mathbf{y}} \Pr(\mathbf{Y}_N = \mathbf{y}^*). \end{aligned}$$

To prove consistency, we must establish (3.9) for the p_1 model on $\{0,1\}^{n\times n}$ and the distribution induced by selection sampling from the p_1 model on $\{0,1\}^{N\times N}$. We can see this for $n = N-1$ and $\mathbf{y} \in \{0,1\}^{(N-1)\times(N-1)}$ by computing

$$\begin{aligned} &\Pr(\mathbf{S}_{N-1,N}\mathbf{Y}_N = \mathbf{y}; \rho, \theta, \alpha, \beta) = \\ &= \sum_{\mathbf{y}^* \in \{0,1\}^{N\times N}: \mathbf{y}^*|_{[N-1]} = \mathbf{y}} \frac{\exp\left\{\rho \sum_{1\le i<j\le N} y^*_{ij}y^*_{ji} + \theta y^*_{\bullet\bullet} + \sum_{i=1}^N \alpha_i y^*_{i\bullet} + \sum_{j=1}^N \beta_j y^*_{\bullet j}\right\}}{\prod_{1\le i<j\le N} \eta_{ij}} \\ &= \sum_{(y^*_{iN}, y^*_{Ni}) \in \{0,1\}\times\{0,1\}, i=1,\ldots,N-1} \Big(\Pr(\mathbf{Y}_{N-1} = \mathbf{y}; \rho, \theta, \alpha|_{[N-1]}, \beta|_{[N-1]}) \times \\ &\quad \times \exp\left\{\rho \sum_{i=1}^{N-1} y^*_{iN}y^*_{Ni} + \theta\Big(\sum_{i=1}^{N-1}(y^*_{iN} + y^*_{Ni})\Big) + \sum_{i=1}^{N-1} \alpha_i y^*_{iN} + \alpha_N y^*_{N\bullet} + \right. \\ &\quad \left. + \sum_{j=1}^{N-1} \beta_j y^*_{Nj} + \beta_N y^*_{\bullet N}\right\} \Big/ \prod_{i=1}^{N-1} \eta_{iN}\Big) \\ &= \sum_{(y^*_{iN}, y^*_{Ni}) \in \{0,1\}\times\{0,1\}, i=1,\ldots,N-1} \Big(\Pr(\mathbf{Y}_{N-1} = \mathbf{y}; \rho, \theta, \alpha|_{[N-1]}, \beta|_{[N-1]}) \times \\ &\quad \times \prod_{i=1}^{N-1} e^{\rho y^*_{iN}y^*_{Ni} + \theta(y^*_{iN} + y^*_{Ni}) + (\alpha_i + \beta_N)y^*_{iN} + (\beta_i + \alpha_N)y^*_{Ni}} \Big/ \prod_{i=1}^{N-1} \eta_{iN}\Big) \\ &= \Pr(\mathbf{Y}_{N-1} = \mathbf{y}; \rho, \theta, \alpha|_{[N-1]}, \beta|_{[N-1]}) \times \\ &\quad \times \left(\prod_{i=1}^{N-1} \sum_{y^*_{iN}, y^*_{Ni} \in \{0,1\}} e^{\rho y^*_{iN}y^*_{Ni} + \theta(y^*_{iN} + y^*_{Ni}) + (\alpha_i + \beta_N)y^*_{iN} + (\beta_i + \alpha_N)y^*_{Ni}}\right) \Big/ \prod_{i=1}^{N-1} \eta_{iN} \\ &= \Pr(\mathbf{Y}_{N-1} = \mathbf{y}; \rho, \theta, \alpha|_{[N-1]}, \beta|_{[N-1]}) \times \\ &\quad \times \left(\prod_{i=1}^{N-1} (1 + e^{\theta + \alpha_i + \beta_N} + e^{\theta + \alpha_N + \beta_i} + e^{\rho + 2\theta + \alpha_i + \alpha_N + \beta_i + \beta_N})\right) \Big/ \prod_{i=1}^{N-1} \eta_{iN} \\ &= \Pr(\mathbf{Y}_{N-1} = \mathbf{y}; \rho, \theta, \alpha|_{[N-1]}, \beta|_{[N-1]}), \end{aligned} \tag{3.10}$$

establishing that

$$\Pr(\mathbf{S}_{N-1,N}\mathbf{Y}_N = \mathbf{y}; \rho, \theta, \alpha, \beta) = \Pr(\mathbf{Y}_{N-1} = \mathbf{y}; \rho, \theta, \alpha|_{[N-1]}, \beta|_{[N-1]})$$

for all $\mathbf{y} \in \{0,1\}^{(N-1)\times(N-1)}$, where $\alpha|_{[N-1]} = (\alpha_1, \ldots, \alpha_{N-1})$ is the restriction of α to its first $N-1$ elements, and similarly for $\beta|_{[N-1]}$, as defined above. Thus, $\mathbf{Y}_N$ from the p_1 model with parameter $(\rho, \theta, \alpha, \beta)$ and $\mathbf{Y}_{N-1}$ from the p_1 model with parameter $(\rho, \theta, \alpha|_{[N-1]}, \beta|_{[N-1]})$ are consistent under selection (Definition 3.1) for all $N \geq 2$. For arbitrary $n \leq N$, we deduce that $\mathbf{Y}_N$ and $\mathbf{Y}_n$ are consistent under selection by noticing that the composition of selection maps is again a selection map: since $[m] \subseteq [n] \subseteq [N]$ for all $m \leq n \leq N$, the composition $\mathbf{S}_{m,n} \circ \mathbf{S}_{n,N}$ corresponding to first selecting $[n]$ from $[N]$ and then selecting $[m]$ from $[n]$ is equivalent to selecting $[m]$ from $[N]$, i.e.,

$$\mathbf{S}_{m,n} \circ \mathbf{S}_{n,N} = \mathbf{S}_{m,N} \quad \text{for all } m \leq n \leq N.$$

In particular, each $\mathbf{S}_{n,N}$ decomposes as

$$\mathbf{S}_{n,N} = \mathbf{S}_{n,n+1} \circ \mathbf{S}_{n+1,n+2} \circ \cdots \circ \mathbf{S}_{N-1,N},$$

and the calculation for N and $N-1$ in (3.10) is enough to establish consistency of the p_1 model for all $N \geq n \geq 1$.

Exercise 3.1 *Prove consistency under selection of the p_1 model 'without calculation'. (Hint: Use independence.)*

3.3 Significance of sampling consistency

The role of sampling consistency in statistical inference is closely related to the discussion of Section 3.1, in which the relationship between the candidate distributions for the population and those for the sampled data proved critical when extending inferences about the observed data to the population. In Section 3.2.1, the population structure $\mathbf{Y}_N$ obeys the p_1 model with parameter $(\rho, \theta, \alpha, \beta)$ ranging over all permissible values. If we want to infer the population parameter ρ based on an observation $\mathbf{Y}_n$ obtained from $\mathbf{Y}_N$ by selection sampling, then for each possible choice of $(\rho, \theta, \alpha, \beta)$ the calculation in (3.10) shows that $\mathbf{Y}_n = \mathbf{S}_{n,N}\mathbf{Y}_N$ follows the p_1 model with parameter $(\rho, \theta, \alpha|_{[n]}, \beta|_{[n]})$. This calculation establishes the relationship between the parameter ρ governing the data $\mathbf{Y}_n$ and the parameter ρ governing the population $\mathbf{Y}_N$. The consistency under selection property of the p_1 model ensures that the parameters marked 'ρ' in the models for $\mathbf{Y}_n$ and $\mathbf{Y}_N$ have a common meaning. Since the meaning of ρ, as a reciprocity parameter, is maintained under selection sampling from $\mathbf{Y}_N$, an estimate $\hat{\rho}_n$ for ρ based on $\mathbf{Y}_n$ can also be used to estimate the population parameter ρ in the distribution of $\mathbf{Y}_N$.

This same logic cannot be applied for inferences from models that lack consistency. By now it is well known, for example, that the class of ERGMs with natural parameter θ and canonical sufficient statistic T as in (2.8) is consistent under selection only if T has *separable increments*. In essence, the separable increments property limits the subclass of ERGMs that are consistent under selection to those models whose sufficient statistics only account for local (i.e., pairwise) or global structure. Sufficient statistics that account for intermediate-range properties, such as clustering, do not have separable increments and, therefore, cannot be incorporated into a consistent family of ERGMs. (See [138, p. 513] for further details about ERGMs and the separable increments property.) For example, suppose $\mathbf{Y}_N$ is modeled by the ERGM on $\{0,1\}^{N\times N}$ given by

$$\Pr(\mathbf{Y}_N = \mathbf{y}; \theta) \propto \exp\{\theta \Delta_N(\mathbf{y})\}, \quad \mathbf{y} \in \{0,1\}^{N\times N}, \tag{3.11}$$

for $\theta \in (-\infty, \infty)$ and sufficient statistic

$$\Delta_N(\mathbf{y}) = \sum_{1\leq i\neq j\neq k\leq N} y_{ij} y_{jk} y_{jk}, \quad \mathbf{y} \in \{0,1\}^{N\times N}, \tag{3.12}$$

which counts the number of triangles in $\mathbf{y}$. Suppose also that $\mathbf{Y}_n$, $n < N$, is modeled by the ERGM having the same form on $\{0,1\}^{n\times n}$, so that

$$\Pr(\mathbf{Y}_n = \mathbf{y}; \theta) \propto \exp\{\theta \Delta_n(\mathbf{y})\}, \quad \mathbf{y} \in \{0,1\}^{n\times n}. \tag{3.13}$$

And further assume that the data $\mathbf{Y}_n$ is obtained from $\mathbf{Y}_N$ by selection sampling, as defined in (3.6). Since the subsampled observation $\mathbf{S}_{n,N}\mathbf{Y}_N$ is obtained by sampling from $\mathbf{Y}_N$, whose distribution is parameterized by θ, it follows that the distribution of $\mathbf{S}_{n,N}\mathbf{Y}_N$ is also parameterized by θ. But since the sufficient statistic in (3.12) does not satisfy the separable increments condition, the main theorems in [138] imply that $\mathbf{S}_{n,N}\mathbf{Y}_N \neq_{\mathscr{D}} \mathbf{Y}_n$. Without a clearly articulated connection between the parameter θ governing $\mathbf{Y}_N$ and that governing $\mathbf{Y}_n$, there is no logical way to relate inferences about θ based on the sample $\mathbf{Y}_n$ to inferences about θ for the population $\mathbf{Y}_N$.

This lack of sampling consistency is not unique to the ERGM. It arises often in network modeling, but has rarely been given much attention, especially in mathematical statistics, where there is a tendency to conflate sampling scheme (as a description of the way in which the observed data relates to the population) with asymptotic regime (as a description of the theoretical growth rate as a function of sample size). The distinction is important, both when formulating an appropriate model for a given application and when interpreting theoretical results in the context of such an application. This lack of clarity in the relationship between asymptotic regimes studied in theoretical work and sampling schemes that arise in practice inhibits the use of the theory for gaining reliable practical insights. In defense of ERGMs, one may argue that the sociometric setting in which ERGMs were initially introduced and have historically been used is not primarily concerned with inferences from subsampled networks. From this perspective, the lack of sampling consistency is not a poignant criticism of the ERGM, but is rather grounds to criticize the use of the ERGM in applications to which it is not well-suited.

3.3.1 *Toward a coherent framework for network modeling*

Though I have spoken in some depth here about selection sampling, I do not mean to suggest that consistency under selection is the be all and end all of sound network analysis. In many applications, selection sampling is far from a realistic description of the sampling mechanism, and thus the logical relationship between population and sample established through the selection map, though theoretically precise, may not accurately reflect the relationship between population and sample in the real world. In such cases, the assumed logical connection between population and sample may not be useful for gleaning insights about the intended application, precisely because such inferences will have been drawn under the false premise that selection sampling accurately models the sampling mechanism.

As we see below, selection is just one of many possible ways to subsample network data, and rarely is it the best sampling description for any given context. Because selection sampling of vertices is untenable for most networks applications, the study of sampling properties for ERGMs, as in [138], and the discussion of selection sampling or projectivity properties found elsewhere in the literature have little bearing on the most pressing practical and theoretical matters facing present-day statistical network analysis. Such results provide little if any guide for how to analyze network data, design network models that properly account for sampling, or glean worthwhile insights in the bulk of situations to which network analysis is now applicable.

The remainder of this and the coming two chapters is dedicated to developing a coherent framework for network analysis with the following three essential observations in mind:

(i) sampling is an indispensible part of network modeling,

(ii) the relationship between observed and unobserved data established by the sampling mechanism is crucial for statistical inference, and

(iii) the nature of this relationship and the reason why it is important have not been properly emphasized in the developments of network analysis to date.

The next two sections describe two specific scenarios in which selection sampling is inadequate.

3.4 Selection from sparse networks

For N very large, assume that $\mathbf{Y}_N = (Y_{ij})_{1\le i,j\le N}$ represents a network with N vertices and $\sum_{1\le i,j\le N} Y_{ij} \approx \varepsilon N$ edges, for some constant $\varepsilon > 0$ that does not depend on N. Suppose that the N vertices are labeled $1,\ldots,N$ according to a uniform random assignment and $\mathbf{Y}_n$ is obtained by selecting a relatively small number of $n \ll N$ vertices from $\mathbf{Y}_N$. (For this example, we assume that the diagonal is 0, i.e., $Y_{ii} \equiv 0$.)

Since the vertices are labeled uniformly, the distribution of $\mathbf{Y}_N = (Y_{ij})_{1\le i,j\le N}$ is exchangeable with respect to rearranging its rows and columns, i.e.,

$$\mathbf{Y}_N^{\sigma} = (Y_{\sigma(i)\sigma(j)})_{1\le i,j\le N} =_{\mathscr{D}} \mathbf{Y}_N \quad \text{for all permutations } \sigma : [N] \to [N],$$

and, moreover, the sample $\mathbf{Y}_n = \mathbf{S}_{n,N}\,\mathbf{Y}_N = (Y_{ij})_{1\le i,j\le n}$ is exchangeable with respect to all permutations $\sigma : [n] \to [n]$.[4] (A thorough treatment of such exchangeable network models can be found in Chapter 6.) By this symmetry, we can compute the

[4] For any $n \ge 1$, a *permutation* of $[n]$ is a one-to-one and onto (i.e., bijective) map $\sigma : [n] \to [n]$. For

marginal probability that there is an edge between the vertices labeled 1 and 2 as

$$\Pr(Y_{12}=1) \approx \varepsilon N/(N(N-1)) \approx \varepsilon/N. \qquad (3.14)$$

(Uniform labeling of vertices makes Y_{12} a uniformly chosen off-diagonal entry of $\mathbf{Y}_N$. There are $N(N-1)$ total off-diagonal entries, of which approximately εN are nonzero.) Also by exchangeability, $Y_{ij} =_{\mathscr{D}} Y_{12}$ (marginally) for all $1 \leq i \neq j \leq n$, meaning that the same calculation in (3.14) applies to each Y_{ij}. Putting these calculations together allows us to bound the probability that $\mathbf{Y}_n$ has at least 1 nonzero entry by

$$\Pr\left(\bigcup_{1\leq i\neq j\leq n} \{Y_{ij}=1\} \right) \leq \sum_{1\leq i\neq j\leq n} \Pr(Y_{ij}=1) \approx n^2\varepsilon/N. \qquad (3.15)$$

Under the assumption that $n \ll \sqrt{N}$, so that $n^2 \ll N$, the probability that $\mathbf{Y}_n$ is non-empty must be close to 0, giving high probability to an empty (and therefore uninformative) observation.

Exercise 3.2 *Before moving on, the reader is encouraged to ponder the practical implications of the calculation in* (3.15). *(Hint: I discuss one possible issue in the next paragraph.)*

From a practical point of view, the trouble with calculation (3.15) is that many networks observed in the real world are 'sparse'[5] in the sense of having few edges relative to the number of vertices (i.e., $\sum_{1\leq i,j\leq N} Y_{ij} \approx \varepsilon N$ for small $\varepsilon > 0$) but are nevertheless well-connected by some complex arrangement of their edges. In such situations, the observed network is 'sparse' but non-empty—had we observed an empty network, we would have no data to model in the first place. We are tasked with modeling the sparse structure in $\mathbf{Y}_n$ for the purpose of drawing inferences about $\mathbf{Y}_N$. But in the setting assumed above, the network obtained by selection sampling is empty with high probability, making any observation '$\mathbf{Y}_n = \mathbf{y}$', for $\mathbf{y}$ non-empty, an event with negligible probability under any assumed model.

This analysis can be carried out prior to seeing the data: we know *a priori* that any network we analyze will be non-empty—for otherwise there would be nothing to analyze—and we know (in many cases) that the population network from which the data will be sampled is 'sparse' (in the above sense of having on the order of εN edges). Thus, without even looking at the data, we can determine that any model for $\mathbf{Y}_n$ induced by selection sampling will assign almost all of its probability to the nonsense event that the observed network is empty. Under the conventional logic of hypothesis testing, this observation would be enough to reject the hypothesis that the

our purposes, any permutation σ represents a relabeling of the units. In this section, the units are vertices. Later, they will be edges, hyperedges, paths, etc. See Section 3.7 for further discussion about statistical units in network analysis.

[5]The term 'sparse' is being used in a loose, imprecise sense in this section. This heuristic notion of 'sparse', which is commonly used in applications of network analysis, interprets the asymptotic property of sparsity as meaning that a (large) finite network has 'few' edges relative to the number of vertices. The technical definition of *sparse* involves a limiting statement based on a network with infinitely many vertices and/or edges. See Chapter 4.2 for further discussion of sparsity.

data $\mathbf{Y}_n$ follows the assumed model. But since the above calculation holds for any assumed distribution of the sparse population network $\mathbf{Y}_N$—even a distribution which assigns probability 1 to the 'true network' $\mathbf{Y}_N$—the assumption that $\mathbf{Y}_n$ has been obtained by selection sampling must be called into question, and a theory of network modeling that accounts for more realistic sampling schemes must be developed.

3.5 Scenario: Ego networks in high school friendships

Staying in the context of the high school friendship network from Section 2.4, suppose that the population network $\mathbf{Y}_N$ is modeled by the Erdős–Rényi–Gilbert distribution with parameter $0 \leq \theta \leq 1$ on $\{0,1\}^{N\times N}$ and 0 diagonal:

$$\Pr(\mathbf{Y}_N = \mathbf{y}; \theta) = \prod_{1 \leq i \neq j \leq N} \theta^{y_{ij}}(1-\theta)^{1-y_{ij}}, \quad \mathbf{y} \in \{0,1\}^{N\times N}. \tag{3.16}$$

[6]

Suppose that the observed data $\mathbf{Y}^*$ is obtained by first sampling 1 student v^* uniformly at random and then observing $\mathbf{Y}^* = \mathbf{Y}_N|_S$ for a set $S \subseteq [N]$ containing v^* along with all other students v such that $Y_{v^*v} = 1$ or $Y_{vv^*} = 1$. (Foreshadowing Section 3.6.2, we call this the *one-step snowball sampling* operation.) In this case, the number of observed vertices is a random quantity which depends on the number of friends of a randomly chosen vertex v^*. As it is immediately clear that this sampling operation differs from selection, we should therefore expect the model for $\mathbf{Y}^*$ to differ from that for $\mathbf{S}_{n,N}\mathbf{Y}_N$.

For any $\theta \in [0,1]$, the distribution of $\mathbf{Y}^*$ is induced by applying the one-step snowball sampling operation to $\mathbf{Y}_N$ distributed according to (3.16). In this case, $\mathbf{Y}^*$ is a random graph with a random number of vertices $V = 1 + B$, where B is a Binomial random variable with $N-1$ trials and success probability θ, i.e.,

$$\Pr(B = k; \theta) = \binom{N-1}{k} \theta^k (1-\theta)^{N-1-k}, \quad k = 0, \ldots, N-1.$$

We represent the observed network $\mathbf{Y}^*$ by labeling the distinguished vertex v^* by 0 and, on the event '$B \geq 1$', labeling the other B vertices arbitrarily $1, \ldots, B$. In this way, the observation $\mathbf{Y}^*$ is given by the symmetric array $(Y_{ij})_{0 \leq i,j \leq B}$ with $Y_{i0} = Y_{0i} = 1$ for $i = 1, \ldots, B$ and each Y_{ij}, $1 \leq i \neq j \leq B$, given by an i.i.d. draw from the Bernoulli distribution with parameter θ.

Exercise 3.3 *Formally derive the distribution of* $\mathbf{Y}^*$.

The straightforward description of this sampling mechanism, with v^* chosen uniformly at random and all edges determined by i.i.d. draws from the Bernoulli distribution, permits the explicit description of the model for $\mathbf{Y}^*$ given above. But if the

---

[6] The Erdős–Rényi–Gilbert model is so classical in the study of random graphs that it is sometimes referred to simply as the 'random graph model'. The distribution in (3.16) describes a random graph for which the edge between each pair of vertices $1 \leq i \neq j \leq N$ is determined by the toss of an independent coin flip with heads probability θ. This model was initially introduced in the late 1950s [68, 81] and has since been studied in some detail, see, e.g., [21]. For our purposes, its simple description is sufficient for demonstrating the most salient aspects of network modeling. This same simplicity, however, makes the model poorly suited to modeling most networks encountered in practice.

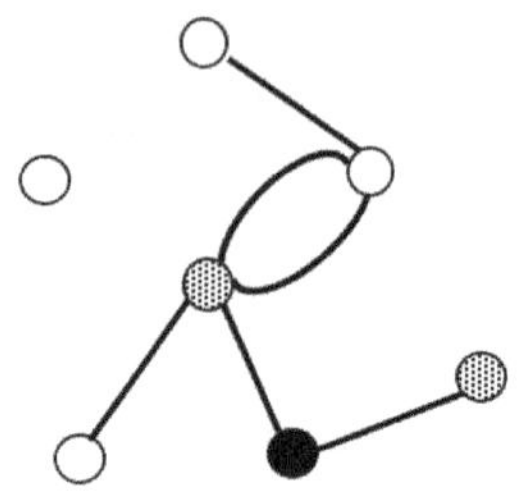

Figure 3.1 *Depiction of one-step snowball sampling operation in Section 3.5. The solid filled vertex (bottom right) corresponds to the randomly chosen vertex v^* and those partially filled with dots are its one-step neighborhood. Only the solidly filled and dotted vertices together with their corresponding edges are observed under this scheme.*

observation mechanism were even slightly more complicated, e.g., if v^* were chosen from a degree-biased distribution or if $\mathbf{Y}^*$ were obtained by two-step snowball sampling from v^*, then such a straightforward description of $\mathbf{Y}^*$ may not be possible.

Research Problem 3.1 *Let $\mathbf{Y}_r^*$ be the graph obtained by applying the r-step snowball sampling operation to a randomly chosen vertex in $\mathbf{Y}_N$. (See Section 3.6.2 for a formal definition of snowball sampling.) First compute the distribution of $\mathbf{Y}_r^*$ under the assumption that v^* is chosen uniformly at random from $\mathbf{Y}_N$ following the Erdős–Rényi–Gilbert distribution in* (3.16). *After warming up with this preliminary case, assume a different model for $\mathbf{Y}_N$, e.g., the graphon model from Chapter 6 or the Hollywood model from Chapter 9, and let v^* be chosen from a non-uniform distribution on the vertices, e.g., degree-biased. Once again compute the distribution of $\mathbf{Y}_r^*$ for $r \geq 1$. If computing the distribution of $\mathbf{Y}_r^*$ is intractable, then describe the distribution of $\mathbf{Y}_r^*$ in any way possible, e.g., by computing the moments or distribution of certain network statistics.*

3.6 Network sampling schemes

We have already seen that some models, namely the Erdős–Rényi–Gilbert and p_1 models, are consistent under selection sampling, while others, e.g., ERGMs whose sufficient statistics lack separable increments, are not. But since selection sampling is rarely an adequate description of the actual sampling mechanism in most networks applications, the question arises: how can inferences be drawn from network data obtained under other, more realistic, sampling schemes? I treat this question in depth in Chapter 5. For the remainder of this chapter, I survey some common network sampling schemes, all of which can be handled within the paradigm of Chapter 5.

As I mentioned in Section 1.7.3, understanding the impact of sampling on network analysis is a topic of great importance to future developments in the field. The sampling schemes surveyed here are just a few of the many network sampling mechanisms that arise in practical applications. But for the sake of highlighting the main challenges and laying down the foundational principles of statistical network modeling, the sampling methods given below should suffice. See Section 3.10 for additional references on network sampling.

3.6.1 Relational sampling

Many network datasets are constructed by sampling relationships and interactions among individuals in a population. In such cases, the primary unit of observation is the interaction/relation (i.e., edge) and the main object of interest is the structure induced by the observed interactions or relations. It is important to note that the characteristics of networks observed by such relational sampling can differ substantially from networks obtained by vertex sampling, as in Section 2.4. If, for example, a network is sampled by intercepting phone calls that pass through a switchboard, then the resulting network structure will be biased toward individuals who interact more frequently than others (because they are more likely to appear in the sample). This example is just one special case of *relational sampling*, which I use here as a generic term for network data obtained by directly sampling *relations* among individuals instead of sampling the individuals first and then observing the relations between them (as in selection sampling). We have previously observed several other examples of networks sampled in this way, e.g., in Sections 1.6.4–1.6.5.

Our observation here that in many networks applications the relations are primitive and the vertices are derivative (i.e., a byproduct of the relational sampling scheme) goes against conventional wisdom in the statistical networks literature. As Handcock and Gile [85, p. 7] write, "Note that in most network samples, the unit of sampling is the actor or node, while the unit of analysis is typically the dyad." To the contrary, in the applications mentioned above, the 'dyads' are the primary units of sampling while the entire structure determined by the dyads is the unit of analysis. In addition to phone call networks, relational sampling is the natural observation mechanism for networks constructed from email or social media communications, professional collaborations, and paths between Internet servers. A basic theory for modeling such networks has been developed in [53, 54] and will be summarized further in Chapters 9 and 10.

3.6.1.1 Edge sampling

Consider a database of telephone calls in which each entry contains a unique identifier (say, a phone number) for the caller and receiver along with other information about the interaction, such as time of call, topic discussed, etc. An illustration of such a database is given in Table 3.1. By ignoring all information except caller and receiver, a sample of n calls from this database results in a sequence $(C_1,R_1),\ldots,(C_n,R_n)$, with C_i identifying the caller of the ith sampled call and R_i the receiver. In a typical network representation of this data, the set of sampled elements $\{C_1,R_1,\ldots,C_n,R_n\}$ determines the vertices and each (C_i,R_i) determines a directed edge from the vertex corresponding to C_i to that corresponding to R_i. (Figure 3.2 illustrates the network constructed by sampling the first four rows from Table 3.1, with vertices labeled *a-e* as indicated in the table.)

Table 3.1 *Database of phone calls. Each row contains information about a single phone call: caller and receiver (identified by phone number), time of call, topic discussed, etc.*

Caller	Receiver	Time of Call	Topic Discussed	...
555-7892 (a)	555-1243 (b)	15:34	Business	...
550-9999 (c)	555-7892 (a)	15:38	Birthday	...
555-1200 (d)	445-1234 (e)	16:01	School	...
555-7892 (a)	550-9999 (c)	15:38	Sports	...
555-1243 (b)	555-1200 (d)	16:17	Business	...
⋮	⋮	⋮	⋮	⋱

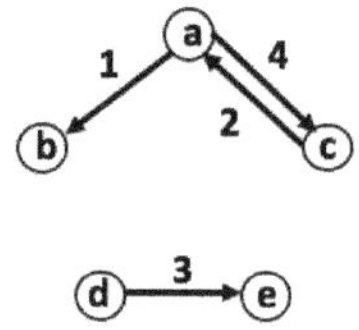

Figure 3.2 *Network depiction of phone call sequence of caller-receiver pairs (a,b), (c,a), (d,e), (a,c) as in the first four rows of Table 3.1. Edges are labeled in correspondence with the order in which the corresponding calls were observed.*

The network in Figure 3.2 differs from the binary arrays discussed so far in several important respects.

- Whereas the high school social network in Section 2.4 was obtained by observing all relationships among $n \geq 1$ selected students (i.e., vertices), the network in Figure 3.2 corresponds to a sample of 4 phone calls (i.e., edges) and the vertices involved in those sampled calls. In the former scenario, the sample size is the number of vertices; in the latter, it is the number of edges.
- In the high school network, it is assumed that the presence or absence of all possible edges between sampled vertices is observed. In this way, the observation of two vertices i and j without an edge between them indicates that students i and j are not friends. But in the phone call dataset, the lack of an edge between two observed vertices does not mean that the two have not communicated or that they do not appear in the database together; it only indicates that any such interactions have not been sampled. In Table 3.1, for example, the fifth row contains a phone call between vertices b and d. But even though there is an edge between b and d in the population database, e.g., in row 5 of Table 3.1, no such call was among the four entries sampled, and thus no such interaction is reflected in the observed network structure of Figure 3.2.
- Since the phone call database can, and likely does, contain multiple entries of different calls between the same caller-receiver pairs, multiple observations of the

Table 3.2 *Database of movies and actors. Each row contains the set of actors in the corresponding movie.*

Movie title	Starring cast
Rocky	Sylvester Stallone, Bert Young, Carl Weathers, ...
Rounders	Matt Damon, Ed Norton, John Malkovich, John Turturro, ...
Groundhog Day	Bill Murray, Andie McDowell, Chris Elliott, ...
A Bronx Tale	Robert DeNiro, Chazz Palminteri, Joe Pesci, ...
Over the Top	Sylvester Stallone, Robert Loggia, ...
⋮	⋮

same interaction are possible, as we see in the appearance of two edges between vertices a and c in Figure 3.2.

Since each phone call (i.e., edge) involves exactly two individuals (i.e., vertices), this example does not seem to stray too far from the prominent 'networks-as-graphs' mindset that pervades the networks literature (see Section 1.2). Nevertheless, the change in observation mechanism brought about by edge sampling makes a substantial difference in the observed network characteristics and proper modeling approach. This scenario and the change of perspective it demands of network analysis motivates the framework of edge exchangeability discussed in Chapter 9.

3.6.1.2 Hyperedge sampling

The phone call database of the previous section is a specific kind of *interaction data.* Many other interaction datasets, which record, e.g., email communications and professional collaborations, arise similarly, and thus exhibit many of the same essential characteristics as the network shown in Figure 3.2. But unlike the database of phone calls, these other interactions need not be restricted to pairs of individuals. Consider, for example, the Internet Movie Database (IMDb),[7] for which each entry contains information about a different movie. For the time being we disregard all information about each movie except its cast of actors, as shown in Table 3.2. Sampling from this database results in a sequence of movie casts $M_1,\dots,M_n$, where each M_i is the set of actors who appear in the ith movie sampled. For each i, M_i may be either an ordered or unordered set: ordered $(M_{i,1},\dots,M_{i,R_i})$ if actors are listed, say, according to the standing of their role (lead role, second role, etc.) and unordered $\{M_{i,1},\dots,M_{i,R_i}\}$ otherwise. In either case, R_i is the (random) number of roles in movie i.

Sampling academic articles from a research repository, such as arXiv,[8] bioRxiv,[9] Social Science Research Network (SSRN),[10] or the Philosophy of Science Archive (PhilSci-Archive),[11] results in a similar data structure to the one constructed by sampling from the IMDb. (See Table 3.3 for illustration.) In this case, each article is

[7] http://www.imdb.com/
[8] http://www.arxiv.org/
[9] http://www.biorxiv.org/
[10] http://www.ssrn.com/
[11] http://philsci-archive.pitt.edu/

Table 3.3 *Database of statistics articles. Each row contains the set of authors of the corresponding article.*

Article title	Authors
A nonparametric view of network models ...	Bickel, Chen
Edge exchangeable models for interaction networks	Crane, Dempsey
Snowball sampling	Goodman
Latent space approaches to social network analysis	Hoff, Raftery, Handcock
⋮	⋮

identified by its set of authors $A_1,\ldots,A_n$. Note that the authors of any given article may be listed in a meaningful order, as in those scientific fields for which the relative order of authors reflects their contribution to the project, or in alphabetical order, as is the norm in mathematics and law. These conventions result in a dataset consisting of ordered or unordered sets of authors, respectively.

In an email database, each email has a unique sender S_i and a set of recipients R_i. Recipients may also be classified further according to whether they were cc'd or bcc'd, so that each sampled email (S_i,R_i,C_i,B_i) consists of sender S_i, set of recipients R_i, set of cc'd recipients C_i, and set of bcc'd recipients B_i.

Each of the above examples describes a specific instance of *hyperedge sampling*. Though not as amenable to visualization, the sample of movie collaborations (or coauthorships or email exchanges) can be conceptualized as an edge-labeled hypergraph (akin to Figure 3.2) with each sampled movie (or article or email) represented as a labeled hyperedge. This representation is directly analogous to the edge-labeled graph representation of phone calls in Figure 3.2, and therefore can be treated within the same conceptual framework. See Section 10.3 for further discussion of networks constructed in this way.

I mention here, and stress again in Chapters 9 and 10, that the structural integrity of network data ought to be maintained to every extent possible, meaning that each hyperedge corresponding to a movie, article, or email should be treated as a single entity by the postulated statistical model. In particular, a hyperedge consisting of three vertices, say $\{a,b,c\}$, ought to be treated in the analysis as a single hyperedge with three vertices. Without a compelling reason to do so, such an interaction should not be decomposed into three binary edges $\{a,b\}$, $\{b,c\}$, and $\{a,c\}$. This basic point is violated far and wide in conventional network analysis. See Sections 9.10 and 10.2 for further discussion.

3.6.1.3 *Path sampling*

For the purpose of this discussion, the (physical) Internet consists of servers (i.e., vertices) and connections between servers along which messages are transmitted (i.e., edges). A guiding motivation in the earliest days of network science was to determine what the Internet network 'looks like' by analyzing the paths traversed when sending information from one part of the Internet to another. Traceroute is one specific sampling method used for this purpose.

```
traceroute to                    (128.135.10.17), 64 hops max, 72
byte packets
 1  fios_quantum_gateway (192.653.22.69) 2.557 ms  3.073 ms  3.881
ms
 2  lo0-100.nyp-sec-4513 -319.comcast-xxg .net ( 158.19.2130 )  5.677
ms  15.916 ms  15.397 ms
 3  b3319.  nyp-sèc -21.comcast-xxg .net (100.41.209.120)  16.386
ms  10.418 ms  16.390 ms
 4  * * *
 5  0.ae3.br2.nyc4.alter.net (140.222.231.133)  13.368 ms  9.816
ms  13.792 ms
 6  204.255.168.110 (204.255.168.110)  15.426 ms  28.583 ms
10.595 ms
 7  be2061.ccr42.jfk02.atlas.cogentco.com (154.54.3.69)  13.331 ms
15.715 ms  15.677 ms
 8  be2890.ccr22.cle04.atlas.cogentco.com (154.54.82.245)  34.393
ms  30.023 ms  26.264 ms
 9  be2718.ccr42.ord01.atlas.cogentco.com (154.54.7.129)  32.745
ms  29.979 ms  28.970 ms
10  be2522.agr21.ord01.atlas.cogentco.com (154.54.81.62)  38.000
ms  32.219 ms  36.152 ms
11  te0-0-2-0.nr11.b010917-1.ord01.atlas.cogentco.com
(154.24.4.38)  48.702 ms  33.046 ms  29.155 ms
12  38.104.103.10 (38.104.103.10)  28.439 ms  35.973 ms  31.425 ms
13  192.170.192.19 (192.170.192.19)  34.475 ms  31.105 ms  28.312
ms
14  192.170.192.27 (192.170.192.27)  29.062 ms  71.833 ms  28.353
ms
15  * * *
16                    (128.135.10.17)  31.506 ms  30.992 ms
35.041 ms
```

Figure 3.3 *Example of traceroute path between IP addresses 192.653.22.69 and 128.135.10.17.*

Given a source s and target t, each identified uniquely by its Internet Protocol (IP) address, traceroute returns the path (i.e., "traces the route") of servers visited in accessing t from s. Thus, for each pair (s,t), traceroute returns an ordered tuple $\text{path}(s,t) = (s, v_1, \ldots, v_k, t)$, which indicates a path from s to v_1, then to v_2, and so on until reaching t. Traceroute also includes other information, such as the time required to traverse the path, which we ignore for this discussion; see Figure 3.3 for an illustration of traceroute output between two IP addresses, or consult [2] for more details on traceroute sampling.

If interested in the structure of the Internet network based on traceroute output, we can imagine seeding the algorithm with a collection of sources and targets $(s_1,t_1),\ldots,(s_n,t_n)$ and then sampling n paths $\text{path}(s_1,t_1),\ldots,\text{path}(s_n,t_n)$ by traceroute, where $\text{path}(s_i,t_i)$ denotes the path traced from s_i to t_i. Assuming that $\text{path}(s,t)$ is deterministic for any given source s and target t reduces any randomness in the sample to the randomness in how the sources and targets have been chosen. This in turn raises conceptual questions regarding the nature of the information contained in the observed network. For example, if given two sources s,s' and a random sample of targets $T_1,\ldots,T_n$, how does the structure determined by the sample of paths $\{\text{path}(s,T_i)\}_{1\leq i\leq n}$ compare to that determined by $\{\text{path}(s',T_i)\}_{1\leq i\leq n}$? Answers to these questions depend on how the respective vertices s and s' are situated

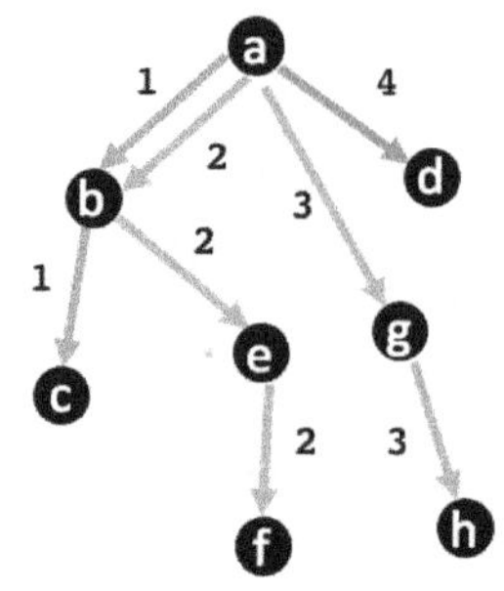

Figure 3.4 *Path-labeled network constructed from sequence* $path(a,c) = (a,b,c)$, $path(a,f) = (a,b,e,f)$, $path(a,h) = (a,g,h)$, *and* $path(a,d) = (a,d)$. *Edges are labeled according to which path they belong. For example, the three edges labeled '2' should be regarded as comprising a single path, namely* $path(a,f) = (a,b,e,f)$, *and not as three distinct edges* $(a,b),(b,e),(e,f)$.

within the network, which by the nature of the investigation is likely to be poorly understood. Another question might be: How does the manner in which the targets $T_1,\ldots,T_n$ are 'randomly' chosen affect the structure determined by the sampled paths $\{\text{path}(s,T_i)\}_{1\leq i\leq n}$?

These basic questions about networks built from path sampling motivate some of the most fundamental considerations in network science [154]. Several critical questions arise from the tendency to overlook the ways in which sampling can distort observed network properties and other unintended consequences of the conventional 'networks-as-graphs' representation. The collection of paths $\{\text{path}(s,T_i)\}_{1\leq i\leq n}$ obtained by traceroute sampling from a vertex s to a sample of target vertices $T_1,\ldots,T_n$ can be represented as a network with edges labeled according to which path they belong, as in Figure 3.4. As mentioned in Section 3.6.1.2 in connection to hyperedges, it is important to treat each path as a single entity, so that the collection of all edges with a given label comprise a single path in the network representation. For example, the path (a,b,e,f) labeled '2' in Figure 3.4 should not be decomposed into its constituent edges (a,b), (b,e), and (e,f). In particular, the structure of the path $a \to b \to e \to f$ cannot be properly analyzed without taking into account that the first traversal $a \to b$ depends on the target vertex f. I discuss other aspects of modeling path-sampled networks in Chapter 10.

3.6.2 *Snowball sampling*

Returning to the high school friendship example in Section 2.4, consider now the sampling procedure which first chooses a student s at random and then observes that student's *ego network* of radius r, for some fixed $r = 1,2,\ldots$. (Figure 3.1 illustrates this sampling scheme for $r = 1$.) More formally, we observe a network of friendships by initializing $N(s,0) = \{s\}$ and first sampling the set of students $N(s,1) = \{s' : Y_{ss'} = 1\}$, i.e., those students whom s identifies as a friend. If $r > 1$, then we sample friends of friends to obtain

$$N(s,2) = \bigcup_{s' \in N(s,1)} \{s'' : Y_{s's''} = 1\} - N(s,1),$$

and in general

$$N(s,k) = \bigcup_{s' \in N(s,k-1)} \{s'' : Y_{s's''} = 1\} - \bigcup_{j=0}^{k-1} N(s,j).^{12}$$

We define $\texttt{snow}(s,r)$ as the set of neighborhoods $\{N(s,k)\}_{1 \le k \le r}$, so that $s' \in N(s,k)$ implies that the shortest friendship path between s and s' has length k.

Similar to path sampling, snowball sampling proceeds by seeding an initial source vertex s along with a radius $r = 1,2,\dots$ describing the width of the neighborhood to be sampled. A snapshot of the network can be obtained by piecing together $\texttt{snow}(S_1,R_1),\dots,\texttt{snow}(S_n,R_n)$ for a sample of sources and radii $(S_1,R_1),\dots,(S_n,R_n)$. In this sense, snowball sampling can be regarded as yet another special case of relational sampling.

3.7 Units of observation

In statistical analysis, the sampling scheme is closely related to the observational *units*, a concept which arises often in the experimental design literature. In a designed experiment, the *units* are defined as the smallest entity to which different treatments can be assigned. (To capture this idea of 'indivisibility' with respect to the treatment, some authors use the term 'atom' instead of 'unit'.) In network analysis, we regard the units as the basic entities of observation, i.e., the 'atomic elements' from which the network structure is constructed. For example, in the scenario of Section 2.4, a social network of high school students is obtained by observing friendships among n selected students. In that case, the units are the vertices, and the observation associated to each unit is its friendships with other units (as represented by the edges of a graph). But a moment's reflection on the different ways that real-world networks arise should make clear that the vertices are not the units in many applications of contemporary interest. In the scenario of Section 3.6.1.1, for example, the observed network is constructed from a sample of binary interactions, i.e., edges, and thus the basic units of observation are the edges. In hyperedge sampling, each observation is a hyperedge, i.e., a multi-way interaction among vertices, and thus the hyperedges are the units. In path sampling, the paths are the units. And so on.

---

[12]For two sets A and B, $B - A$ denotes the set containing all elements of B which are not in A. This is sometimes written as $B \setminus A$ and is defined by

$$B - A = B \setminus A = B \cap A^c,$$

where A^c is the complement of A.

The network sampling schemes described in Section 3.6 highlight the many ways in which entities other than the vertices can serve as the units, thus contradicting the aforementioned claim of Handcock and Gile that "in most network samples, the unit of sampling is the actor or node" [85, p. 7]. This misconception is perpetuated by the rote application of vertex-centric models (i.e., random graph models) found throughout the networks literature. (See Sections 1.2–1.5 for further elaboration on this point.) But while identifying the units is an important starting point for any application, it is not enough to simply acknowledge what the units are. The units must also be properly integrated into the model specification. If, for instance, the observational units are the edges (as in Section 3.6.1.1) but the formal model setup and subsequent analysis implicitly treats the vertices as the units, then any ensuing inferences will have been conducted under an (unspoken) false assumption and subsequent conclusions are prone to be interpreted in the wrong context.

Implicit and explicit units

The fact that the units specified (implicitly) by the model may be different from the units identified (explicitly) by the observation mechanism marks a subtle distinction between *explicit* and *implicit* units which warrants careful consideration in network analysis. The *explicit units* are the 'real' or 'actual' units of a given application, in the sense of being the basic entities of the (real-world) observation process. Because the explicit units are determined by the scenario under which the data is actually observed, and thus are not affected by the model setup or subsequent analysis, I often refer to the explicit units simply as 'the units'. For example, the explicit units in Section 2.4 are vertices, in the phone call scenario of Section 3.6.1.1 are edges, in the coauthorship scenario of Section 3.6.1.2 are hyperedges, in the traceroute sampling scenario of Section 3.6.1.3 are paths, and in the snowball sampling scenario of Section 3.6.2 are neighborhoods of radius r.

On the other hand, there is a notion of units that is *implicit* in any formulation of a network model. Most often, when a network is represented as a graph with vertex set V and edge set $E \subseteq V \times V$, or equivalently as a $\{0,1\}$-valued array with vertices corresponding to the rows and columns, the vertices are the *implicit units*, in the sense that the resulting 'random graph model' implicitly treats the vertices as the basic unit of observation. (I discuss this further in Chapter 6.) Just as the real-world context under which the data is observed determines the explicit units, the theoretical context in which the model is specified determines the implicit units. Because the objective of modeling is to describe the real world in such a way that inferences based on the model are as meaningful as possible, it is a basic requirement of sound statistical modeling that the implicit and explicit units should align. This observation motivates the paradigm of Chapter 5 and the modeling frameworks in Chapters 7–11.

3.8 What is the sample size?

The question of sample size is one of the oldest in statistical network analysis, and yet it remains poorly understood. It was once common to think of an observation of network data as 'a sample of size 1'. (In fact, this point of view is still held by some networks researchers.) The inevitable next question arises: how is one to conduct reliable statistical inference based on a sample of size one? It is clear from the numerous successful applications of statistical methods to network data that such inferences are possible. Just as clear, after reflection on the contents of this chapter, and in particular the preceding section, is that the sample size of network data is not the number of *networks* which have been observed (most often 1), but rather the number of *units* which have been observed in constructing the network. So while the sample size is 1 in certain situations, it is not 1 across the board. As I have emphasized in previous sections (e.g., Section 1.4), context plays a major role.

To appreciate the absurdity of the 'sample size 1' viewpoint in many applications, it is helpful to consider the more standard setup involving an i.i.d. sequence $\mathbf{X} = (X_1, X_2, \ldots)$, from which the initial segment $\mathbf{X}_n = (X_1, \ldots, X_n)$ is observed. Here it seems uncontroversial that the sample size of $\mathbf{X}_n$ is the number of observed measurements, namely n. But if applying the same rationale which takes network data as a 'sample of size 1' (because only *one* network is observed), then $\mathbf{X}_n$ could also be viewed as a sample of size 1, since only *one* sequence $(X_1, \ldots, X_n)$ has been observed. Of course, the reason why $(X_1, \ldots, X_n)$ is a sample of size n is that each X_i, $i = 1, \ldots, n$, is an independent draw from a common distribution, and the observation $\mathbf{X}_n = (X_1, \ldots, X_n)$ consists of n such observations.

Conversely, suppose that $\mathbf{X}_n = (X_1, \ldots, X_n)$ instead represents the result of a single draw from an urn with balls labeled $1, \ldots, n$. If the chosen ball has label i, then the components of $\mathbf{X}_n$ are given by $X_i = 1$ and $X_j = 0$ for all $j \neq i$. In particular, if the chosen ball is labeled 1, then $\mathbf{X}_n = (1, 0, 0, \ldots, 0)$; if the chosen ball is labeled 2, then $\mathbf{X}_n = (0, 1, 0, \ldots, 0)$; and so on. As in the previous example, the sequence $\mathbf{X}_n$ has length n. But since $\mathbf{X}_n$ reflects just a single draw from the urn, this length n sequence represents a sample of size 1.

As these two examples make clear, the sample size is intrinsic to the application. It cannot be manipulated or altered by arbitrary choices of how the data is represented. In the previous paragraph, a single draw from the urn could be represented either as a length n sequence, as described above, or as a single observation, say, $Y = i$ if the chosen ball has label i. Each represents the same observation, and thus both correspond to a sample of size 1. In the earlier sample of n i.i.d. observations $X_1, \ldots, X_n$, both a sequence $\mathbf{X}_n = (X_1, \ldots, X_n)$ and an array $\mathbf{Y}_n = (Y_{ij})_{1 \leq i,j \leq n}$, with $Y_{ii} = X_i$ and $Y_{ij} = 0$ for $i \neq j$, represent the same observation, and thus both correspond to a sample of size n.

Stated most simply:

the sample size is the number of observed units.

So while it is true in many cases that the data consists of one network, this single network is most often the result of repeated draws from the observation process. And it is this repetition which makes reliable inferences possible. To parallel the discussion of units from the previous section, the sample size in the scenario of Section 2.4 is the number of sampled vertices, in the phone call scenario of Section 3.6.1.1 is the number of sampled phone calls (or edges), in the coauthorship scenario of Section 3.6.1.2 is the number of sampled articles (or hyperedges), in the traceroute scenario of Section 3.6.1.3 is the number of sampled paths, and in the snowball sampling scenario of Section 3.6.2 is the number of sampled neighborhoods.

Exercise 3.4 *Describe a scenario under which a non-trivial network (i.e., non-empty with $n \geq 5$ vertices) can be rightly regarded as a sample of size 1.*

3.9 Consistency under subsampling

The above discussion of units and sample size, especially in light of the different sampling schemes of Section 3.6, suggests the following refinement to the concept of consistency under selection (Definition 3.1). Specializing again to networks represented as $\{0,1\}$-valued arrays, we interpret any injection $\psi : [n] \to [N]$, $1 \leq n \leq N$, as a vertex sampling scheme described by subsampling vertices $\psi(1), \dots, \psi(n)$ along with their incident edges as follows. Given $\psi : [n] \to [N]$, define the operation of *ψ-selection from* $\{0,1\}^{N\times N}$ by the action

$$\begin{aligned} \mathbf{S}^{\psi}_{n,N} : &\{0,1\}^{N\times N} \to \{0,1\}^{n\times n} \\ &\mathbf{y} \mapsto \mathbf{y}^{\psi} = (y_{\psi(i)\psi(j)})_{1\leq i,j\leq n}. \end{aligned} \tag{3.17}$$

Notice that the selection map $\mathbf{S}_{n,N}$ defined in (3.6) coincides with ψ-selection for $\psi : [n] \to [N]$ given by the inclusion map $\psi(i) = i$ for $i = 1, \dots, n$. For example, with $n = 3$, $N = 4$, and $\mathbf{y}$ given by

$$\mathbf{y} = \begin{pmatrix} 0 & 1 & 0 & 1 \\ 1 & 0 & 1 & 1 \\ 0 & 1 & 0 & 0 \\ 1 & 1 & 0 & 0 \end{pmatrix},$$

$\psi : [3] \to [4]$ with $\psi(1) = 2$, $\psi(2) = 4$, and $\psi(3) = 1$ acts on $\mathbf{y}$ by

$$\mathbf{y}^{\psi} = \begin{array}{c} \psi(1) \\ \psi(2) \\ \psi(3) \end{array} \overset{\begin{array}{ccc} \psi(1) & \psi(2) & \psi(3) \end{array}}{\begin{pmatrix} 0 & 1 & 1 \\ 1 & 0 & 1 \\ 1 & 1 & 0 \end{pmatrix}}.$$

With $\mathbf{S}^{\psi}_{n,N}$ as in (3.17), define

$$\mathscr{S}_{n,N} = \{\mathbf{S}^{\psi}_{n,N} \mid \psi : [n] \to [N] \text{ an injection}\}$$

as the set of all ψ-selection maps, which we interpret as the set consisting of all ways to sample n units from $[N]$ according to the above mechanism. A *random sampling scheme* $\Sigma_{n,N}$ is a ψ-selection map chosen randomly from $\mathscr{S}_{n,N}$. To reflect that real-world networks are often sampled in a way that depends on the population network (e.g., in relational sampling), we allow the distribution of $\Sigma_{n,N}$ to depend on $\mathbf{Y}_N$. Thus, for any $\mathbf{Y}_N$ in $\{0,1\}^{N\times N}$ and a random sampling scheme $\Sigma_{n,N}$, $\Sigma_{n,N}\mathbf{Y}_N$ denotes an array obtained by applying a randomly chosen ψ-selection map to $\mathbf{Y}_N$. The distribution induced on $\{0,1\}^{n\times n}$ is computed by

$$\begin{aligned}
\Pr(\Sigma_{n,N}\mathbf{Y}_N=\mathbf{y}) &= \\
&= \sum_{\mathbf{y}^*\in\{0,1\}^{N\times N}}\sum_{\psi:[n]\to[N]}\Pr(\Sigma_{n,N}=\mathbf{S}^{\psi}_{n,N}\mid\mathbf{Y}_N=\mathbf{y}^*)\Pr(\mathbf{Y}_N=\mathbf{y}^*)\mathbf{1}(\mathbf{y}^{*\psi}=\mathbf{y}) \\
&= \sum_{\psi:[n]\to[N]}\sum_{\mathbf{y}^*\in\{0,1\}^{N\times N}:\mathbf{y}^{*\psi}=\mathbf{y}}\Pr(\Sigma_{n,N}=\mathbf{S}^{\psi}_{n,N}\mid\mathbf{Y}_N=\mathbf{y}^*)\Pr(\mathbf{Y}_N=\mathbf{y}^*), \quad (3.18)
\end{aligned}$$

where

$$\mathbf{1}(\mathbf{y}^{*\psi}=\mathbf{y})=\begin{cases}1, & \mathbf{y}^{*\psi}=\mathbf{y},\\ 0, & \text{otherwise},\end{cases}$$

denotes the indicator function.

Definition 3.2 (Consistency under subsampling) *Let* $\mathbf{Y}_n$ *and* $\mathbf{Y}_N$, $n\le N$, *be random* $\{0,1\}$*-valued arrays and let* $\Sigma_{n,N}$ *be a random sampling operation in* $\mathscr{S}_{n,N}$. *Then* $\mathbf{Y}_n$ *and* $\mathbf{Y}_N$ *are* consistent under subsampling from $\Sigma_{n,N}$, *or alternatively* consistent with respect to $\Sigma_{n,N}$ *or* $\Sigma_{n,N}$-consistent, *if*

$$\Sigma_{n,N}\mathbf{Y}_N=_{\mathscr{D}}\mathbf{Y}_n, \quad (3.19)$$

for $\Sigma_{n,N}\mathbf{Y}_N$ *as distributed in* (3.18).

Notice that *consistency under selection* (Definition 3.1) is a special case of Definition 3.2 for a random sampling operation $\Sigma_{n,N}$ with degenerate distribution at $\mathbf{S}_{n,N}$, i.e., $\Pr(\Sigma_{n,N}=\mathbf{S}_{n,N})=1$. When surveying the literature, however, the reader should note that the terms *consistency under selection* and *consistency under subsampling* are often used interchangeably, and treated as synonymous with the term *projectivity*. Here I reserve the more general term *consistency under subsampling* for the property in Definition 3.2.

With few exceptions, e.g., [52], the networks literature is mostly confined to the more limited definition of consistency under selection from Section 3.2. As the above discussion emphasizes the relevance of sampling schemes other than selection, the implications of more generic sampling mechanisms for network analysis is an important topic for future research. To my knowledge, Definition 3.2 has not appeared previously in the networks literature or otherwise. The mere statement of this definition opens a number of possible avenues of research which curious readers are encouraged to explore.

Subsampling in arbitrary networks

Here I have defined consistency under subsampling only in the special case of networks represented as $\{0,1\}$-valued arrays. The variety of sampling schemes discussed in Section 3.6 emphasizes the need to expand this definition even further to handle networks represented, for example, as edge-labeled, hyperedge-labeled, or path-labeled graphs. But until we encounter these specific cases in Chapters 9 and 10, it is safe to proceed with the following generic, albeit vague, understanding of ψ-selection for arbitrary networks.

Let the units of observation be well-defined and let $\mathcal{N}_n$ denote the set of all networks with n units labeled $1,\ldots,n$. Except in pathological cases, there should be a well-defined notion of *restriction to* $\mathcal{N}_m$ (analogous to the operation in (3.6) when the vertices are units) which acts on $\mathbf{y} \in \mathcal{N}_n$ by projecting to the network labeled by units $1,\ldots,m$. If this action is well-defined, then so is ψ-selection: for any injection $\psi : [m] \to [n]$, relabel units $\psi(1),\ldots,\psi(m)$ as $1,\ldots,m$, respectively, and then apply the restriction map. If this exposition was too quick, the reader need not worry. Further details on generic sampling schemes are given as needed.

3.10 Further reading

Statistical investigations into network sampling have mostly been confined to the literature on social network analysis, for which the works of Frank [75, 76, 77] are good references. To the best of my knowledge, there has been little discussion of statistical units and sample size in more general treatments of network analysis, with one notable exception being Krivitsky and Kolaczyk's early work on effective sample size in networks [110]. Kolaczyk [106, Chapter 5] also discusses sampling in his broader coverage of network analysis. As this book was nearing completion, a more recent study of network sampling and invariance principles from an algorithmic perspective appeared in [127]. Comparing the results of [127] to the forthcoming discussion in Chapters 5–10 could prove worthwhile for the reader interested in these foundational issues.

Although sampling issues in network analysis have received increasing attention in the relevant conferences and workshops of late, the topic remains marginalized or ignored in the broader literature. There have been some efforts by computer scientists and physicists, e.g., [4, 112, 154], to understand the impact of sampling on observed network structure. But aside from a few recent efforts to better understand the effects of sampling and missing data on network analysis, e.g., [52, 85, 102, 105, 114, 138, 162], the implications of network sampling have been mostly overlooked by statisticians. I especially highlight [102, 114], and references therein, for relevant recent work on network-driven and respondent-driven sampling. I single out [112, 154] as essential reading about the potentially substantial effects of sampling on observed network attributes.

3.11 Solutions to exercises

3.11.1 Exercise 3.1

Consistency under selection for the p_1 model can be proven 'without calculation' by noting that the p_1 model is a special case of the dyad independence model (Section 2.2). In the dyad independence model, the pairs (Y_{ij}, Y_{ji}), $1 \leq i < j \leq N$, in $\mathbf{Y}_N$ are independent, and thus the dyads (Y_{ij}, Y_{ji}), $1 \leq i < j \leq N-1$, that determine the distribution of the sampled array $\mathbf{S}_{N-1,N}\mathbf{Y}_N$ are independent of (Y_{iN}, Y_{Ni}), $1 \leq i \leq N-1$. It follows that the distribution of $\mathbf{Y}_{N-1}$ constructed from the independent dyads (Y_{ij}, Y_{ji}), $1 \leq i < j \leq N-1$, coincides with the distribution of $\mathbf{S}_{N-1,N}\mathbf{Y}_N$

obtained by first taking independent dyads (Y_{ij}, Y_{ji}), $1 \leq i < j \leq N$, and then disregarding (Y_{iN}, Y_{Ni}), for all $1 \leq i \leq N-1$. This proves the more general result that the dyad independence model is consistent under selection. Consistency of the p_1 model follows by noting that it is a special case of the dyad independence model.

3.11.2 Exercise 3.2

This exercise is open-ended. The observation that selection sampling from a sparse graph produces an empty graph with high probability has a number of practical implications. An immediate, and very serious, implication is discussed in the text following the statement of this exercise. The reader is encouraged to identify other issues by thinking more deeply about this question.

3.11.3 Exercise 3.3

The network $\mathbf{Y}^*$ in this exercise is obtained by first generating $\mathbf{Y}_N$ according to the Erdős–Rényi–Gilbert distribution with parameter θ on N vertices, as in (3.16), and then sampling one vertex v^* uniformly at random along with the subgraph consisting of all vertices adjacent to v^* in $\mathbf{Y}_N$. To express the distribution of $\mathbf{Y}^*$ formally, we condition first on the random number B of vertices in the one-step neighborhood of v^*. Since v^* is chosen independently of $\mathbf{Y}_N$, B is equal in distribution to the number of edges adjacent to any given vertex in the Erdős–Rényi–Gilbert distribution, and thus follows the binomial distribution with success probability θ on $N-1$ trials. Suppose $B = k \geq 1$ and label the vertices adjacent to v^* by $i_1, \ldots, i_k$. Then, by the independence assumption of the Erdős–Rényi–Gilbert distribution, each edge $Y_{i_r i_s}$ between vertices i_r and i_s, $1 \leq r, s \leq k$, is an independent draw from the Bernoulli distribution with parameter θ. Combining these observations together by conditioning on the value of B and using the law of total probability gives the distribution of $\mathbf{Y}^*$ by

$$
\begin{aligned}
\Pr(\mathbf{Y}^* = \mathbf{y}) &= \sum_{k=0}^{N-1} \Pr(B = k) \times \Pr(\mathbf{Y}^* = \mathbf{y} \mid B = k) \\
&= \sum_{k=0}^{N-1} \binom{N-1}{k} \theta^k (1-\theta)^{N-1-k} \times \Pr(\mathbf{Y}^* = \mathbf{y} \mid B = k),
\end{aligned}
$$

for

$$
\Pr(\mathbf{Y}^* = \mathbf{y} \mid B = k) = \prod_{1 \leq i \neq j \leq k} \theta^{y_{ij}} (1-\theta)^{1-y_{ij}} \times \prod_{i=1}^{k} y_{0i} y_{i0}, \quad \mathbf{y} = (y_{ij})_{0 \leq i,j \leq k},
$$

where the product

$$
\prod_{1 \leq i \neq j \leq k} \theta^{y_{ij}} (1-\theta)^{1-y_{ij}}
$$

is the usual Erdős–Rényi–Gilbert probability governing the connections of the vertices adjacent to v^* and

$$
\prod_{i=1}^{k} y_{0i} y_{i0}
$$

evaluates to 1 only if all of the vertices $1,\ldots,k$ are adjacent to 0 in $\mathbf{y}$ (i.e., $y_{i0} = y_{0i} = 1$), as is required by the one-step snowball sampling scheme with v^* assigned label 0. In the above expression, we adopt the convention $\prod_{1\leq i\neq j\leq 0} \theta^{y_{ij}}(1-\theta)^{1-y_{ij}} = 1$ and $\prod_{i=1}^{0} y_{i0}y_{0i} = 1$ in the event that $B = 0$.

3.11.4 Exercise 3.4

There are a number of possible answers to this exercise. One possible solution is below. The reader is encouraged to come up with other examples.

Consider a network obtained by sampling a high school uniformly at random from among all U.S. high schools with more than 5 students. Having chosen a specific high school, let $\mathbf{Y} = (Y_{ij})_{1\leq i,j\leq N}$ be the binary array representing all friendships among the N students in that school. Then $\mathbf{Y}$ is a binary $N \times N$ array reflecting a single observation from the sampling process, which chooses 1 high school and observes the social network among students within that high school. The sampling unit in this case is the high school, and $\mathbf{Y}$ is a sample of size 1 from this process.

Chapter 4

Generative models

The sampling schemes of the previous chapter arise most naturally when modeling a partially observed network, i.e., for $\mathbf{Y}_n$ obtained by sampling from a population network $\mathbf{Y}_N$. Generative models are relevant in the complementary context in which the observation $\mathbf{Y}_n$ is assumed to be evolving according to some generating process. Thus, instead of specifying the model by describing how observations of smaller size are obtained by sampling from one of larger size, generative models describe how networks evolve (i.e., grow larger) according to a random mechanism.

4.1 Specification of generative models

In parallel to the formulation of sampling models, which we specified in Chapter 3 by describing the *sampling mechanism* according to which an observation $\mathbf{Y}_n$ is obtained from a population network $\mathbf{Y}_N$, a generative model is specified by a *generating mechanism* that describes network evolution. For $n \leq N$, call $P : \{0,1\}^{n\times n} \to \{0,1\}^{N\times N}$ an *evolution map* if

$$P(\mathbf{y})|_{[n]} = \mathbf{y} \quad \text{for all } \mathbf{y} \in \{0,1\}^{n\times n}. \tag{4.1}$$

In words, an evolution map is an operation by which $\mathbf{y} \in \{0,1\}^{n\times n}$ 'evolves' into $P(\mathbf{y}) \in \{0,1\}^{N\times N}$ by holding fixed that part of the network which already exists, namely $\mathbf{y}$. For $n \leq N$, let $\mathscr{P}_{n,N}$ be the set of all evolution maps $\{0,1\}^{n\times n} \to \{0,1\}^{N\times N}$ and define a *generating scheme* as a random map $\Pi_{n,N}$ chosen randomly according to some probability distribution on $\mathscr{P}_{n,N}$. We allow the distribution of $\Pi_{n,N}$ to depend on the input $\mathbf{Y}_n$.

Given a random array $\mathbf{Y}_n$ and a generating scheme $\Pi_{n,N}$, we write $\Pi_{n,N}\mathbf{Y}_n$ to denote the random element of $\{0,1\}^{N\times N}$ obtained by applying the generating scheme $\Pi_{n,N}$ to a realization of $\mathbf{Y}_n$. More precisely, $\Pi_{n,N}\mathbf{Y}_n$ is the network with N vertices obtained by first generating $\mathbf{Y}_n$ and, given $\mathbf{Y}_n = \mathbf{y}$, putting $\Pi_{n,N}\mathbf{Y}_n = P(\mathbf{y})$, for $P \in \mathscr{P}_{n,N}$ chosen according to the conditional distribution of $\Pi_{n,N}$ given $\mathbf{Y}_n = \mathbf{y}$. The distribution of $\Pi_{n,N}\mathbf{Y}_n$ is computed by

$$\Pr(\Pi_{n,N}\mathbf{Y}_n = \mathbf{y}) = \sum_{P\in\mathscr{P}_{n,N}} \Pr(\Pi_{n,N} = P \mid \mathbf{Y}_n = \mathbf{y}|_{[n]}) \Pr(\mathbf{Y}_n = \mathbf{y}|_{[n]}) \mathbf{1}(P(\mathbf{y}|_{[n]}) = \mathbf{y}), \tag{4.2}$$

where $\mathbf{1}(\cdot)$ is the indicator function.

Definition 4.1 (Generative consistency) *Let $\mathbf{Y}_n$ and $\mathbf{Y}_N$ be random $\{0,1\}$-valued arrays and let $\Pi_{n,N}$ be a generating scheme. Then $\mathbf{Y}_n$ and $\mathbf{Y}_N$ are* consistent with respect to $\Pi_{n,N}$ *if*

$$\Pi_{n,N}\mathbf{Y}_n =_{\mathscr{D}} \mathbf{Y}_N, \tag{4.3}$$

for $\Pi_{n,N}\mathbf{Y}_n$ defined by the distribution in (4.2).

Among the most attractive features of generative models is that they allow for the distributions of all finite sample networks $(\mathbf{Y}_n)_{n\geq 1}$ to be defined inductively in a way that guarantees generative consistency. For example, if given a generating scheme $\Pi_{n,N}$ and a network $\mathbf{Y}_n$, then $\mathbf{Y}_N$ can be constructed so that it automatically satisfies (4.3) by defining its distribution as in (4.2). This built-in consistency property at least partially explains the popularity of generative models in Bayesian nonparametrics and machine learning. But, like sampling models, a generative model is only as useful as its assumed generating mechanism is a realistic description of network evolution.

For any $\mathbf{Y}_n$ and generating mechanism $\Pi_{n,N}$, define $\mathbf{Y}_N$ by $\mathbf{Y}_N = \Pi_{n,N}\mathbf{Y}_n$. Then by the defining property (4.1) of an evolution map, $\mathbf{Y}_n$ and $\mathbf{Y}_N$ enjoy the relationship

$$\mathbf{S}_{n,N}\mathbf{Y}_N = \mathbf{S}_{n,N}\Pi_{n,N}\mathbf{Y}_n = \mathbf{Y}_n \quad \text{with probability 1;} \tag{4.4}$$

that is, $\mathbf{Y}_n$ and $\Pi_{n,N}\mathbf{Y}_n$ are consistent under selection by default. (Note that (4.4) follows from (i) $(\Pi_{n,N}\mathbf{Y}_n)|_{[n]} = \mathbf{Y}_n$ with probability 1 by the definition of an evolution map (4.1) and (ii) $\mathbf{S}_{n,N}\mathbf{y} = \mathbf{y}|_{[n]}$ by definition of the selection map in (3.6).) Thus, when a generative model is treated instead as a sampling model, it is automatically consistent under selection. But it is important not to misconstrue this mathematical consequence as a generic endorsement or justification of selection as a way to model how real networks are sampled. The choice to use a generative model in place of a sampling model reflects the perspective from which the network is being analyzed. Generative models treat the observed network as evolving, while sampling models treat the observed network as having been sampled from a population network. In the former case, the nature of network evolution must be incorporated into the inference; in the latter case, interest lies in drawing inferences about the population based on a sample.

4.2 Generative model 1: Preferential attachment model

The Barabási–Albert preferential attachment model [14], abbreviated here as the BA model, was proposed to explain the prevalence of specific empirical properties, namely sparsity and power law degree distribution, which have been found in real-world networks from a range of disciplines. Its generating dynamics are based on Simon's preferential attachment scheme [139] for producing heavy tailed (i.e., power law) distributions. In a nutshell, under the dynamics of the BA model, a network evolves by adding one new vertex at each step, with each new vertex attaching to existing vertices *preferentially* according to their degree. In this way, high-degree vertices tend to attract more connections, in a phenomenon sometimes called the 'rich get richer' or the 'Matthew effect'.[1]

The formal description of the BA model takes $m \geq 1$ (integer) and $\delta > -m$ (real number) so that at each step a new vertex appears and attaches randomly to m existing vertices with probability proportional to the degree of the vertex offset by the parameter δ. The process is initiated at a graph $\mathbf{y}_0$ with $n_0 \geq 1$ vertices, which then evolves successively into $\mathbf{y}_1, \mathbf{y}_2, \ldots$ by, at each step, connecting a new vertex to the existing graph according to the following rule. For any symmetric $\{0,1\}$-valued adjacency matrix $\mathbf{y} = (y_{ij})_{1 \leq i,j \leq n}$ and for every $i = 1, \ldots, n$, define the *degree* of i in $\mathbf{y}$ to be the number of edges incident to i,

$$\deg_{\mathbf{y}}(i) = \sum_{j \neq i} y_{ij}.$$

At step $n \geq 1$, a new vertex v_n attaches to $m \geq 1$ vertices in $\mathbf{y}_{n-1}$, with each of the m vertices v' chosen independently without replacement with probability proportional to $\deg_{\mathbf{y}_{n-1}}(v') + \delta/m$, where $\deg_{\mathbf{y}_{n-1}}(v')$ is the *degree* of v' in $\mathbf{y}_{n-1}$, i.e., the number of edges involving v' in $\mathbf{y}_{n-1}$.[2] In keeping with the notation of Section 4.1, let $\Pi_{k,n}^{\delta,m}$, $k \leq n$, denote the generating mechanism for the process parameterized by $m \geq 1$ and $\delta > -m$.

By letting the parameters $n_0 \geq 1$, $m \geq 1$, and $\delta > -m$ vary over all permissible values and treating the initial conditions $\mathbf{y}_0$ and n_0 as fixed, the above generating mechanism determines a family of distributions for each finite sample size $n \geq 1$, where n is the number of vertices that have been added to $\mathbf{y}_0$. (In particular, an observation $\mathbf{y}$ with n vertices corresponds to a sample of size $n - n_0$, since the n_0 initial vertices are not assumed to be part of the observation process.)[3] For each $n \geq 1$, this process gives a collection of distributions $\mathcal{M}_n$ indexed by (m, δ), and each distribution in $\mathcal{M}_k$ indexed by (m, δ) is related to a distribution in $\mathcal{M}_n$, $n \geq k$, with the same parameters through the preferential attachment scheme $\Pi_{k,n}^{\delta,m}$ associated to the model. For any choice of parameter (δ, m), we express the relationship between $\mathbf{Y}_k$ and $\mathbf{Y}_n$, $n \geq k$, by

$$\mathbf{Y}_n =_{\mathcal{D}} \Pi_{k,n}^{\delta,m} \mathbf{Y}_k.$$

[1]The Matthew effect is so named because its behavior coincides with a principle from the Gospel of Matthew: "For to everyone who has will more be given, and he will have an abundance. But from the one who has not, even what he has will be taken away." (Matthew 25:29, The Bible, English Standard Version, 2001.)

[2]For simplicity, I have described a version of the model by which all $m \geq 1$ edges involving v_n are chosen independently according to the same distribution, and therefore the resulting graph can, and with high probability will, have multiple edges. There are several variants of this scheme, e.g., whereby the m edges are chosen sequentially so that the vertex degrees in the sampling distribution are updated within each step and the m vertices connected to v_n are sampled without replacement to avoid multiple edges. For discussion of these variations see [36, 146].

[3]To be more precise, the n_0 initial vertices are observed, but we do not assume that they reflect properties of the generating mechanism, and therefore we do not regard them as observational units.

Empirical properties

A major selling point of the BA model is its ability to replicate sparsity and power law degree distributions through its easily interpretable preferential attachment generating mechanism. Sparsity and power law are properties of the degree distribution of an infinitely large or growing sequence of networks defined as follows. For each $n \geq 1$, let $\mathbf{y}^{(n)} = (y_{ij}^{(n)})_{1 \leq i,j \leq n}$ be a graph with n vertices, and consider a sequence $\mathbf{y} = (\mathbf{y}^{(n)})_{n \geq 1}$ of graphs that are growing in size. The collection $\mathbf{y}$ is called *sparse* if

$$\lim_{n\to\infty} \frac{1}{n(n-1)} \sum_{1 \leq i \neq j \leq n} y_{ij}^{(n)} = 0. \tag{4.5}$$

By the generating dynamics of the BA model, there are exactly $m \geq 1$ new edges added at each step, so that after n steps there are exactly $mn + k_0$ edges in $\mathbf{Y}_n$, where k_0 is the fixed number of edges in the initial graph $\mathbf{y}_0$. From this we easily see that the rescaled edge density satisfies

$$\frac{1}{n(n-1)}(mn + n_0) \to 0 \quad \text{as } n \to \infty$$

and the sequence of graphs $(\mathbf{Y}_n)_{n\geq 1}$ generated from the BA model is sparse with probability 1.

The *degree distribution of* $\mathbf{y}$ counts the relative proportion of vertices with each integer degree:

$$p_{\mathbf{y}}(k) = \sum_{i=1}^{n} \mathbf{1}(\deg_{\mathbf{y}}(i) = k), \quad k = 0, 1, \ldots.$$

A sequence of networks $\mathbf{y} = (\mathbf{y}^{(n)})_{n\geq 1}$ exhibits *power law degree distribution with exponent* γ if its degree distributions satisfy

$$p_{\mathbf{y}^{(n)}}(k) \sim k^{-\gamma} \quad \text{for all large } k \text{ as } n \to \infty, \tag{4.6}$$

for some $\gamma > 1$, where $a(k) \sim b(k)$ indicates that $a(k)/b(k) \to 1$ as $k \to \infty$. More precisely, $(\mathbf{y}^{(n)})_{n\geq 1}$ has power law with exponent γ if

$$\lim_{k\to\infty} \lim_{n\to\infty} \frac{p_{\mathbf{y}^{(n)}}(k)}{k^{-\gamma}} = 1.$$

With a little more work, one can also see that the generating mechanism $\Pi_{k,n}^{\delta,m}$ gives rise to a sequence of networks $(\mathbf{Y}_n)_{n\geq 1}$ whose degree distributions satisfy (4.6) with exponent $3 - \delta/m$. See [146] for a formal derivation.

Both sparsity and power law are asymptotic properties defined for an infinite collection of networks $(\mathbf{y}^{(n)})_{n\geq 1}$. Although the components of $(\mathbf{y}^{(n)})_{n\geq 1}$ need not be related to one another in any obvious way (e.g., $\mathbf{y}_m = \mathbf{S}_{m,n}\mathbf{y}_n$ need not hold for $m \leq n$), these properties are best understood by envisioning $(\mathbf{y}^{(n)})_{n\geq 1}$ as the finite components of an 'infinite size' population network $\mathbf{y} = (y_{ij})_{i,j\geq 1}$. From this population network, we define each $\mathbf{y}^{(n)}$ as the restriction of $\mathbf{y}$ to its first n labeled vertices,

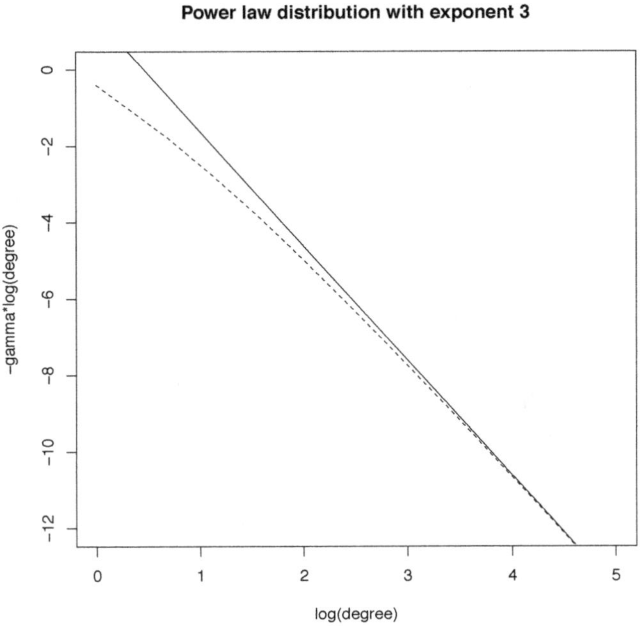

Figure 4.1 *Dotted line shows log-log plot of the Yule–Simon distribution in* (4.8) *for* $\gamma = 3$. *Solid line shows the linear approximation of* (4.8) *by approximating* $\Gamma(\gamma)/\Gamma(k+\gamma) \sim \gamma^{-k}$, *which holds asymptotically for large values of k.*

so that $\mathbf{y}^{(n)} = \mathbf{y}\,|_{[n]}$ for each $n \geq 1$. In this way, the sparsity and power law conditions in (4.5) and (4.6), respectively, can be interpreted as empirical properties of a limiting population network.

Also, since sparsity and power law are asymptotic properties, neither can be verified from any finite observation of a network. In practice, however, the limiting statements in (4.5) and (4.6) are often interpreted to hold for networks that are 'large' (as opposed to infinite). A 'large' network $\mathbf{y} = (y_{ij})_{1 \leq i,j \leq n}$ is often called *sparse* if its edge density is judged to be 'small' relative to its size, i.e.,

$$n^{-2} \sum_{1 \leq i,j \leq n} y_{ij} \approx 0.$$

Note that 'large' and 'small' have no precise meaning here. They must instead be interpreted heuristically. Similarly, the power law property (4.6) is often judged by comparing the empirical degree distribution of a network $\mathbf{y} = (y_{ij})_{1 \leq i,j \leq n}$ to a plot of its degree distribution on the log-log scale. From the expression in (4.6), a power law degree distribution satisfies

$$\log p_{\mathbf{y}}(k) \sim -\gamma \log(k) \quad \text{for all large } k, \tag{4.7}$$

giving an immediate heuristic check of the power law property (i.e., whether the log-log plot shows a negative linear relationship for large values).

For a concrete illustration of this last point, the Yule–Simon distribution on the positive integers assigns probabilities

$$\Pr(K = k; \gamma) = \frac{(\gamma - 1)\Gamma(k)\Gamma(\gamma)}{\Gamma(k+\gamma)}, \quad k \geq 1, \tag{4.8}$$

for $\gamma > 1$, where $\Gamma(t) = \int_0^\infty u^{t-1}e^{-u}du$ is the gamma function. The Yule–Simon distribution satisfies the power law property with exponent $\gamma > 1$ for large $k \geq 1$. Figure 4.1 shows the log-log plot for the Yule–Simon distribution and its linear approximation by (4.7). Notice, in particular, that a power law distribution can deviate from the line $-\gamma \log(k)$ for small values of k, as the condition in (4.6) only requires the relationship to hold for all large $k \geq 1$. See [27, 39] for further discussion on power law distributions, including some caveats about using the relationship in (4.7) as a heuristic check for power law distributions from finite samples.

4.3 Generative model 2: Random walk models

Random walk (RW) models evolve by adding a new edge at each step, instead of a new vertex as in the BA model. Let $\mathbf{y}_0$ be an initial graph and $n_0 \geq 1$ be an initial number of edges. The network $\mathbf{y}_1, \mathbf{y}_2, \ldots$ evolves as follows. At step $n \geq 1$, select a vertex v_n in $\mathbf{y}_{n-1}$ randomly according to a probability distribution F_n (which can depend on $\mathbf{y}_{n-1}$). Next draw a random nonnegative integer L_n from a probability distribution (which is also allowed to depend on $\mathbf{y}_{n-1}$). Given v_n and $L_n = \ell$, perform a simple random walk on $\mathbf{y}_{n-1}$ for ℓ steps starting at vertex v_n.[4] If after the ℓth step the random walk is at a vertex $v^* \neq v_n$, then add an edge between v^* and v_n; otherwise, add a new vertex v^{**} to the network and insert an edge between v^{**} and v_n.

The random walk dynamics of this model refine several properties of the BA model. Let $\mathbf{y} = (y_{ij})_{1 \leq i,j \leq n}$ be any undirected, connected graph[5] and generate a random sequence of states $(X_t)_{t=0,1,\ldots}$ taking values in the vertex set of $\mathbf{y}$ with distribution

$$X_0 \sim G \quad \text{for any distribution } G \text{ on } 1, \ldots, n \quad \text{and}$$

$$\Pr(X_{t+1} = v' \mid X_t = v) = \begin{cases} 1/\deg_{\mathbf{y}}(v), & y_{vv'} = 1, \\ 0, & \text{otherwise.} \end{cases}$$

The process $\mathbf{X}$ is called a simple random walk on $\mathbf{y}$ with initial distribution G. The marginal distribution of X_t converges to the degree-biased distribution on $\mathbf{y}$ as $t \to \infty$:

$$\lim_{t\to\infty} \Pr(X_t = i) = \frac{\deg_{\mathbf{y}}(i)}{\sum_{1 \leq j \leq n} \deg_{\mathbf{y}}(j)}, \quad i = 1, \ldots, n. \tag{4.9}$$

[4] A simple random walk on a graph is one that moves along the edges of $\mathbf{y} = (y_{ij})_{1 \leq i,j \leq n}$ as follows: given that the walk is at vertex v, the next state is chosen uniformly from among all v' for which $y_{vv'} = 1$.

[5] A connected graph is one for which there is a path (i.e., sequence of edges) connecting any two of its vertices. More precisely, $\mathbf{y} = (y_{ij})_{1 \leq i,j \leq n}$ is connected if for any $i, j = 1, \ldots, n$ there exists a sequence of vertices $i = i_0, i_1, \ldots, i_k = j$ such that $y_{i_r, i_{r+1}} = 1$ for $r = 0, 1, \ldots, k-1$.

Returning to the RW model, we see that the edge added between v_n and v^* at step n features a vertex v_n chosen according to a distribution F and v^* chosen by taking L_n steps in a simple random walk on $\mathbf{y}_{n-1}$. The analysis in (4.9) shows that if the number of steps L_n is large, then v^* is approximately chosen from the degree-biased distribution, as in the BA model. Because its generating dynamics are more complicated than the BA model, inference from the RW model tends to be more challenging. For example, if the order in which edges arrive is not observed, then it must be imputed during inference. See [20] for many more details about this model. In addition to the model in [20], a number of variations on the preferential attachment model have been proposed and studied, including preferential attachment-type models for networks that grow by sequential addition of edges instead of vertices. See [22, 80, 133, 149] and references therein for further details.

4.4 Generative model 3: Erdős–Rényi–Gilbert model

In addition to its description in terms of selection sampling, the Erdős–Rényi–Gilbert model in (3.16) admits the following generative description. For any $\theta \in [0,1]$, define $\Pi^{\theta}_{n,N}$ as the generating scheme which acts on $\{0,1\}^{n\times n}$ by

$$\mathbf{y} \mapsto \Pi_{n,N}(\mathbf{y})$$

$$\mathbf{y} \mapsto \begin{pmatrix} & & & B_{1,n+1} & \cdots & B_{1,N} \\ & \mathbf{y} & & \vdots & \ddots & \vdots \\ & & & B_{n,n+1} & \cdots & B_{n,N} \\ B_{n+1,1} & \cdots & B_{n+1,n} & 0 & \cdots & B_{n+1,N} \\ \vdots & \ddots & \vdots & \vdots & \ddots & \vdots \\ B_{N,1} & \cdots & B_{N,n} & B_{N,n+1} & \cdots & 0 \end{pmatrix},$$

which fixes the upper $n \times n$ submatrix to be $\mathbf{y}$ and fills in the rest of the off-diagonal entries with i.i.d. Bernoulli random variables $(B_{ij})_{1\leq i\neq j\leq N}$ with success probability θ. From this description, $\Pi^{\theta}_{n,N}\mathbf{Y}_n$ is distributed according to (3.16) with parameter θ as long as $\mathbf{Y}_n$ is distributed according to (3.16) with parameter θ on $\{0,1\}^{n\times n}$. This example illustrates the duality between sampling under selection and generative consistency, as expressed in (4.4).

4.5 Generative model 4: General sequential construction

Each of the above examples begins with a base case $\mathbf{Y}_0$, from which a family of networks $\mathbf{Y}_1, \mathbf{Y}_2, \ldots$ is constructed inductively according to a random scheme. Thus, a generic way to specify a generative network model is to specify, for every $n \geq 1$, a conditional distribution for $\mathbf{Y}_n$ given $\mathbf{Y}_{n-1}$ such that $\mathbf{Y}_n|_{[n-1]} = \mathbf{Y}_{n-1}$ with probability 1. In this way, the conditional distribution $\Pr(\mathbf{Y}_n = \cdot \mid \mathbf{Y}_{n-1})$ determines the distribution of a random generating mechanism $\Pi_{n-1,n}$ in $\mathscr{P}_{n-1,n}$ and $\mathbf{Y}_n$ can be expressed as $\mathbf{Y}_n = \Pi_{n-1,n}\mathbf{Y}_{n-1}$ for every $n \geq 1$.

Composing these actions for successive values of n determines the generating mechanism $\Pi_{n,N}$, $n \leq N$, by the law of iterated conditioning. In particular, for $n \leq N$, let $\Pi_{n,n+1}, \ldots, \Pi_{N-1,N}$ be the generating mechanisms determined by the conditional distributions of $\mathbf{Y}_m$ given $\mathbf{Y}_{m-1}$ for $m = n+1, \ldots, N$. Then, given $\mathbf{Y}_n$, construct $\mathbf{Y}_N = \Pi_{n,N}\mathbf{Y}_n$ by

$$\mathbf{Y}_N = \Pi_{N-1,N}(\Pi_{N-2,N-1}(\cdots(\Pi_{n,n+1}\mathbf{Y}_n)));$$

that is, $\mathbf{Y}_N$ is constructed by iterated application of the random functions $\Pi_{n,n+1}, \ldots, \Pi_{N-1,N}$, each with distribution determined by the conditional distribution of $\mathbf{Y}_{m+1}$ given $\mathbf{Y}_m$, for $m = n, \ldots, N-1$. The conditional distribution of $\mathbf{Y}_N$ given $\mathbf{Y}_n$ can be computed as

$$\begin{aligned}
&\Pr(\mathbf{Y}_N = \mathbf{y}^* \mid \mathbf{Y}_n = \mathbf{y}^*|_{[n]}) = \\
&= \Pr(\mathbf{Y}_N = \mathbf{y}^* \mid \mathbf{Y}_{N-1} = \mathbf{y}^*|_{[N-1]}) \times \Pr(\mathbf{Y}_{N-1} = \mathbf{y}^*|_{[N-1]} \mid \mathbf{Y}_n = \mathbf{y}^*|_{[n]}) \\
&= \prod_{i-1}^{N-n} \Pr(\Pi_{N-i,N-i+1}(\mathbf{y}^*|_{[N-i]}) = \mathbf{y}^*|_{[N-i+1]} \mid \mathbf{Y}_{N-i} = \mathbf{y}^*|_{[N-i]}).
\end{aligned}$$

4.6 Further reading

In Section 4.1, I discussed the automatic consistency properties of generative models, both in terms of generative consistency and consistency under selection. Because of this connection (shown explicitly in (4.4)), the language of generative models is often invoked when modeling real-world networks, even in the absence of any natural interpretation for the generating mechanism in the given context. So even though the connection between generative and sampling models may be helpful for developing mathematical theory or computational tools in network analysis, one should exercise caution when interpreting results from models whose initial specification (in terms of the generating process) lacks a meaningful interpretation vis-à-vis the application at hand.

The discussion of generative models in this chapter has been brief compared to the more in-depth discussion of sampling models in Chapter 3. With this abbreviated presentation, I do not mean to downplay the importance of generative models in network analysis. But by comparison to generative models, network modeling in the presence of sampling is poorly understood and has received limited attention from statisticians. Thus, although generative models are of interest for predictive modeling and machine learning applications, my emphasis on sampling models reflects the intention to focus here on those essential elements of network analysis which have not been given due attention elsewhere. The reader looking for additional discussion of generative network models will have no trouble finding it. See [7, 18, 36, 64, 67, 123, 146] and references therein for further details.

Chapter 5

Statistical modeling paradigm

Chapters 3 and 4 highlight two primary contexts of statistical network analysis:

- Chapter 3 focuses on network data that is assumed to have been sampled from a larger population network, with the goal of drawing inferences about the population based on whatever information can be extracted from the sample; and
- Chapter 4 focuses on evolving networks, with the goal of better understanding the generating process.

From these two cases, we have the following immediate observations:

- The concept of 'network' should not be conflated with the mathematical notion of 'graph' (Section 1.2).
- For networks obtained by sampling, the sampling mechanism plays a crucial role in model specification and statistical inference (Chapter 3).
- The statistical units for a given application are determined by the way in which the data is observed (Section 3.7).
- The explicit and implicit units should be aligned so that model-based inferences are compatible with their intended interpretation (Section 3.7).

Throughout this chapter, we write $\mathbf{Y}$ to generically denote 'network data', with $\mathbf{Y}_n$ indicating 'network data of size n'. Although the discussion so far has focused on the case in which $\mathbf{Y}_n$ is a $\{0,1\}$-valued array, the framework is meant to include networks which may be better represented as something other than a $\{0,1\}$-valued array. Chapters 9 and 10 discuss a class of models for networks with such an alternative representation, but unfortunately most of the discussion throughout this book remains confined to the conventional networks-as-graphs setting. As discussed in Chapter 1, the limited scope of the book reflects the current limitations of the field. I nevertheless attempt to lay down principles that will transfer to alternative approaches, as they arise in the future.

In order to speak as generally as possible, we write $\mathcal{N}_n$ to denote the set of all *networks of size n*, where the concepts of 'network' and 'size' are understood based on context. In general, the 'size' refers to the number of observed units, with each element of $\mathcal{N}_n$ corresponding to a mathematical representation of a network of size n within the prevailing context; refer back to Sections 3.7–3.8 for further discussion about sample size and units. For example, in the setting of Section 2.4, $\mathcal{N}_n = \{0,1\}^{n\times n}$, a 'network' is a graph (or binary relation), and the 'size' is the number of sampled students (i.e., vertices). For interaction networks, such as the network representation of phone call activity in Section 3.6.1.1, the 'network' is the structure induced by the interactions, as in Figure 3.2, and the 'size' is the number of interactions. For networks obtained by sampling paths (Section 3.6.1.3), the 'network' is the structure induced by those paths, as in Figure 3.4, and the 'size' is the number of sampled paths.

5.1 The quest for coherence

Keeping with the scenario of Section 2.4, suppose we are interested in reciprocity and differential attractiveness based on the observed friendships among a sample of high school students. We opt to model the population network $\mathbf{Y}_N$ by the p_1 model in (2.6), for which the parameterization by $(\rho, \theta, \alpha, \beta)$ affords the natural interpretation of ρ and β as the main parameters of interest for inferring reciprocity and differential attractiveness, respectively. How should the partially observed data $\mathbf{Y}_n$ be modeled so that valid inferences can be drawn about $\mathbf{Y}_N$?

A common, seemingly intuitive, approach is to estimate ρ and β by first fitting (2.6) to the observation $\mathbf{Y}_n = \mathbf{y} \in \{0,1\}^{n \times n}$ and then using the estimates for ρ and β based on $\mathbf{Y}_n$ as the estimates for ρ and β governing the population $\mathbf{Y}_N$. To be more explicit, assume that $\mathbf{Y}_n$ is distributed according to (2.6) for some unknown parameters $\rho, \theta \in (-\infty, \infty)$, $\alpha = (\alpha_1, \ldots, \alpha_n)$, and $\beta = (\beta_1, \ldots, \beta_n)$. Given an observation $\mathbf{Y}_n = \mathbf{y}$, we obtain estimates $\hat{\rho}_n, \hat{\theta}_n, \hat{\alpha}_n, \hat{\beta}_n$ for $\rho, \theta, \alpha, \beta$, respectively, by maximizing the likelihood function

$$L(\rho, \theta, \alpha, \beta; \mathbf{y}) = \Pr(\mathbf{Y}_n = \mathbf{y}; \rho, \theta, \alpha, \beta)$$

jointly with respect to $\rho, \theta, \alpha, \beta$, for $\Pr(\cdot; \rho, \theta, \alpha, \beta)$ as given in (2.6).[1] In what way are $\hat{\rho}_n$ and $\hat{\beta}_n$ informative about reciprocity and differential attractiveness in the population?

This 'intuitive' approach to inference is so standard in classical statistics that it may seem unnecessary to discuss in much detail here. For example, in a standard statistical application with $X_1, X_2, \ldots$ assumed to be i.i.d. from a distribution P_θ, for $\theta \in \Theta$, it is taken for granted that an estimator $\hat{\theta}_n$ for θ based on $X_1, \ldots, X_n$ also serves as an estimator for the parameter θ governing the entire sequence $X_1, X_2, \ldots$. However, the opening example of Section 3.1 and followup discussion throughout Chapter 3 advises caution about blindly using $\hat{\theta}_n$ as an estimator for θ in the population model. In a nutshell, the parameter 'θ' estimated by $\hat{\theta}_n$ may be related to the parameter 'θ' governing the population $X_1, X_2, \ldots$ without having the same meaning or interpretation. In essence, the 'θ' parameterizing the distribution of the sample $X_1, \ldots, X_n$ and the 'θ' parameterizing the distribution of the population $X_1, X_2, \ldots$

[1] Since estimation techniques are not the focus of the book, we often default to the most standard methods, e.g., maximum likelihood (frequentist) or maximum posterior (Bayesian) inference. Here I have chosen maximum likelihood estimation for illustration only. The modeling principles discussed here do not discriminate between these or any other inferential approaches one might prefer, such as Martin–Liu inferential models [118].

may not be the 'same θ', as in the example from Section 3.1. And while it is commonly assumed in standard applications that the observed data is representative of the population in a straightforward way, such situations are the exception, rather than the rule, in many networks applications.

In many complex data problems, there is no clearly specified logical relationship between the parameters governing the data $\mathbf{Y}_n$ and those governing the population $\mathbf{Y}_N$; or, in cases where this relationship is clearly specified, the logical relationship between $\mathbf{Y}_n$ and $\mathbf{Y}_N$ established by the model may not be compatible with the actual relationship between data and population, as determined by the real-world observation process. In the former case, when $\mathbf{Y}_n$ and $\mathbf{Y}_N$ have no logical relationship, it is meaningless to use parameter estimates for $\mathbf{Y}_n$ to draw conclusions about $\mathbf{Y}_N$. In the latter case, when the model implies a relationship between $\mathbf{Y}_n$ and $\mathbf{Y}_N$ in a way that contradicts their actual relationship, conclusions about $\mathbf{Y}_N$ based on parameter estimates for $\mathbf{Y}_n$ are bound to be spurious. This latter situation happens, for example, when a network obtained by degree-biased sampling is modeled as if obtained by selection sampling. These considerations underlie the condition of *model coherence*, which avoids potential ambiguity in statistical inferences by making sure that the models specified for the population and sample are compatible with their assumed sampling/generative relationship.

Rationale behind coherence

For an imperfect analogy, suppose you are deciding on whether to eat dinner at a new restaurant that has just opened in your neighborhood. Let θ represent your utility (on the scale $[0,1]$) of dining at this restaurant, with 0 being the lowest and 1 being the highest. Since you haven't yet eaten at this restaurant, θ is unknown, and so you try to estimate it by sampling the opinion of a friend who has eaten there. Supposing your friend reports to you his utility of $\tilde{\theta}$, how would you use $\tilde{\theta}$ as a proxy to estimate your own utility θ if

- the restaurant serves Chinese food?
- the restaurant is expensive?

As these questions suggest, the relationship between $\tilde{\theta}$ and θ depends on the context, which differs under the two scenarios described above. For example, if your friend has poor taste in Chinese food, then you might discount his high rating $\tilde{\theta}$ and estimate your utility θ to be much lower or much more uncertain. Or, if your friend is cheap, then you might choose to disregard a low rating $\tilde{\theta}$ given on the grounds that the restaurant is too expensive. Or, if you and your friend have similar taste in restaurants, then you might take $\tilde{\theta}$ as a good proxy for your own utility θ.

Just as we regularly engage in such contextual reasoning when making everyday decisions, such as our expected utility of eating at a certain restaurant, we ought to do the same in formal statistical inference. The upcoming framework is intended to formalize this logic by accounting for *context* in the specification of a statistical model. In this example about rating a new restaurant, the context is given by the eating preferences of your friend and how those preferences relate to your own. More generally, the context is given by all of the relevant circumstances under which the data has been observed, including a description of the sampling and/or generating mechanism, if applicable.

5.2 An incoherent model

Before formally defining coherence, we first discuss some potentially adverse consequences of incoherence. Assume a hypothetically infinite population network represented as a $\{0,1\}$-valued array $\mathbf{Y} = (Y_{ij})_{i,j\geq 1}$ with 0 on the diagonal. For each $n \geq 1$, assume that a network of size n is obtained by selection, so that $\mathbf{Y}_n = \mathbf{Y}\,|_{[n]} = (Y_{ij})_{1\leq i,j\leq n}$. In this way, all finite observations $\mathbf{Y}_m$ and $\mathbf{Y}_n$, $m \leq n$, are related to one another by $\mathbf{Y}_m = \mathbf{S}_{m,n}\mathbf{Y}_n$, for $\mathbf{S}_{m,n} : \{0,1\}^{n\times n} \to \{0,1\}^{m\times m}$ defined as in (3.6).

As I alluded in Section 3.4 and will discuss again in later chapters, much of modern network science is geared towards analyzing networks that are 'sparse'. In the setting assumed here, the population network $\mathbf{Y}$ is said to be *sparse* if the edge density of its finite components vanishes as the sample size grows. More precisely, the edge density of $\mathbf{Y}_n = (Y_{ij})_{1\leq i,j\leq n}$ is the ratio $n^{-1}(n-1)^{-1}\sum_{1\leq i\neq j\leq n} Y_{ij}$ of edges to the total number of possible edges in $\mathbf{Y}_n$. The population network $\mathbf{Y}$ is *sparse* if these edge densities converge to 0 as $n \to \infty$, i.e.,

$$\lim_{n\to\infty} \frac{1}{n(n-1)} \sum_{1\leq i\neq j\leq n} Y_{ij} = 0. \tag{5.1}$$

For now, let us take for granted that (5.1) is a known property of the population network $\mathbf{Y}$. How might we draw inferences about the sparse population network $\mathbf{Y}$ based on data $\mathbf{Y}_n = \mathbf{Y}\,|_{[n]}$ obtained by selection?

One preliminary approach to modeling sparse $\mathbf{Y}$ is to assume that it evolves under a *sparse regime*, meaning that $\mathbf{Y}$ results from the evolution of finite sample networks $(\mathbf{Y}_n)_{n\geq 1}$, each of which is described by a family of distributions such that the sparsity condition (5.1) holds with probability 1. To this end, let us assume that each $\mathbf{Y}_n$ is modeled by a set of candidate distributions $\mathcal{M}_n$ consisting of the n-scaled Erdős–Rényi–Gilbert distributions, i.e., for $0 \leq \theta \leq 1$,

$$\Pr(\mathbf{Y}_n = \mathbf{y}; \theta) = \prod_{1\leq i\neq j\leq n} (\theta/n)^{y_{ij}}(1-\theta/n)^{1-y_{ij}}, \quad \mathbf{y} \in \{0,1\}^{n\times n}. \tag{5.2}$$

With this specification, each $\mathbf{Y}_n$ is assumed to follow one of the Erdős–Rényi–Gilbert distributions in (3.16) with parameter in the range $[0,1/n]$. Since θ is bounded in the range $[0,1]$ and the edges of $\mathbf{Y}_n$ from (5.2) are i.i.d. Bernoulli, the strong law of large numbers implies that

$$\frac{1}{n(n-1)} \sum_{1\leq i\neq j\leq n} Y_{ij} \approx \theta/n \to 0 \quad \text{as } n\to\infty.$$

We see at once that a sequence $(\mathbf{Y}_n)_{n\geq 1}$ with each $\mathbf{Y}_n$ distributed according to (5.2) satisfies (5.1).

But under the assumed selection sampling scheme, the family of models $\{\mathscr{M}_n\}_{n\geq 1}$ is incoherent in the following sense. Let $\mathbf{Y}_n$ be modeled by (5.2) and let $\mathbf{S}_{m,n}\mathbf{Y}_n$ be the network obtained by selection sampling. By independence of the edges, the induced distribution of $\mathbf{S}_{m,n}\mathbf{Y}_n$ is again Erdős–Rényi–Gilbert with parameter θ/n on $\{0,1\}^{m\times m}$. In this specification, the set of candidate distributions for $\mathbf{Y}_m = \mathbf{S}_{m,n}\mathbf{Y}_n$ is given by the subset of distributions

$$\Pr(\mathbf{S}_{m,n}\mathbf{Y}_n = \mathbf{y};\theta) = \prod_{1\leq i\neq j\leq m} \theta^{y_{ij}}(1-\theta)^{1-y_{ij}}, \quad \mathbf{y}\in\{0,1\}^{m\times m}, \tag{5.3}$$

for all $\theta \in [0,1/n]$. But the model $\mathscr{M}_m$ specified in (5.2) consists of distributions in (5.3) with $\theta \in [0,1/m]$.

Here we see that the explicitly specified model, i.e., the model explicitly defined for $\mathbf{Y}_m$ by

$$\mathscr{M}_m = \{\Pr(\mathbf{Y}_m = \cdot;\theta) : \theta \in [0,1/m]\},$$

and the implicitly induced model, i.e., the model induced by the assumed sampling scheme and the model for $\mathbf{Y}_n$ by

$$\mathbf{S}_{m,n}\mathscr{M}_n = \{\Pr(\mathbf{Y}_m = \cdot;\theta) : \theta \in [0,1/n]\}, \tag{5.4}$$

differ for all $n > m$. Fixing $m \geq 1$ and letting $n > m$ vary over all possible finite sample sizes results in infinitely many distinct sets of candidate distributions for $\mathbf{Y}_m$, i.e.,

$$\mathscr{M}_m \neq \mathbf{S}_{m,m+1}\mathscr{M}_{m+1} \neq \cdots \neq \mathbf{S}_{m,n}\mathscr{M}_n \neq \cdots, \tag{5.5}$$

raising the question about which of these inequivalent models ought to be used as the set of candidate distributions for an observation $\mathbf{Y}_m$. We would observe a similar incompatibility to that in (5.5) if we instead specified the model in terms of the Erdős–Rényi–Gilbert generating scheme from Section 4.4.

We call a model *incoherent* whenever it fails to give a single (coherent) description of the circumstances under which the data has been observed. Specifically, the model is incoherent if the set of candidate distributions $\mathscr{M}_m$ for a given observation $\mathbf{Y}_m$ (as in (5.2)) differs from the set of candidate distributions for $\mathbf{Y}_m$ that is induced by the assumed context of the model (as in (5.4)). I formalize this notion below.

5.3 What is a statistical model?

Under the traditional textbook definition, a statistical model is merely

"a set of probability distributions on the sample space" [120, p. 1225].

But we have seen in Chapters 3 and 4 that many networks applications involve more than a set of distributions and more than one space. In the scenario of Section 2.4, for example, we are interested in a population network for N students from which we observe only a sample of size $n < N$. The population network of interest thus occupies a different space than the observed network, and the model entails not only sets of candidate distributions for population and sample but also a description of the sampling mechanism relating the two. In this case, the spaces occupied by the population and sampled networks seem to be related to one another, but how can we specify a model in a way that makes this precise? As alluded earlier, the precise relationship between different sample spaces depends on the context of application, as determined by the manner in which the network for n students has been sampled. We consider two possible ways to specify this model.

5.3.1 Population model

Assuming that there is a well-defined sample space $\mathcal{N}_N$ for the population network, the standard textbook definition implies that a statistical model can be specified as a set of distributions on $\mathcal{N}_N$. In the above example, $\mathcal{N}_N = \{0,1\}^{N\times N}$ is the set of all adjacency arrays representing binary relations among N high school students. The model for the population network is a set of candidate distributions $\mathcal{M}_N$, and it is our objective to identify which candidate distribution best explains the friendship patterns among all students in the school based on a partial observation of friendships $\mathbf{Y}_n$. Since the data consists only of friendships among $n < N$ sampled students, $\mathbf{Y}_n$ occupies a different space, namely $\mathcal{N}_n = \{0,1\}^{n\times n}$, and thus the candidate distributions in $\mathcal{M}_n$ differ from those in $\mathcal{M}_N$. If $\mathbf{Y}_n$ is to be of any use for inference, the statistical model cannot be just the set of distributions $\mathcal{M}_N$. It must also account for the *context* in which the data and population are related to one another. In particular, the context relates the distributions in $\mathcal{M}_n$ to those in $\mathcal{M}_N$ so that inferences based on $\mathcal{M}_n$ are compatible with inferences about the population.

In the scenario of Section 2.4, we assume that $\mathbf{Y}_n$ is the friendship network of n students chosen uniformly at random without replacement. In the notation of Section 3.9, we define $\Sigma_{n,N}$, for $1 \le n \le N$, as the sampling mechanism corresponding to this action, i.e.,

$$\Pr(\Sigma_{n,N} = \mathbf{S}^{\psi}_{n,N}) = 1/N^{\downarrow n}, \quad \psi : [n] \to [N],$$

for $N^{\downarrow n} = N(N-1)\cdots(N-n+1)$ and notation $\mathbf{S}^{\psi}_{n,N}$ as defined in (3.17). (Since there are $N^{\downarrow n}$ total injections $\psi : [n] \to [N]$, uniform random vertex sampling corresponds to choosing each such injection with equal probability and observing $\mathbf{Y}_n = \mathbf{S}^{\psi}_{m,n}\mathbf{Y}_N$.) Given $\Sigma_{n,N}$ and any probability distribution $P \in \mathcal{M}_N$, we can compute the distribution of $\Sigma_{n,N}\mathbf{Y}_N$ for $\mathbf{Y}_N \sim P$ by

$$(\Sigma_{n,N}P)(\mathbf{Y}_n = \mathbf{y}) = \Pr(\Sigma_{n,N}\mathbf{Y}_N = \mathbf{y}), \quad \mathbf{y} \in \{0,1\}^{n\times n},$$

as in (3.18). Together the set $\mathcal{M}_N$ and the sampling mechanism $\Sigma_{n,N}$ induce a model $\mathcal{M}_n$ for $\mathbf{Y}_n$ by

$$\mathcal{M}_n = \Sigma_{n,N}\mathcal{M}_N = \{\Sigma_{n,N}P : P \in \mathcal{M}_N\}. \tag{5.6}$$

We call $\Sigma_{n,N}\mathcal{M}_N$ the $\Sigma_{n,N}$-*induced model* for $\mathbf{Y}_n$. In this case, it is enough to specify the statistical model by $(\mathcal{M}_N, \{\Sigma_{n,N}\}_{1\le n\le N})$, from which each of the finite sample models $\mathcal{M}_n$, $n \le N$, can be deduced via (5.6).

5.3.2 Finite sample models

In many practical situations, the population space is unknown or undefined, or it may be technically difficult or impossible to specify $\mathcal{M}_N$ directly on $\mathcal{N}_N$, as we assumed in Section 5.3.1. For example, if the population size is unknown or unbounded, then it is common to take $N = \infty$ and define the population model on the space of networks for a countable population. In this case, the population sample space tends to be infinite (or even uncountable), making it most natural to specify the finite sample models $\mathscr{M}_n$ explicitly for each possible finite sample size $1 \leq n \leq N$. Thus, instead of defining one set of distributions $\mathscr{M}_N$ on a population space and deducing the finite sample models $\mathscr{M}_n$ as in (5.6), the model is specified at the outset as the set of finite sample models $\{\mathscr{M}_n\}_{n\geq 1}$ for all finite sample sizes $n \geq 1$. Defined in this way, it is immediate that the statistical model is not *a set of probability distributions on the population space*, but rather involves *a set of sets of probability distributions defined on a family of sample spaces*, with $\mathscr{M}_n$ regarded as the set of candidate distributions for data $\mathbf{Y}_n$ residing in $\mathscr{N}_n$. But, as in Section 5.3.1, the collection $\{\mathscr{M}_n\}_{n\geq 1}$ of finite sample models alone does not constitute a *statistical* model unless these models can be interpreted in a common statistical context.

With this observation in mind, I define here a statistical model as a family $\{\mathscr{M}_n\}_{n\geq 1}$ of *finite sample models*, where each $\mathscr{M}_n$ is the set of candidate distributions on the sample space of observations of size n, along with a *context* $\mathscr{C}$ in which the candidate distributions are to be interpreted. We will later see how the context $\mathscr{C}$ can be described by a system of sampling mechanisms $\{\Sigma_{n,N}\}_{1\leq n\leq N}$ or generating mechanisms $\{\Pi_{n,N}\}_{1\leq n\leq N}$, from Chapters 3 and 4, respectively.

Definition 5.1 (Statistical model) *A* statistical model *is a pair* $(\{\mathscr{M}_n\}_{n\geq 1}, \mathscr{C})$, *where*

$\mathscr{M}_n$ *is a set of candidate distributions for observations of size n and*

$\mathscr{C}$ *is the context in which these models are interpreted.*

Referring back to Section 1.5, $\{\mathscr{M}_n\}_{n\geq 1}$ is the *descriptive* component of the statistical model, with each $\mathscr{M}_n$ describing a sample of size n in terms of its candidate distributions, and $\mathscr{C}$ is the *inferential* component of the statistical model, which provides the necessary link for interpreting inferences about $\mathscr{M}_n$ within the broader context of all finite sample models $\{\mathscr{M}_n\}_{n\geq 1}$.

Table 5.1 partly explains the need for context $\mathscr{C}$ in addition to the candidate distributions $\{\mathscr{M}_n\}_{n\geq 1}$ in network modeling. In the table, $\mathscr{M}_N = \{P_\theta : \theta \in \Theta\}$ models the population network $\mathbf{Y}_N$, and $\mathbf{Y}_n$ is obtained by sampling from $\mathbf{Y}_N$ in an unspecified manner. Without knowing the relationship between population and sample (i.e., the *context*) we cannot specify a model for $\mathbf{Y}_n$ in a way that is compatible with the model for $\mathbf{Y}_N$. The incomplete specification of the model in Table 5.1 is reflected by the expression '???' in place of the model for $\mathbf{Y}_n$. Furthermore, as each set of candidate models $\mathscr{M}_n$ pertains only to data of size n, the set $\mathscr{M}_n$ alone treats data *out of context*. Coherent inference from a model requires that the data can be analyzed and conclusions stated in the appropriate context, necessitating the additional component $\mathscr{C}$ in the model specification. Definition 5.1 thus affords the interpretation of a statistical model $(\{\mathscr{M}_n\}_{n\geq 1}, \mathscr{C})$ as 'candidate models + context', decomposing a

Table 5.1 *Illustration of population network (left) and sampled network (right), along with population model $\{P_\theta : \theta \in \Theta\}$ for $\mathbf{Y}_N$. As presented, the sampling mechanism by which $\mathbf{Y}_n$ is obtained from $\mathbf{Y}_N$ is unspecified, leaving the model for the sampled network $\mathbf{Y}_n$ unspecified, as represented by '???' in the above graphic.*

	Population	Observed network (sample)
	$\mathbf{Y}_N$	$\mathbf{Y}_n$
	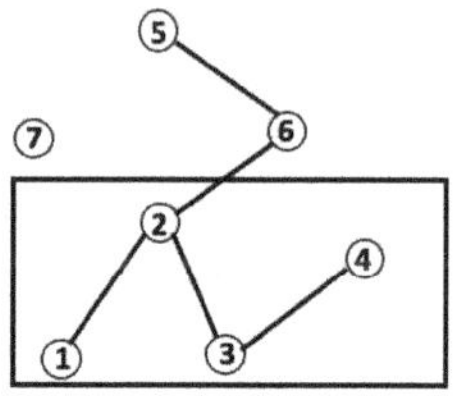	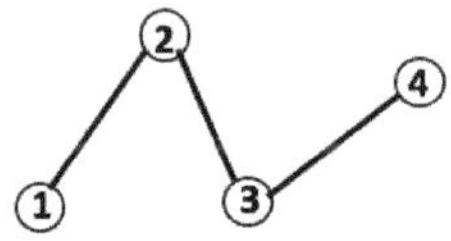
Model	$\{P_\theta : \theta \in \Theta\}$	???

statistical model into a *descriptive* component $\{\mathscr{M}_n\}_{n\geq 1}$ and an *inference* component $\mathscr{C}$, as foreshadowed in Section 1.5.

Note that a population model $(\mathscr{M}_N, \{\Sigma_{n,N}\}_{1\leq n\leq N})$, as defined in the previous section, can be reformulated as a statistical model $(\{\mathscr{M}_n\}_{1\leq n\leq N}, \mathscr{C})$ as in Definition 5.1 by setting

$$\mathscr{M}_n = \Sigma_{n,N}\mathscr{M}_N \quad \text{for each } n = 1, \ldots, N \text{ and}$$
$$\Pr(\Sigma_{m,n} = \mathbf{S}^{\psi}_{m,n}) = 1/n^{\downarrow m}, \quad \mathbf{S}^{\psi}_{m,n} : [m] \to [n], \quad \text{for all } 1 \leq m \leq n \leq N,$$

and defining the context by $\mathscr{C} = \{\Sigma_{m,n}\}_{1\leq m\leq n\leq N}$.

Remark 5.1 (Dependence on context) *I note the following deficiency of Definition 5.1 with respect to specifying models whose inferential component (i.e., context) depends on its descriptive component (i.e., finite sample models). For example, in the Barabási–Albert model $(\{\mathscr{M}_n\}_{n\geq 1}, \{\Pi_{m,n}\}_{n\geq m\geq 1})$ parameterized by (m, δ) as in Section 4.2, the generating mechanism $\Pi_{m,n}$ is implicitly assumed to depend on the distribution of $\mathbf{Y}_m$ on which it acts, in the sense that the generating mechanism $\Pi_{m,n}$ that acts on $\mathbf{Y}_m$ from the BA model with parameter (m, δ) is also parameterized by (m, δ), denoted $\Pi_{m,n}\mathbf{Y}_m = \Pi^{\delta,m}_{m,n}\mathbf{Y}_m$ in the notation of Section 4.2. For the most part this dependence does not affect the specific cases discussed below, and so I often suppress it in Definition 5.1 and throughout the rest of this and subsequent chapters. The reader should nevertheless be aware of this possibility when studying network models that are not covered in this text.*

5.4 Coherence

On their own, the components $\{\mathscr{M}_n\}_{n\geq 1}$ and $\mathscr{C}$ in Definition 5.1 are necessary but not sufficient to ensure that inferences based on $(\{\mathscr{M}_n\}_{n\geq 1}, \mathscr{C})$ make logical sense.

I introduce here the condition of *coherence*, which captures the additional logical requirement that $(\{\mathscr{M}_n\}_{n\geq 1},\mathscr{C})$ be self-consistent in the sense that the descriptions $\mathscr{M}_n$, $n \geq 1$, make sense (i.e., are 'coherent') within a single context $\mathscr{C}$. Incoherent models, as in Section 5.2, may give rise to incoherent inferences, i.e., inferences which cannot be interpreted within the same context. Referring back to the Boxian trope from Section 1.1—"all models are wrong, but some are useful"—it is this conception of 'making sense' (via coherence) that I posit here as the first necessary step for a model to be 'useful'. I first introduce the concept of coherence for sampling models, with context given by a system of sampling mechanisms $(\mathscr{C} = \{\Sigma_{m,n}\}_{n\geq m\geq 1})$. The definition for generative models, in which the context is given by a system of generating mechanisms $(\mathscr{C} = \{\Pi_{m,n}\}_{n\geq m\geq 1})$, follows in an analogous way and is mentioned briefly in Section 5.4.2.

5.4.1 Coherence in sampling models

For a sampling model $(\{\mathscr{M}_n\}_{1\leq n\leq N},\{\Sigma_{m,n}\}_{1\leq m\leq n\leq N})$ with each $\mathscr{M}_n$ specifying a set of candidate distributions on $\mathscr{N}_n$ and $\Sigma_{m,n}$ describing a (possibly random) sampling mechanism $\mathscr{N}_n \to \mathscr{N}_m$, we write $\Sigma_{m,n}\mathscr{M}_n$ to denote the set of distributions *induced* on $\mathscr{N}_m$ by $\mathscr{M}_n$ through $\Sigma_{m,n}$ as in (5.6). In particular, for any candidate distribution $P \in \mathscr{M}_n$, $\Sigma_{m,n}P$ denotes the distribution of $\mathbf{Y}_m$ obtained according to the scheme:

$$\begin{aligned}\mathbf{Y}_n &\sim P\\ \mathbf{Y}_m &= \Sigma_{m,n}\mathbf{Y}_n.\end{aligned}$$

Thus, $\Sigma_{m,n}P$ is the distribution of a random network $\Sigma_{m,n}\mathbf{Y}_n$ of size m. (An explicit calculation of $\Sigma_{m,n}P$ in the special case of $\mathscr{N}_n = \{0,1\}^{n\times n}$ is shown in (3.18).) The set of all possible distributions obtained in this way determines the *$\Sigma_{m,n}$-induced model*

$$\Sigma_{m,n}\mathscr{M}_n = \{\Sigma_{m,n}P : P \in \mathscr{M}_n\} \tag{5.7}$$

as in (5.6).

To illustrate this action further, let each $\mathscr{M}_n$ be the set of Erdős–Rényi–Gilbert distributions (3.16) parameterized by $\theta \in [0,1]$ on $\{0,1\}^{n\times n}$. For $n \geq m \geq 1$, define $\Sigma_{m,n}$ as the uniform random sampling mechanism given by

$$\Pr(\Sigma_{m,n} = \mathbf{S}^{\psi}_{m,n}) = 1/n^{\downarrow m}, \quad \psi : [m] \to [n].$$

For fixed $\theta \in [0,1]$, notice that the distribution of $\Sigma_{m,n}\mathbf{Y}_n$, for $\mathbf{Y}_n$ distributed as in (3.16) with parameter θ, is again Erdős–Rényi–Gilbert with parameter θ; whence, for every θ, the distribution of $\Sigma_{m,n}\mathbf{Y}_n$ is given by (3.16) on $\{0,1\}^{m\times m}$ and

$$\Sigma_{m,n}\mathscr{M}_n = \{\Pr(\mathbf{Y}_m = \cdot;\theta) : \theta \in [0,1]\}, \tag{5.8}$$

i.e., the Erdős–Rényi–Gilbert model defined on $\{0,1\}^{m\times m}$. In this case, the model for $\mathbf{Y}_m$ as *specified by $\mathscr{M}_m$* agrees with (i.e., is *coherent with*) the model for $\Sigma_{m,n}\mathbf{Y}_n$ *induced by $\mathscr{M}_n$* through the sampling scheme $\Sigma_{m,n}$. We call such a model $(\{\mathscr{M}_n\}_{1\leq n\leq N},\{\Sigma_{m,n}\}_{1\leq m\leq n\leq N})$ *coherent*.

Definition 5.2 (Coherence (Sampling Model)) *A statistical model with finite sample models* $\{\mathscr{M}_n\}_{1\leq n\leq N}$ *and sampling context* $\mathscr{C} = \{\Sigma_{m,n}\}_{1\leq m\leq n\leq N}$ *is* coherent *if*

(C) $\Sigma_{m,n}\mathscr{M}_n = \mathscr{M}_m$ *for all* $1 \leq m \leq n \leq N$.

Remark 5.2 *In principle, a model can be specified to automatically satisfy coherence by defining* $(\{\mathscr{M}_n\}_{n\geq 1}, \{\Sigma_{m,n}\}_{n\geq m\geq 1})$ *as the models induced by sampling from a population model* $\mathscr{M}$ *on a hypothetical population space for an infinite population, as in Section 5.3.1 and equation* (5.6). *But in practice,* deductive definitions *of coherent models are hard to describe since they depend on the initial specification of* $\mathscr{M}$ *on a sample space for a potentially infinite population network.*

The concept of *model coherence* is not widespread in the statistical literature, but it is a critical component to statistical modeling and inference of all kinds. According to Definition 5.2, a model is coherent as long as the candidate distributions for any observation $\mathbf{Y}_m$ of size m does not depend on the manner in which $\mathbf{Y}_m$ may have been obtained by subsampling from a larger sample or population. The rationale underlying this definition is that the assumed behavior of observations of all sizes, i.e., $\mathbf{Y}_n$ for all possible sample sizes $1 \leq n \leq N$, ought to 'fit together' in the sense of determining a single 'coherent model' within the context of the given application. We saw in Section 5.2 that finite sample models need not fit together as in (5.5), and that such incoherence could have adverse implications for statistical inference. In Section 5.5, I discuss the significance of this 'fitting together', or lack thereof.

5.4.2 Coherence in generative models

Coherence for generative models follows the same recipe as in Definition 5.2. As before we let $\mathscr{M}_n$ be a set of finite sample models for each $n \geq 1$, but now we write $\Pi_{m,n}$, $m \leq n$, to denote the family of generating mechanisms. (Refer to Chapter 4 for an explanation of this notation and terminology.) For any $P \in \mathscr{M}_m$, $\Pi_{m,n}P$ denotes the distribution of a random network $\mathbf{Y}_n$ obtained by applying $\Pi_{m,n}$ to $\mathbf{Y}_m \sim P$, i.e.,

$$\mathbf{Y}_m \sim P$$
$$\mathbf{Y}_n = \Pi_{m,n}\mathbf{Y}_m\,.$$

Aggregating over all candidate distributions $P \in \mathscr{M}_m$ gives the $\Pi_{m,n}$-*induced model* on $\mathscr{N}_n$,

$$\Pi_{m,n}\mathscr{M}_m = \{\Pi_{m,n}P : P \in \mathscr{M}_m\}.$$

Definition 5.3 (Coherence (Generative Model)) *A generative statistical model with finite sample models* $\{\mathscr{M}_n\}_{1\leq n\leq N}$ *and generative context* $\mathscr{C} = \{\Pi_{m,n}\}_{1\leq m\leq n\leq N}$ *is* coherent *if*

(C′) $\Pi_{m,n}\mathscr{M}_m = \mathscr{M}_n$ *for all* $1 \leq m \leq n \leq N$.

Remark 5.3 *For the precise interpretation of (C′) for context-dependent generating schemes, see Remark 5.1.*

Though it is often difficult to specify a coherent sampling model deductively from a population model $\mathscr{M}$—see the discussion following Definition 5.2—generative models often admit a straightforward *inductive definition* from a base model $\mathscr{M}_0$ on a hypothetical population of size 0 and a system of generating mechanisms $\{\Pi_{n-1,n}\}_{n\geq 1}$ by putting

$$\mathscr{M}_n = \Pi_{n-1,n}\mathscr{M}_{n-1}, \quad n \geq 1, \quad \text{and} \tag{5.9}$$

$$\Pi_{m,n} = \Pi_{n-1,n} \circ \cdots \circ \Pi_{m,m+1}, \quad m \leq n, \tag{5.10}$$

where $\Pi_{m,n} \circ \Pi_{\ell,m}$ denotes the composition of independent random operators $\Pi_{m,n}$ and $\Pi_{\ell,m}$. All of the generative models discussed in Chapter 4 were defined in this way. We can easily verify that a model $(\{\mathscr{M}_n\}_{n\geq 1}, \{\Pi_{m,n}\}_{n\geq m\geq 1})$ defined by (5.9) and (5.10) is coherent in the sense of Definition 5.3.

Exercise 5.1 *Prove that any generative model* $(\{\mathscr{M}_n\}_{n\geq 1}, \{\Pi_{m,n}\}_{n\geq m\geq 1})$ *defined as in* (5.9) *and* (5.10) *is coherent in the sense of Definition 5.3.*

In addition to coherence, generative models offer other practical and computational benefits, such as

- facilitating numerical analysis by simulation,
- providing intuition and heuristic justification for the model by relating the presumed generating mechanism to the way in which the network is actually observed, and
- allowing for predictive inferences by computing how the observed network is expected to evolve if more units were observed according to the inferred generating mechanism.

The main conceptual difference between the sampling and generative cases is that coherence in sampling models requires compatibility between $\mathscr{M}_m$ and the model $\Sigma_{m,n}\mathscr{M}_n$ induced by sampling from a model for a larger population or sample, while coherence for generative models requires compatibility between $\mathscr{M}_n$ and the model $\Pi_{m,n}\mathscr{M}_m$ induced by evolving from a smaller network. It is in this sense that (C) and (C′) can be seen as dual notions of coherence.

Notice also the subtle distinction between *coherence* (Definitions 5.2 and 5.3) and *consistency under subsampling* (Definitions 3.1 and 3.2). Whereas coherence is a property of a *model* $(\{\mathscr{M}_n\}_{n\geq 1}, \mathscr{C})$, consistency is a property of *distributions* P_m and P_n. Moreover, while consistency establishes an exact equality of two distributions as in (3.19), coherence is a condition for two *sets* of distributions, as in (C) and (C′). Put another way, we can regard consistency as a probabilistic condition and coherence as a statistical condition.

For the rest of this chapter, I mostly specialize to coherence for sampling models, as in Definition 5.2, but many of the same observations also hold for generative models, as in Definition 5.3.

5.5 Statistical implications of coherence

To understand why coherence is a logical requirement for model-based statistical inference, consider how sampling models $(\{\mathscr{M}_n\}_{1\leq n\leq N}, \{\Sigma_{m,n}\}_{1\leq m\leq n\leq N})$ figure into two basic kinds of inference.

(I) Out of sample/predictive inference: Given data $\mathbf{Y}_m$ of size $m \geq 1$, let $\hat{P}_m \subseteq \mathcal{M}_m$ be an inference for the optimal candidate distribution(s) describing $\mathbf{Y}_m$ and infer the optimal distribution(s) for a larger sample/population of size $n > m$ by

$$\hat{P}_n = \bigcup_{\tilde{P} \in \hat{P}_m} \{P \in \mathcal{M}_n : \Sigma_{m,n} P = \tilde{P}\}.^2 \tag{5.11}$$

The inference $\hat{P}_n$ is the set of all candidate distributions in $\mathcal{M}_n$ which are compatible with the optimal $\hat{P}_m$ through the assumed sampling description $\Sigma_{m,n}$. If each $\mathcal{M}_n = \{P_\theta : \theta \in \Theta\}$ is parameterized by Θ, then an inference $\hat{\theta}_m \subseteq \Theta$ based on $\mathbf{Y}_m$ gives rise to

$$\hat{\theta}_n = \bigcup_{\tilde{\theta} \in \hat{\theta}_m} \{\theta \in \Theta : \Sigma_{m,n} P_\theta = P_{\tilde{\theta}}\} \tag{5.12}$$

for the model of $\mathbf{Y}_n$. If the model is identifiable then $\hat{P}_n$ (respectively, $\hat{\theta}_n$) are singleton sets whenever $\hat{P}_m$ (resp., $\hat{\theta}_m$) are singletons. (I allow the inference $\hat{P}_m$ (resp., $\hat{\theta}_m$) to be a set in order to allow for inferences, such as confidence regions, which may not correspond to a unique point estimate. The case in which $\hat{P}_m$ (resp., $\hat{\theta}_m$) is a singleton set corresponds to a point estimate.)

(II) Within sample inference: Given data $\mathbf{Y}_n$ of size $n \geq 1$, infer the optimal candidate distribution(s) $\hat{P}_n \subseteq \mathcal{M}_n$ and deduce the optimal distribution(s) for a subsample of size $m < n$ from $\mathbf{Y}_n$ by

$$\hat{P}_m = \Sigma_{m,n} \hat{P}_n, \tag{5.13}$$

where the action of $\Sigma_{m,n}$ on the set $\hat{P}_n$ is as defined in (5.6),

$$\Sigma_{m,n} \hat{P}_n = \{\Sigma_{m,n} \tilde{P} : \tilde{P} \in \hat{P}_n\}.$$

Notice how these two kinds of inference match up with the two fundamental implications of condition (C), namely $\mathcal{M}_m = \Sigma_{m,n} \mathcal{M}_n$ (as sets) if and only if $\mathcal{M}_m \subseteq \Sigma_{m,n} \mathcal{M}_n$ and $\mathcal{M}_m \supseteq \Sigma_{m,n} \mathcal{M}_n$:

(I′) $\mathcal{M}_m \subseteq \Sigma_{m,n} \mathcal{M}_n$: every distribution in $\mathcal{M}_m$ can be described as the distribution induced by sampling from $\mathbf{Y}_n \sim P$ via $\Sigma_{m,n}$, for some $P \in \mathcal{M}_n$. Related to point (I) above, the inference $\hat{P}_n$ (resp., $\hat{\theta}_n$) in (5.11) (resp., (5.12)) is guaranteed to be non-empty since every $\tilde{P} \in \hat{P}_m$ (resp., $\tilde{\theta} \in \hat{\theta}_m$) corresponds to at least one candidate distribution in $\mathcal{M}_n$ through $\Sigma_{m,n}$.

(II′) $\mathcal{M}_m \supseteq \Sigma_{m,n} \mathcal{M}_n$: every distribution in $\mathcal{M}_n$ corresponds to a distribution in $\mathcal{M}_m$ through $\Sigma_{m,n}$. This relates to the inference in (II) since it guarantees that $\hat{P}_m$ in (5.13) is a subset of the candidate distributions in $\mathcal{M}_m$.

These observations clarify how coherence makes it possible to extend inferences for $\mathcal{M}_m$ to inferences about $\mathcal{M}_n$ through the relationship established by the sampling

[2] As noted before, we do not assume a preferred way of estimating the optimal distribution(s) $\hat{P}_m$. We are instead concerned about what can be done with this estimate once it has been obtained.

context $\{\Sigma_{m,n}\}_{1\le m\le n\le N}$. If (C) is not satisfied, then there are 'rogue' distributions with no logical relationship to some candidate distribution of a different sample size. If such a rogue distribution were inferred from $\mathbf{Y}_m$, then there would be no logical way to extend the inference to $\mathscr{M}_n$. For instance, consider the incoherent family of models from Section 5.2 (i.e., each $\mathscr{M}_n$ consists of Erdős–Rényi–Gilbert distributions with parameter $0 \le \theta \le 1/n$) and suppose that inference based on $\mathbf{Y}_m$ modeled by (5.2) gives a point estimate in the range $1/n < \hat{\theta}_m \le 1/m$. Since such parameter values are not in the model $\mathbf{S}_{m,n}\mathscr{M}_n$ induced by selection sampling, how should one extend $\hat{\theta}_m$ to an inference about a network of size n?

For an estimate $\hat{\theta}_m$ based on $\mathbf{Y}_m$, the inference rule in (5.12) gives

$$\hat{\theta}_n = \{\theta' : \mathbf{S}_{m,n}\Pr(\mathbf{Y}_n = \cdot;\theta') = \Pr(\mathbf{Y}_m = \cdot;\hat{\theta}_m)\} = \{(n/m)\hat{\theta}_m\} \cap [0,1].$$

Since $\hat{\theta}_m$ can take values in the range $[0,1]$, the estimate $\hat{\theta}_n = (n/m)\hat{\theta}_m$ derived from $\hat{\theta}_m$ can take values in the range $[0,n/m]$. Since $n \ge m$, there are estimates $\hat{\theta}_m$ which put $\hat{\theta}_n$ outside the parameter space $[0,1]$ of $\mathscr{M}_n$, making it unclear how to extend $\hat{\theta}_m > 1/n$ to an inference for $\mathscr{M}_n$. Such uncertainty about what to do about some parameter values might also cause confusion about how to justify the extension of estimates $\hat{\theta}_m$ lying within $[0,1/n]$. In light of Definition 5.2, this example demonstrates the precise sense in which the model of Section 5.2 is incoherent.

5.6 Examples

By Definition 5.1, a statistical model has two key components:

(M1) a set $\mathscr{M}_n$ of candidate distributions for every possible sample size n and

(M2) a context $\mathscr{C}$ within which observations of different sample sizes are to be interpreted.

The first requirement (M1) is fulfilled by defining, for each possible sample size $n = 1,2,\ldots,N$,[3] a set of probability distributions $\mathscr{M}_n$ on the observation space $\mathscr{N}_n$. The constituents of $\mathscr{M}_n$ are the candidate distributions for data $\mathbf{Y}_n$ observed on n units. The second step (M2) is achieved by articulating the relationship between observations of size $n \le N$ and size $m \le n$. Most commonly the context is described by a sampling scheme or generating mechanism, and we shall restrict attention to those cases here. The further consideration of coherence (Definitions 5.2 and 5.3) is necessary to make sure that inferences drawn from the model are sound, as discussed in Section 5.5. The next few examples demonstrate the specification of a network model and the determination of coherence, or lack thereof, when the vertices are the units and $\mathscr{N}_n = \{0,1\}^{n\times n}$.

5.6.1 *Example 1: Erdős–Rényi–Gilbert model under selection sampling*

Suppose that the population is of known finite size $N < \infty$ and each $\mathscr{M}_n$ is given by

$$\mathscr{M}_n = \{\Pr_n(\mathbf{Y}_n = \cdot;\theta) : \theta \in [0,1]\}, \quad 1 \le n \le N,$$

[3] If there is no well-defined upper bound on sample size, we can take $N = \infty$ and define $\mathscr{M}_n$ for all finite sample sizes $n = 1,2,\ldots$.

with

$$\Pr_n(\mathbf{Y}_n = \mathbf{y}; \theta) = \prod_{1 \le i \ne j \le n} \theta^{y_{ij}} (1-\theta)^{1-y_{ij}}, \tag{5.14}$$

for all $\mathbf{y} = (y_{ij})_{1 \le i,j \le n} \in \{0,1\}^{n \times n}$ with 0 along the diagonal. For $1 \le m \le n \le N$, define $\Sigma_{m,n} = \mathbf{S}_{m,n}$ as the selection map

$$\begin{aligned} \mathbf{S}_{m,n} : \{0,1\}^{n \times n} &\to \{0,1\}^{m \times m} \\ \mathbf{y} &\mapsto \mathbf{S}_{m,n}\mathbf{y} = (y_{ij})_{1 \le i,j \le m}, \end{aligned} \tag{5.15}$$

as given previously in (3.6). The sampling maps $\{\Sigma_{m,n}\}_{n \ge m \ge 1}$ describe the context for network data modeled by $\{\mathscr{M}_n\}_{n \ge 1}$. Together the pair $(\{\mathscr{M}_n\}_{1 \le n \le N}, \{\Sigma_{m,n}\}_{1 \le m \le n \le N})$ is a completely specified statistical model. Since the Erdős–Rényi–Gilbert model is consistent under selection, i.e., $\mathbf{Y}_n$ distributed in (5.14) with parameter θ implies that $\mathbf{S}_{m,n}\mathbf{Y}_n$ is distributed according to (5.14) on $\{0,1\}^{m \times m}$ with the same parameter θ, it follows that $\mathbf{S}_{m,n}\mathscr{M}_n = \mathscr{M}_m$ for all $1 \le m \le n \le N$ and the model is coherent.

5.6.2 *Example 2: ERGM under selection sampling*

Let $N < \infty$ be finite and for each $n = 1, \ldots, N$ define $\mathscr{M}_n$ as the set of distributions given by the ERGM in (2.8) with sufficient statistic $\Delta_n(\mathbf{y}) = \sum_{1 \le i < j < k \le n} y_{ij} y_{jk} y_{ki}$ that counts the number of triangles in an undirected graph $\mathbf{y} \in \mathscr{N}_n$. For $m \le n$, define $\Sigma_{m,n} = \mathbf{S}_{m,n}$ as in (5.15) above. With this specification, the $\mathbf{S}_{m,n}$-induced model $\mathbf{S}_{m,n}\mathscr{M}_n$ on $\{0,1\}^{m \times m}$ is given by the distribution of $\mathbf{S}_{m,n}\mathbf{Y}_n$ obtained by selection sampling from the ERGM with some parameter θ and sufficient statistic $\Delta_n(\cdot)$. According to the main theorem of [138], $\mathbf{S}_{m,n}\mathbf{Y}_n$ does not follow the ERGM with parameter θ and sufficient statistic $\Delta_m(\cdot)$ on $\{0,1\}^{m \times m}$. It follows that $\mathbf{S}_{m,n}\mathbf{Y}_n \ne_{\mathscr{D}} \mathbf{Y}_m$ for non-degenerate choices of θ. Some further calculation shows that $\mathbf{S}_{m,n}\mathscr{M}_n \ne \mathscr{M}_m$, and therefore the model is incoherent.

5.6.3 *Example 3: Erdős–Rényi–Gilbert model under edge sampling*

Suppose N is unknown and let $\mathscr{M}_n$ be just as in Section 5.6.1, except now $\mathscr{M}_n$ is defined for all $n \ge 1$. Instead of selection sampling, define each $\Sigma_{m,n}$ as the operation which samples m edges uniformly at random from $\mathbf{y} \in \{0,1\}^{n \times n}$ and discards the rest. If $\mathbf{y}$ has fewer than m edges, then we put $\Sigma_{m,n}\mathbf{y} = \mathbf{y}$. By specifying each $\mathscr{M}_n$ as a set of candidate distributions on the space of graphs, the vertices are (implicitly) treated as the units under $\{\mathscr{M}_n\}_{n \ge 1}$. (See Section 3.7 for further discussion on implicit and explicit units.) On the other hand, $\Sigma_{m,n}$ describes a procedure for sampling m edges from $\mathbf{Y}_n$, suggesting that the edges are the units. Thus, the distributions in $\mathscr{M}_m$ are defined on the space $\{0,1\}^{m \times m}$ of graphs with m vertices, but the outcome $\Sigma_{m,n}\mathbf{Y}_n$ by sampling m edges from $\mathbf{Y}_n$ will generally reside in a different space. This model is incoherent.

5.7 Invariance principles

Section 5.5 explains why the definitions of statistical model (Definition 5.1) and coherence (Definitions 5.2–5.3) are necessary to ensure that inferences in the schematic of (5.11) can be interpreted within a well-articulated context. The reader might have noticed that network models, and statistical models more generally, are almost never specified according to Definition 5.1, i.e., as a collection of probability distributions together with (a family of) sampling or generating mechanisms. More often, the connection between observed (data) and unobserved (population) is either undetermined or defined implicitly by a symmetry or invariance principle. The former case occurs, for example, when the finite sample models are given by a family of ERGMs whose sufficient statistics lack the separable increments property, cf. [138]. The latter case is common in classical statistics where, for example, the i.i.d. assumption serves as the ultimate symmetry condition linking sample and population. In time series analysis, *stationarity* is often assumed to relate observations across time. And when modeling discrete structures, such as networks, the principle of *exchangeability* plays a central role, and is the focus of the next several chapters.

In any statistical analysis, there is always a tradeoff between principle and practice. The model must strike a balance between computational tractability, empirical properties, and theoretical considerations, and all the while maintain the integrity of the intended application and uphold the principles of statistical inference. Among the sacrifices that might be made in the course of striking this balance, coherence cannot be among them. Exchangeability is often cited as a way to achieve coherence without sacrificing tractability. But exchangeability often imposes more symmetry than may be warranted.

Speaking generally, a network model is *exchangeable* if it assigns equal probability to any two networks that are equivalent up to relabeling of the units (in some well-defined sense). For network data represented as a graph with labeled vertices, exchangeability is a distributional invariance with respect to arbitrary relabeling of vertices. This condition together with the conventional networks-as-graphs perspective (Section 1.2) explains why the most common form of exchangeability studied in the networks literature to date has been *vertex exchangeability* (Chapter 6). (Because vertex exchangeability was, until recently, the only kind of exchangeability studied in the networks literature, many authors refer to these models simply as *exchangeable*.)

As a general principle, exchangeability assumes that the observed network structure is homogeneous with respect to its context. For vertex exchangeability, this homogeneity reflects a probabilistic symmetry with respect to vertex properties, implying that the observed vertices are representative of the population of all vertices. Because this assumption is plainly violated in many modern networks applications, vertex exchangeability often takes a backseat to newly developed invariance principles, such as relative exchangeability, edge exchangeability, and relational exchangeability.

Relative exchangeability (Chapter 8) expresses the probabilistic symmetries of a network in terms of the symmetries of some underlying structure on the population of vertices. Canonical examples include stochastic blockmodels, latent space models, and random graphs whose symmetries are determined by another graph. I discuss these further in Chapter 8.

The vertex-centric point of view assumed by vertex exchangeable and relatively exchangeable models is incongruous with applications for which the vertices are not the units or the observed vertices are not a representative sample of the population. *Edge exchangeability* (Chapter 9) takes an initial step beyond the networks-as-graphs perspective by instead representing network data as an edge-labeled graph and assuming a model which assigns equal probability to any two edge-labeled graphs that are isomorphic up to relabeling of their edges. Edge exchangeable models make up a canonical class of statistical models for interaction networks, such as those discussed in Sections 1.6.4–1.6.5 and 3.6.1. See Chapter 9 and [54] for further discussion of these models.

Out of edge exchangeability, and the change in perspective it inspires, emerges the more general principle of *relational exchangeability* (Chapter 10), which is better suited to networks constructed from a sample of higher-order relations, such as descriptions of the Internet network by repeated path sampling. Relational exchangeability was introduced by Crane and Dempsey [53] shortly after their development of edge exchangeability [54]. An immediate upshot of edge and relational exchangeability is its ability to model sparse, scale-free structure within a sound statistical framework. I discuss several additional consequences of this approach in Chapters 9 and 10.

5.8 Further reading

The concept of model coherence introduced in this chapter is a novelty of the network modeling framework proposed in [52]. Though naturally arising by consideration of how networks and other complex data structures are modeled, coherence is a staple of any statistical analysis. The closest known discussion to the present one on coherence can be found in McCullagh's treatment of statistical modeling [120], though the connection may be hard to see for readers unfamiliar with basic category theory.

I have focused mostly on coherence in sampling models $(\{\mathscr{M}_n\}_{n\geq 1}, \{\Sigma_{m,n}\}_{n\geq m\geq 1})$ for which each $\Sigma_{m,n}$ is a (random) sampling mechanism. In general, however, the sampling mechanism may not be known precisely, raising the possibility that the sampling mechanism should also be described by a set of candidate sampling schemes $\mathscr{G}_{m,n}$. In this case, the model is specified by $(\{\mathscr{M}_n\}_{n\geq 1}, \{\mathscr{G}_{m,n}\}_{n\geq m\geq 1})$ for sets of candidate distributions $\mathscr{M}_n$ and sets of sampling mechanisms $\mathscr{G}_{m,n}$. The $\Sigma_{m,n}$-induced model defined in (5.7) can be refined to the *$\mathscr{G}_{m,n}$-induced model* defined by

$$\mathscr{G}_{m,n}\mathscr{M}_n = \bigcup_{\Sigma_{m,n}\in\mathscr{G}_{m,n}} \Sigma_{m,n}\mathscr{M}_n = \{\Sigma_{m,n}P_n : \Sigma_{m,n} \in \mathscr{G}_{m,n} \text{ and } P_n \in \mathscr{M}_n\},$$

where $\Sigma_{m,n}\mathscr{M}_n$ is as defined in (5.7). Coherence (C) is modified accordingly by replacing the singleton $\Sigma_{m,n}$ by the set $\mathscr{G}_{m,n}$, i.e., (C) $\mathscr{G}_{m,n}\mathscr{M}_n = \mathscr{M}_m$ for all $n \geq m \geq 1$. From such a model, the inferred $\hat{P}_m \subseteq \mathscr{M}_m$ based on $\mathbf{Y}_m$ produces an inference of sampling scheme-distribution pairs $(\hat{\Sigma}, \hat{P})$ for $\hat{\Sigma} \in \mathscr{G}_{m,n}$ and $\hat{P}_n \in \mathscr{M}_n$ satisfying

$$\hat{\Sigma}\hat{P}_n \in \hat{P}_m.$$

This extra generality adds no conceptual difficulty, but we streamline the presentation above and below by restricting to singleton sampling schemes $\mathscr{G}_{m,n} = \{\Sigma_{m,n}\}$. An analogous extension is possible for generative models.

Given the varied circumstances under which network data arise, there are a number of other practical matters which one must take into account when specifying a model in a specific application. Considerations such as computational feasibility and interpretability of estimates bear on the usefulness of a statistical model, and such aspects are important when applying a model to data. If, for example, the candidate models $\mathscr{M}_n$ provide a poor description of the variation in and uncertainty about the actual network, or the assumed sampling scheme $\Sigma_{m,n}$ or generating mechanism $\Pi_{m,n}$ is artificial or inaccurate, then a coherent model will be of limited practical use. Because these kinds of judgments are application-specific, such matters lie far beyond the scope of the 'Probabilistic Foundations of Statistical Network Analysis' covered here. Practical considerations about goodness of fit and model validation in network analysis remain underdeveloped, but the reader can consult [93, 107] for discussion and further references.

5.9 Solutions to exercises

5.9.1 Exercise 5.1

From Definition 5.3, we need to show that

$$(\mathrm{C}') \quad \Pi_{m,n}\mathscr{M}_m = \mathscr{M}_n \quad \text{for all } n \geq m \geq 1,$$

for any $(\{\mathscr{M}_n\}_{n\geq 1}, \{\Pi_{m,n}\}_{n\geq m\geq 1})$ defined as in (5.9) and (5.10). By (5.9), (C$'$) holds by definition for all $m = n-1$ and $n \geq 2$. For arbitrary $n \geq 1$ and $m < n$, we have

$$\Pi_{m,n}\mathscr{M}_m = \Pi_{m+1,n} \circ \Pi_{m,m+1}\mathscr{M}_m = \Pi_{m+1,n}(\Pi_{m,m+1}\mathscr{M}_m) = \Pi_{m+1,n}\mathscr{M}_{m+1},$$

by (5.10). This is enough to set up a double induction on m and n which establishes (C$'$).

Chapter 6

Vertex exchangeable models

Over the next several chapters I will discuss the implications of the exchangeability assumption for statistical network analysis, with special emphasis on the connection between exchangeability, sampling, and statistical inference. I begin here with *vertex exchangeable random graph models*, which are specifically tailored to network data represented as a graph (V,E) with vertices labeled in a set V and edges $E \subseteq V \times V$. Such networks are expressed mathematically as an adjacency array $\mathbf{y} = (y_{ij})_{i,j \in V}$ with

$$y_{ij} = \begin{cases} 1, & (i,j) \in E, \\ 0, & \text{otherwise}, \end{cases}$$

as in (2.1). When edges are undirected (i.e., $(i,j) \in E$ if and only if $(j,i) \in E$ for all $i,j \in V$) I often write $ij \in E$ (instead of $(i,j) \in E$ or $(j,i) \in E$) to indicate that there is an edge between i and j. See Figure 6.1 for illustration.

6.1 Preliminaries: Formal definition of exchangeability

Exchangeability can be understood informally as a probabilistic invariance with respect to arbitrary relabeling of the statistical units. Throughout this chapter, the units are assumed to be the vertices of a graph represented as an array $\mathbf{y} = (y_{ij})_{i,j \in V}$. For any such array, the *relabeling of* $\mathbf{y}$ *by permutation* $\sigma : V \to V$ is defined as the graph $\mathbf{y}^\sigma$ obtained by relabeling the vertices of $\mathbf{y}$ according to σ, i.e.,

$$\mathbf{y}^\sigma = (y_{\sigma(i)\sigma(j)})_{i,j \in V}, \tag{6.1}$$

where a permutation σ is a bijective function $V \to V$. Figure 6.2 illustrates the action in (6.1).

Definition 6.1 (Vertex exchangeability) *A random array* $\mathbf{Y} = (Y_{ij})_{i,j \in V}$ *is* (vertex) exchangeable *if*

$$\mathbf{Y}^\sigma =_{\mathscr{D}} \mathbf{Y} \quad \textit{for all permutations } \sigma : V \to V. \tag{6.2}$$

In terms of the probability operator Pr, exchangeability (6.2) *is equivalent to*

$$Pr(\mathbf{Y} \in A) = Pr(\mathbf{Y} \in A^\sigma), \textit{ for all permutations } \sigma : V \to V \textit{ and } A \subseteq \{0,1\}^{V \times V}, \tag{6.3}$$

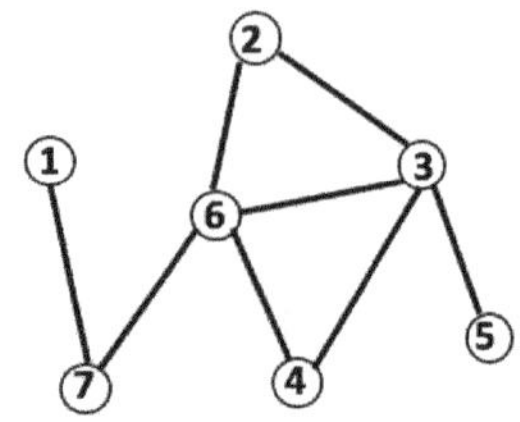

Figure 6.1 *An undirected graph with vertex set* $V = \{1,2,3,4,5,6,7\}$ *and (undirected) edge set* $E = \{17,23,26,34,35,36,46,67\}$.

where $A^{\sigma} = \{\mathbf{y}^{\sigma} : \mathbf{y} \in A\}$ *is the set obtained by relabeling all elements of A according to* σ. *In particular, for any* $\mathbf{y} \in \{0,1\}^{V \times V}$, (6.3) *implies*

$$Pr(\mathbf{Y} = \mathbf{y}) = Pr(\mathbf{Y} = \mathbf{y}^{\sigma}) \quad \textit{for all permutations } \sigma : V \to V.$$

Without loss of generality we restrict to arrays $\mathbf{Y}$ indexed by either $[n] = \{1,\dots,n\}$, if V is a finite set of size n, or $\mathbb{N} = \{1,2,\dots\}$, if V is countably infinite. When the context is clear, we refer to any $\mathbf{Y}$ satisfying (6.2) simply as *exchangeable*.

Exchangeable distributions vs. exchangeable models

Before moving on to more technical aspects of vertex exchangeability, note the distinction between exchangeable *distributions* and exchangeable *models*. Since we focus in this book on statistical analysis, it makes sense that we are primarily interested in network models, as formalized by $(\{\mathscr{M}_n\}_{n\geq 1}, \mathscr{C})$ with each $\mathscr{M}_n$ being a set of distributions on $\{0,1\}^{n\times n}$ interpreted in the context $\mathscr{C}$ as in Chapter 5. In this framework, the condition of exchangeability in (6.3) easily extends to a definition of an *exchangeable model* as any $(\{\mathscr{M}_n\}_{n\geq 1}, \mathscr{C})$ satisfying

(Ex) every $P \in \mathscr{M}_n$ is exchangeable in the sense of (6.3), for every $n \geq 1$.

Since exchangeable models consist entirely of exchangeable distributions, we can restrict our attention throughout most of this and the following chapters to exchangeable distributions, from which the salient properties of exchangeable models follow. We need only speak of exchangeable *models* when considering how exchangeability behaves in a certain *context* $\mathscr{C}$. (As in previous chapters, I specialize here to the case in which $\mathbf{Y}$ is $\{0,1\}$-valued, but note that the discussion applies just as well to multigraphs and weighted graphs, in which case Y_{ij} can take values in the non-negative integers or real numbers, respectively.)

6.2 Implications of exchangeability

In Chapter 3, I discussed how the sampling scheme is a necessary component for converting partial observations of a network into inferences about the unobserved population. In much of statistical work, exchangeability plays this same role, serving

Figure 6.2 *Two isomorphic graphs. The graph on the right can be obtained by relabeling the graph on the left by the permutation* $\sigma : [7] \to [7]$ *with* $\sigma(1) = 3$, $\sigma(2) = 2$, $\sigma(3) = 5$, $\sigma(4) = 1$, $\sigma(5) = 6$, $\sigma(6) = 7$, *and* $\sigma(7) = 4$. *Both graphs are assigned equal probability by any vertex exchangeable model.*

primarily to tie together the observed and unobserved by (implicitly) asserting that the observed units are 'representative' of the unobserved, in a sense made precise by (6.2). Over the course of this and the next several chapters, we examine how exchangeability accomplishes this task on the basis of two implicit assumptions:

(i) the population is homogeneous and

(ii) the sampled units are representative of the population.

With the vertices serving as the units, (i) and (ii) speak of homogeneity and representativeness as viewed from the perspective of the vertices. Subsequent chapters will explore the implications of (i) and (ii) for networks viewed from the perspective of edges (Chapter 9), more general relations (Chapter 10), and edge patterns observed over a fixed duration of time (Section 7.3).

(i) Homogeneous population

Assuming for the moment a population of finite size N, exchangeability of the population network motivates the interpretation of a network which "looks the same" (in distribution) from the perspective of every vertex. To describe this formally, compare the expected behavior of any network statistic $g : \{0,1\}^{N\times N} \to \mathscr{X}$ between $\mathbf{Y}_N$ and $\mathbf{Y}_N^\sigma$, where $\sigma : [N] \to [N]$ is any permutation of $[N]$. (Here $\mathscr{X}$ is a generic space which we tacitly assume to be equipped with a σ-field with respect to which g is measurable.) By the definition of exchangeability in (6.2), we readily deduce

$$\mathbf{Y}_N =_{\mathscr{D}} \mathbf{Y}_N^\sigma \quad \Longrightarrow \quad g(\mathbf{Y}_N) =_{\mathscr{D}} g(\mathbf{Y}_N^\sigma) \tag{6.4}$$

for all permutations $\sigma : [N] \to [N]$. From this, the distribution of any network statistic is unaffected by the perspective from which the data is viewed, whether as $\mathbf{Y}_N$ or $\mathbf{Y}_N^\sigma$, and thus inferences based on any statistic are invariant with respect to any such change in perspective. It is in this sense that a vertex exchangeable model views the network as "looking the same" from the viewpoint of every vertex.

Exercise 6.1 *Formally derive the implication in* (6.4).

(ii) Representative sampling

When fitting an exchangeable model to sampled network data, the homogeneity implied by (6.4) suggests further that the sampled network is "representative" of other networks which could have been observed under the same circumstances. To see this directly, let $\mathbf{S}_{n,N}$ be the selection map $\{0,1\}^{N\times N} \to \{0,1\}^{n\times n}$ defined in (3.6). Observe first that any permutation $\sigma : [n] \to [n]$ of the sampled vertices $[n]$ can be extended to a permutation $\sigma^* : [N] \to [N]$ of the population $[N]$ by putting

$$\sigma^*(i) = \begin{cases} \sigma(i), & 1 \le i \le n, \\ i, & n < i \le N. \end{cases} \tag{6.5}$$

For $\mathbf{Y}_N = (Y_{ij})_{1\le i,j\le N}$ satisfying (6.2), $\mathbf{Y}_n = \mathbf{Y}_N|_{[n]} = (Y_{ij})_{1\le i,j\le n}$, and any permutation $\sigma : [n] \to [n]$, we have

$$\mathbf{Y}_n^{\sigma} = (\mathbf{S}_{n,N}\mathbf{Y}_N)^{\sigma} = \mathbf{S}_{n,N}(\mathbf{Y}_N^{\sigma^*}) =_{\mathscr{D}} \mathbf{S}_{n,N}\mathbf{Y}_N = \mathbf{Y}_n,$$

where the first equality is by definition of $\mathbf{Y}_n = \mathbf{S}_{n,N}\mathbf{Y}_N$, the second is by definition of σ^* in (6.5), and the equality in distribution $\mathbf{S}_{n,N}(\mathbf{Y}_N^{\sigma^*}) =_{\mathscr{D}} \mathbf{S}_{n,N}\mathbf{Y}_N$ follows from exchangeability of $\mathbf{Y}_N$ and (6.4).[1] We see immediately that exchangeability is preserved by selection sampling, and in particular that exchangeability of $\mathbf{Y}_N$ implies exchangeability of each of its finite restrictions $\mathbf{Y}_N|_{[n]} = (Y_{ij})_{1\le i,j\le n}$, $n \ge 1$.

More generally, if $\Sigma_{n,N}$ is any (possibly random) sampling scheme for which n vertices are chosen from $[N]$ in a way that does not depend on $\mathbf{Y}_N$, then $\Sigma_{n,N}\mathbf{Y}_N$ is exchangeable. To see this explicitly, let $\Sigma_{n,N}$ be a random sampling scheme with distribution satisfying

$$\Pr(\Sigma_{n,N} = \mathbf{S}_{n,N}^{\psi} \mid \mathbf{Y}_N = \mathbf{y}) = \Pr(\Sigma_{n,N} = \mathbf{S}_{n,N}^{\psi}) \quad \text{for all } \mathbf{y} \in \{0,1\}^{N\times N}, \quad \psi : [n] \to [N],$$

that is, $\Sigma_{n,N}$ is independent of $\mathbf{Y}_N$. By the law of total probability, the distribution of $\Sigma_{n,N}\mathbf{Y}_N$ can be computed by

$$\begin{aligned}
\Pr(\Sigma_{n,N}\mathbf{Y}_N = \mathbf{y}) &= \\
&= \sum_{\mathbf{y}'\in\{0,1\}^{N\times N}} \Pr(\mathbf{Y}_N = \mathbf{y}') \sum_{\psi:[n]\to[N]} \Pr(\Sigma_{n,N} = \mathbf{S}_{n,N}^{\psi} \mid \mathbf{Y}_N = \mathbf{y}')\mathbf{1}(\mathbf{S}_{n,N}^{\psi}\mathbf{y}' = \mathbf{y}) \\
&= \sum_{\mathbf{y}'\in\{0,1\}^{N\times N}} \Pr(\mathbf{Y}_N = \mathbf{y}') \sum_{\psi:[n]\to[N]} \Pr(\Sigma_{n,N} = \mathbf{S}_{n,N}^{\psi})\mathbf{1}(\mathbf{S}_{n,N}^{\psi}\mathbf{y}' = \mathbf{y}) \\
&= \sum_{\psi:[n]\to[N]} \Pr(\Sigma_{n,N} = \mathbf{S}_{n,N}^{\psi}) \sum_{\mathbf{y}'\in\{0,1\}^{N\times N}} \Pr(\mathbf{Y}_N = \mathbf{y}')\mathbf{1}(\mathbf{S}_{n,N}^{\psi}\mathbf{y}' = \mathbf{y}) \\
&= \sum_{\psi:[n]\to[N]} \Pr(\Sigma_{n,N} = \mathbf{S}_{n,N}^{\psi})\Pr(\mathbf{S}_{n,N}^{\psi}\mathbf{Y}_N = \mathbf{y}),
\end{aligned} \tag{6.6}$$

where the third line follows by independence of $\Sigma_{n,N}$ and $\mathbf{Y}_N$ and the last line follows

[1] Note that since the expression $\mathbf{S}_{n,N}(\mathbf{Y}_N^{\sigma^*})$ only makes sense if σ^* is applied first to $\mathbf{Y}_N$ and then $\mathbf{S}_{n,N}$ is applied to the relabeled structure $\mathbf{Y}_N^{\sigma^*}$, then we could have omitted the parentheses and written $\mathbf{S}_{n,N}\mathbf{Y}_N^{\sigma^*}$ without any potential for confusion. Moving forward, I often adopt this more economical notation.

since

$$\sum_{\mathbf{y}' \in \{0,1\}^{N \times N}} \Pr(\mathbf{Y}_N = \mathbf{y}')\mathbf{1}(\mathbf{S}_{n,N}^{\psi}\mathbf{y}' = \mathbf{y}) = \Pr(\mathbf{S}_{n,N}^{\psi}\mathbf{Y}_N = \mathbf{y}) \tag{6.7}$$

for any fixed $\psi : [n] \to [N]$. By (6.7), the following holds for every permutation $\sigma : [n] \to [n]$ and $\sigma^* : [N] \to [N]$ defined as in (6.5):

$$\begin{aligned}
\Pr((\Sigma_{n,N}\mathbf{Y}_N)^{\sigma} = \mathbf{y}) &= \sum_{\psi:[n]\to[N]} \Pr(\Sigma_{n,N} = \mathbf{S}_{n,N}^{\psi}) \Pr((\mathbf{S}_{n,N}^{\psi}\mathbf{Y}_N)^{\sigma} = \mathbf{y}) \\
&= \sum_{\psi:[n]\to[N]} \Pr(\Sigma_{n,N} = \mathbf{S}_{n,N}^{\psi}) \Pr(\mathbf{S}_{n,N}^{\psi}\mathbf{Y}_N^{\sigma^*} = \mathbf{y}) \\
&= \sum_{\psi:[n]\to[N]} \Pr(\Sigma_{n,N} = \mathbf{S}_{n,N}^{\psi}) \Pr(\mathbf{S}_{n,N}^{\psi}\mathbf{Y}_N = \mathbf{y}) \\
&= \Pr(\Sigma_{n,N}\mathbf{Y}_N = \mathbf{y}),
\end{aligned} \tag{6.8}$$

whence $\Sigma_{n,N}\mathbf{Y}_N$ is exchangeable. A similar calculation allows us to conclude the precise sense in which $\mathbf{Y}_n = (Y_{ij})_{1 \le i,j \le n}$ can be regarded as "representative" of any sampled network $\Sigma_{n,N}\mathbf{Y}_N$ of size n. By the homogeneity of $\mathbf{Y}_N$ (i.e., (6.4)) and the independence of $\Sigma_{n,N}$ and $\mathbf{Y}$, we compute, for any permutation $\sigma : [N] \to [N]$,

$$\begin{aligned}
\Pr(\Sigma_{n,N}\mathbf{Y}_N^{\sigma} = \mathbf{y}) &= \sum_{\psi:[n]\to[N]} \Pr(\Sigma_{n,N} = \mathbf{S}_{n,N}^{\psi}) \Pr(\mathbf{S}_{n,N}^{\psi}\mathbf{Y}_N^{\sigma} = \mathbf{y}) \\
&= \sum_{\psi:[n]\to[N]} \Pr(\Sigma_{n,N} = \mathbf{S}_{n,N}^{\psi}) \Pr(\mathbf{S}_{n,N}^{\psi}\mathbf{Y}_N = \mathbf{y}) \\
&= \Pr(\Sigma_{n,N}\mathbf{Y}_N = \mathbf{y}),
\end{aligned} \tag{6.9}$$

indicating that the distribution of the sampled network is the same regardless of the initial perspective from which the population network $\mathbf{Y}_N$ is viewed, i.e., either as $\mathbf{Y}_N$ or $\mathbf{Y}_N^{\sigma}$ for any permutation $\sigma : [N] \to [N]$.

Exercise 6.2 *Suppose $\mathbf{Y}_N$ is exchangeable and $\Sigma_{n,N}$ is independent of $\mathbf{Y}_N$, for $1 \le n \le N$. Prove that $\Sigma_{n,N}\mathbf{Y}_N =_{\mathscr{D}} \mathbf{S}_{n,N}\mathbf{Y}_N$, i.e.,*

$$Pr(\Sigma_{n,N}\mathbf{Y}_N = \mathbf{y}) = Pr(\mathbf{S}_{n,N}\mathbf{Y}_N = \mathbf{y}) \quad \textit{for all } \mathbf{y} \in \{0,1\}^{n \times n}.$$

Remark 6.1 *Notice that $(\Sigma_{n,N}\mathbf{Y}_N)^{\sigma}$ in* (6.8) *and $\Sigma_{n,N}\mathbf{Y}_N^{\sigma}$ in* (6.9) *differ in the order in which the sampling and relabeling operations are applied, with* (6.8) *giving the distribution of the network obtained by first sampling and then relabeling (i.e., $(\Sigma_{n,N}\mathbf{Y}_N)^{\sigma}$) and* (6.9) *giving the distribution of the network obtained by first relabeling according to $\sigma : [N] \to [N]$ and then sampling (i.e., $\Sigma_{n,N}\mathbf{Y}_N^{\sigma}$). By comparing* (6.8) *and* (6.9) *we see that the relabeling and sampling operations commute for exchangeable models, provided the sampling scheme is independent of the network being sampled from.*

To close this section, I stress that the assumption of vertex exchangeability does *not* correspond to observing a network by simple random vertex sampling. It instead implies that the distribution of the network restricted to every subset of vertices is

Figure 6.3 *The set $\mathscr{U}_3$ of unlabeled graphs with exactly 3 vertices.*

representative of every other subset of the same size, as shown in the above calculations and Exercise 6.2. As long as the sampling scheme is independent of $\mathbf{Y}_N$, the resulting observation is exchangeable. Observing a network by simple random vertex sampling is one of many such sampling schemes. Selection sampling is another.

Implications of exchangeability

Implications (i) and (ii) are worth keeping in mind when considering whether a vertex exchangeable model is appropriate for a given application. Regarding (i), it seems rare in practice to encounter a network for which such a strong homogeneity property is realistic. In fact, modern network analysis is almost exclusively focused on networks that are heterogeneous and complex in how the vertices interact with one another, hence the term 'complex networks'. Regarding (ii), the prevalence of heterogeneity in real networks nullifies the practical use of any representative vertex sampling scheme. Under a representative sampling scheme, inferences apply to any 'typical' part of the network. But if understanding heterogeneity is the goal, then it is precisely the 'atypical' parts of the network that are of primary interest. Representative sampling is of little use for detecting such atypicality. A specific illustration of this is shown in Section 3.4. The further assumption that $\Sigma_{n,N}$ and $\mathbf{Y}_N$ are independent is also unrealistic in most cases. In later chapters we discuss how some of the relational sampling schemes surveyed in Section 3.6 may be more useful.

6.3 Finite exchangeable random graphs

Although the definition of vertex exchangeability in (6.2) applies to both finite and countably infinite arrays, the theory differs between the two cases. Before studying the theory of countable exchangeable models in detail (Sections 6.4–6.6), I briefly discuss exchangeable models for finite networks.

Let $N < \infty$ and assume $\mathbf{Y}_N$ is vertex exchangeable in the sense of (6.2). For the generic description of such *finite exchangeable random graph models*, we let $\mathscr{U}_N$ denote the set of all *unlabeled graphs* with N vertices. Intuitively, such 'unlabeled graphs' are understood as the abstract 'shapes' obtained by removing the vertex labels from ordinary vertex-labeled graphs. The shapes corresponding to unlabeled graphs with 3 vertices are drawn in Figure 6.3.

Formally, the elements of $\mathscr{U}_N$ are the equivalence classes of $\{0,1\}^{N\times N}$ up to relabeling. For $\mathbf{y},\mathbf{y}' \in \{0,1\}^{N\times N}$, write $\mathbf{y} \cong \mathbf{y}'$ to indicate that there exists a permutation

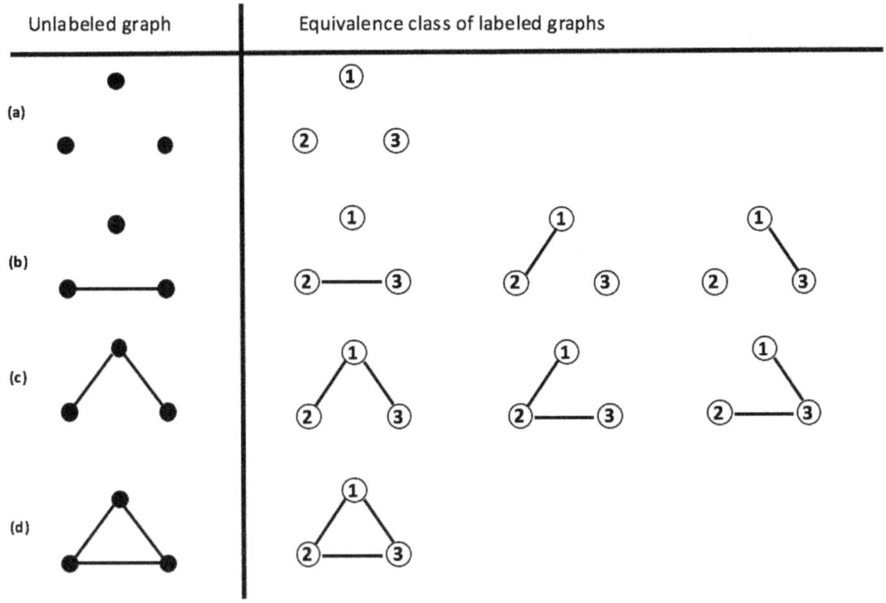

Figure 6.4 *The set of unlabeled graphs $\mathscr{U}_3$ from Figure 6.3 along with their associated equivalence class of labeled graphs as defined in* (6.10). *Notice how the unlabeled graphs on the left describe the 'shape' of the labeled graphs on the right.*

$\sigma : [N] \to [N]$ such that $\mathbf{y}^\sigma = \mathbf{y}'$, i.e., $\mathbf{y}'$ can be obtained from $\mathbf{y}$ by some relabeling σ of its vertices. This relation associates each $\mathbf{y} \in \{0,1\}^{N\times N}$ with the equivalence class

$$\langle \mathbf{y} \rangle_{\cong} = \{\mathbf{y}' \in \{0,1\}^{N\times N} : \mathbf{y}' \cong \mathbf{y}\} \tag{6.10}$$

consisting of all $\mathbf{y}' \in \{0,1\}^{N\times N}$ with the same 'shape' as $\mathbf{y}$. It follows that

$$\langle \mathbf{y} \rangle_{\cong} = \langle \mathbf{y}' \rangle_{\cong} \quad \text{if and only if} \quad \mathbf{y} \cong \mathbf{y}',$$

putting these equivalence classes in correspondence with the elements of $\mathscr{U}_N$. The grouping of elements according to their equivalence under relabeling makes precise our interpretation of the elements in $\mathscr{U}_N$ as abstract 'shapes' of graphs with vertices labeled in $\{1,\dots,N\}$. Figure 6.4 shows this correspondence, with each of the four unlabeled graphs in $\mathscr{U}_3$ associated to its equivalence class via (6.10).

By the definition of exchangeability in (6.3), the distribution of any exchangeable random graph $\mathbf{Y}_N$ satisfies

$$\Pr(\mathbf{Y}_N = \mathbf{y}) = \Pr(\mathbf{Y}_N = \mathbf{y}') \quad \text{for all } \mathbf{y} \cong \mathbf{y}'.$$

From this definition we should expect the probability of the event '$\mathbf{Y}_N = \mathbf{y}$' to depend only on the equivalence class $\langle \mathbf{y} \rangle_{\cong}$. We make this observation precise as follows.

Given any probability distribution $\mathbf{p}$ on $\mathscr{U}_N$, let $\Pr(\mathbf{Y}_N^* = \cdot;\mathbf{p})$ denote the distribution of $\mathbf{Y}_N^*$ generated by first drawing a random unlabeled graph $\mathbf{U} \sim \mathbf{p}$ and, given $\mathbf{U} = \mathbf{u}$, choosing $\mathbf{Y}_N^*$ uniformly from the equivalence class of all $\mathbf{y}' \in \{0,1\}^{N\times N}$ for which $\langle \mathbf{y}'\rangle_{\cong} = \mathbf{u}$. By this description, the distribution of $\mathbf{Y}_N^*$ can be expressed as

$$\Pr(\mathbf{Y}_N^* = \mathbf{y};\mathbf{p}) = \mathbf{p}(\langle \mathbf{y}\rangle_{\cong})/|\langle \mathbf{y}\rangle_{\cong}|, \quad \mathbf{y} \in \{0,1\}^{N\times N}, \tag{6.11}$$

where $\mathbf{p}(\langle \mathbf{y}\rangle_{\cong})$ is the probability assigned to $\langle \mathbf{y}\rangle_{\cong}$ by $\mathbf{p}$ and $|\langle \mathbf{y}\rangle_{\cong}|$ is the cardinality of the equivalence class $\langle \mathbf{y}\rangle_{\cong}$.

Theorem 6.1 *Let $\mathbf{Y}_N$ be an exchangeable random graph on $\{0,1\}^{N\times N}$. Then there exists a unique probability distribution $\mathbf{p}$ on $\mathscr{U}_N$ such that $\mathbf{Y}_N =_{\mathscr{D}} \mathbf{Y}_N^*$, for $\mathbf{Y}_N^*$ with distribution $Pr(\mathbf{Y}_N^* = \cdot;\mathbf{p})$ given in* (6.11).

Exercise 6.3 *Prove Theorem 6.1.*

In many practical situations, the formulation in (6.11) is neither computationally nor theoretically feasible. In particular, the space $\mathscr{U}_N$ is elusive both conceptually and computationally, making the act of defining or inferring a probability distribution on $\mathscr{U}_N$ practically impossible without imposing further constraints. In such cases, we may sometimes assume a more tractable parametric form which preserves exchangeability while expressing the model structure in terms of a small number of parameters and network statistics. Exchangeable ERGMs from Section 2.3 make up one such class of models.

6.3.1 *Exchangeable ERGMs*

Recall the exponential random graph model (ERGM) from Section 2.3: for real-valued parameters $\theta = (\theta_1,\ldots,\theta_k)$ and sufficient statistics $T = (T_1,\ldots,T_k) : \{0,1\}^{N\times N} \to \mathbb{R}^k$, the class of ERGMs on $\{0,1\}^{N\times N}$ with parameter θ and sufficient statistic T assigns probability

$$\Pr(\mathbf{Y}_N = \mathbf{y};\theta,T) = \frac{\exp\left\{\sum_{i=1}^k \theta_i T_i(\mathbf{y})\right\}}{\sum_{\mathbf{y}^*\in\{0,1\}^{N\times N}} \exp\left\{\sum_{i=1}^k \theta_i T_i(\mathbf{y}^*)\right\}} \tag{6.12}$$

to each $\mathbf{y} \in \{0,1\}^{N\times N}$. The distribution in (6.12) assigns the same probability to any $\mathbf{y}$ and $\mathbf{y}'$ for which $\sum_{i=1}^k \theta_i T_i(\mathbf{y}) = \sum_{i=1}^k \theta_i T_i(\mathbf{y}')$. Thus, in order for (6.12) to be exchangeable, the parameters and sufficient statistics must satisfy

$$\sum_{i=1}^k \theta_i T_i(\mathbf{y}) = \sum_{i=1}^k \theta_i T_i(\mathbf{y}^\sigma)$$

for all $\mathbf{y} \in \{0,1\}^{N\times N}$ and all permutations $\sigma : [N] \to [N]$. This condition is ensured as long as every T_i is preserved under permutation, i.e., $T_i(\mathbf{y}) = T_i(\mathbf{y}^\sigma)$ for all permutations $\sigma : [N] \to [N]$. Under this requirement, the sufficient statistics must depend on none or all of the entries of $\mathbf{y}$, not on a subset as in the statistics $y_{i\bullet} = \sum_{j=1}^N y_{ij}$ and $y_{\bullet j} = \sum_{i=1}^N y_{ij}$ of the p_1 model.

Because its parameters and sufficient statistics can favor one labeling of the vertices over another, the ERGM need not be exchangeable. For example, if just a single sufficient statistic $T_1(\mathbf{y}) = \sum_{j=2}^{N} y_{1j}$ counts the out-degree of the vertex labeled 1, then all arrays $\mathbf{y}$ with the same first row sum have equal probability; but the model is not exchangeable since the distribution of $\mathbf{Y}_N$ is not preserved by any σ which permutes the first row with another row. The p_1 model (Chapter 2) fails to be exchangeable since the vertex specific parameters $\alpha_1, \ldots, \alpha_N$ and $\beta_1, \ldots, \beta_N$ allow for skewness in the distribution unless $\alpha_i = \alpha_j$ and $\beta_i = \beta_j$ for all $i \neq j$.

On the other hand, exchangeable ERGMs can account for some types of non-local dependence, such as transitivity and cliques; but these 'non-local' characteristics can only be captured on a 'global' scale, e.g., the number of triangles is invariant under relabeling vertices and can be used to quantify the degree of transitivity in a network. As alluded in Sections 2.3 and 2.5, the viability of ERGMs for modeling sampled network data hinges on whether its sufficient statistics possess the separable increments property; see Section 2.3 and [138] for more details.

Research Problem 6.1 *Consider an ERGM with sufficient statistic T for a random graph $\mathbf{Y}_N$ of size $N \geq 1$. Let $\mathbf{Y}_n = \Sigma_{n,N}\mathbf{Y}_N$ for a sampling scheme $\Sigma_{n,N}$ given by (i) selection, (ii) snowball sampling, (iii) simple random edge sampling, or (iv) any other reasonable sampling method. Derive the distribution of the statistic $T(\mathbf{Y}_n)$. (Hint: this problem is likely to be very difficult. If unable to derive the distribution of $T(\mathbf{Y}_n)$, then try to compute other distributional properties, such as moments. Though I am not aware of prior attempts at this problem, some related work might be found in the probability and statistical physics literature.)*

Research Problem 6.2 *Consider an ERGM for $\mathbf{Y}_N$ whose sufficient statistics do not have separable increments. Is there a necessary and/or sufficient condition on the sufficient statistics (weaker than separable increments) by which selection sampling yields a random graph $\mathbf{S}_{n,N}\mathbf{Y}_N$ still of ERGM-type? If so, can the natural parameter and canonical sufficient statistics be expressed as a function of the parameters and sufficient statistics in the model for $\mathbf{Y}_N$? (Note the question is asking only if $\mathbf{S}_{n,N}\mathbf{Y}_N$ can be expressed in the form of* (2.8) *for* some *natural parameter and canonical sufficient statistic, not necessarily the same parameter and statistic as in the model for $\mathbf{Y}_N$.)*

Following up on our discussion from Chapter 3, if $\mathbf{Y}_N$ is distributed as an ERGM with natural parameter θ and Σ is any (possibly random) sampling operation, then the distribution of $\mathbf{Y}^* = \Sigma\mathbf{Y}_N$ will also be parameterized by θ. But since the nature of this parameterization is generally unknown, inference for θ based on $\mathbf{Y}^*$ is unlikely to be straightforward. With that said, the fact that selection sampling is rarely a relevant sampling scheme in practical applications suggests that alternative approaches to estimating network models ought to be explored.

Research Problem 6.3 *Let $\mathbf{Y}_N$ be distributed as an ERGM with natural parameter θ and canonical sufficient statistic T as in* (6.12)*. Let Σ^* be any (possibly random) sampling operation, including perhaps $\Sigma^* = \mathbf{S}_{n,N}$ for some fixed $n \geq 1$. How can the parameter θ be estimated from an observation $\mathbf{Y}^* = \Sigma^*\mathbf{Y}_N$? In the case when the sufficient statistics have separable increments and $\Sigma^* = \mathbf{S}_{n,N}$, maximum likelihood* *estimation for θ should be sufficient based on the results in [138]. (Verify this.) But I am otherwise aware of no theoretical or computational study of this question.*

Problem 6.3 applies just as well whether $\mathbf{Y}_N$ is exchangeable or not. In particular, if $\mathbf{Y}_N$ is modeled by $\mathscr{M}$ on $\{0,1\}^{N\times N}$ and Σ^* is any sampling operation, how can the optimal distribution of $\mathbf{Y}_N$ be estimated based on an observation of $\Sigma^*\mathbf{Y}_N$? In principle, the general approach in (5.11) remains valid, but in practice it is hard to compute the estimator for arbitrary sampling operations Σ^*. Given the challenge of modeling in the presence of arbitrary sampling Σ^*, I leave this as an important problem for future research.

6.4 Countable exchangeable models

Since in many applications the population is known to be finite but of unknown size, it is often natural to specify a network model by $(\{\mathscr{M}_n\}_{n\geq 1},\mathscr{C})$, without declaring any finite bound on the population size. Note well, however, that this specification does not imply that the population is assumed to be infinite. It only means that the model does not presume that the size of the population is limited by any given finite upper bound. Leaving the population size unspecified imposes further limitations on the class of coherent models due to the constraint of having to satisfy the coherence condition (C) for all $n \geq m \geq 1$.

6.4.1 *Graphon models*

The term 'graphon' originated with the study of limits of graph sequences in the early 2000s; see [115]. These same 'graphon models' were originally introduced under a different name in the late 1970s and early 1980s by Diaconis and Freedman [61], cf. [9, 12, 62]. Below we use the notation ϕ to denote what is commonly called a graphon in the current terminology. This notation follows the convention of Aldous [9, pp. 124–125], whose 'ϕ-processes' correspond to 'graphon processes directed by ϕ' in the forthcoming discussion.

6.4.1.1 *Generative model*

Let $\Phi = \{\phi : [0,1]\times[0,1]\to[0,1]\}$ be the set of all functions $[0,1]\times[0,1]\to[0,1]$ with 0 diagonal (i.e., $\phi(u,u)\equiv 0$ for all $0\leq u\leq 1$) and fix any $\phi\in\Phi$.[2] Any such ϕ parameterizes the distribution of a random array $\mathbf{Y}_N$ as follows. To construct $\mathbf{Y}_N$, first draw $U_1,\ldots,U_N$ i.i.d. Uniform$[0,1]$ and, given $U_1,\ldots,U_N$, assign Y_{ij} conditionally independently with probabilities

$$\begin{aligned}\Pr(Y_{ij}=1\mid U_1,\ldots,U_N;\phi) &= \phi(U_i,U_j) \quad \text{and}\\ \Pr(Y_{ij}=0\mid U_1,\ldots,U_N;\phi) &= 1-\phi(U_i,U_j) \qquad (6.13)\end{aligned}$$

[2] We allow ϕ to be asymmetric, i.e., $\phi(u,v)\neq\phi(v,u)$, so that the resulting random graphs, e.g., in (6.14), are directed. We can restrict the model to undirected graphs by forcing $\phi(u,v)=\phi(v,u)$ and putting $Y_{ji}=Y_{ij}$ for $i<j$. Since these two cases are technically similar, we discuss only the directed case here.

for all $1 \leq i \neq j \leq N$. The random variables $U_1,\ldots,U_N$ can be thought of as latent random effects associated to each vertex so that, for example, all relations involving vertex i share a common dependence on U_i.

From the construction in (6.13), we can express the distribution of $\mathbf{Y}_N = (Y_{ij})_{1\leq i,j\leq N}$ in closed form by

$$\Pr(\mathbf{Y}_N = \mathbf{y};\phi) = \int_{[0,1]^N} \prod_{1\leq i\neq j\leq N} \phi(u_i,u_j)^{y_{ij}}(1-\phi(u_i,u_j))^{1-y_{ij}}\, du_1\cdots du_N, \quad (6.14)$$

for each $\mathbf{y} = (y_{ij})_{1\leq i,j\leq N}$. (The expression in (6.14) can be explained as follows. Conditionally independently given $U_1 = u_1,\ldots,U_N = u_N$, each outcome y_{ij} has probability

$$\phi(u_i,u_j)^{y_{ij}}(1-\phi(u_i,u_j))^{1-y_{ij}}.$$

The integral is obtained by averaging over the $U_1,\ldots,U_N$, which are independent and uniformly distributed in $[0,1]$.)

Definition 6.2 (Graphon distribution) *For any $\phi : [0,1]\times[0,1] \to [0,1]$, the construction in* (6.13) *is called a* graphon process directed by ϕ, *or simply a* ϕ-process, *and the distribution in* (6.14) *is called the* graphon distribution with parameter ϕ.

Generative description

The construction in (6.13) suggests the following generative description of graphon models. Given $\mathbf{Y}_n = \mathbf{y}$, construct $\mathbf{Y}_{n+1}$ from $\mathbf{Y}_n$ by using the same random variables $U_1,\ldots,U_n$ as in the construction of $\mathbf{Y}_n$ plus an additional Uniform$[0,1]$ random variable U_{n+1} that is independent of $U_1,\ldots,U_n$ and of $\mathbf{Y}_n$. A fully generative version of this model, expressed in the framework of Chapter 4, can be written as $(\{\mathscr{M}_n^{\Phi}\}_{n\geq 1},\{\Pi_{m,n}\}_{n\geq m\geq 1})$, with

$$\mathscr{M}_n^{\Phi} = \{\Pr(\mathbf{Y}_n = \cdot;\phi) : \phi \in \Phi\} \quad (6.15)$$

for $\Pr(\mathbf{Y}_n = \cdot;\phi)$ as defined in (6.14), and for each $\phi \in \Phi$ the generating mechanism $\Pi_{m,n} = \Pi_{m,n}^{\phi}$ defined as follows. Given $\mathbf{y} \in \{0,1\}^{m\times m}$, let $\Pi_{m,n}^{\phi}\mathbf{y} = (Y_{ij})_{1\leq i,j\leq n}$ be a random graph obtained by first generating $U_1,\ldots,U_n$ i.i.d. Uniform$[0,1]$ conditional on the event '$\mathbf{Y}_m = \mathbf{y}$', and then putting

$$Y_{ij} = y_{ij} \quad \text{for } 1 \leq i,j \leq m \quad (6.16)$$

and otherwise drawing Y_{ij} as in (6.13), for all $i \neq j$. This description has the net effect of conditioning on $(Y_{ij})_{1\leq i,j\leq m} = \mathbf{y}$ and generating the rest of $\mathbf{Y}$ according to the ϕ-process.

Remark 6.2 *The specification surrounding display* (6.15) *beckons Remark 5.1, in which I noted the possible dependence between context and candidate distributions. Perhaps a better way to state the generative graphon model above would be to express each element of $\mathscr{M}_m^{\Phi}$ as a pair consisting of $Pr(\mathbf{Y}_m = \cdot;\phi)$, for $\phi \in \Phi$, together* *with the system* $\{\Pi^{\phi}_{m,n}\}_{n\geq m}$ *of all generating mechanisms that go along with that candidate distribution. In this way, each candidate distribution in* $\mathscr{M}^{\Phi}_m$, *for* $m \geq 1$, *has its own context, e.g.,* $\{\Pi_{m,n}\}_{n\geq m}$. *The context* $\mathscr{C}$ *in the formal specification of the model (Definition 5.1) would then state additional coherence conditions for these dependent contexts, e.g., for* $\ell \leq m \leq n$ *and* $\phi \in \Phi$, $\mathscr{C}$ *expresses the identity* $\Pi^{\phi}_{\ell,m} \circ \Pi^{\phi}_{m,n} = \Pi^{\phi}_{\ell,n}$, *which would not be apparent otherwise. For present purposes, the concern expressed in this remark is too technical and will be ignored moving forward.*

Exercise 6.4 *For any* $\phi \in \Phi$, *verify that if* $\mathbf{Y}_m$ *is a* ϕ*-process on* $\{0,1\}^{m\times m}$, *then* $\Pi^{\phi}_{m,n}\mathbf{Y}_n$ *is a* ϕ*-process on* $\{0,1\}^{n\times n}$.

Sampling description

From the sequential construction in (6.13) we see that selection $\mathbf{S}_{n,N}\mathbf{Y}_N$ from the ϕ-process is again a ϕ-process on $\{0,1\}^{n\times n}$, cf. Section 4.1 and equation (4.4). We can also show this directly by performing the same calculation used to show consistency of the p_1 model in (3.10). We first consider the case $n = N-1$ and $\mathbf{y} \in \{0,1\}^{(N-1)\times(N-1)}$ and compute

$$
\begin{aligned}
&\Pr(\mathbf{S}_{N-1,N}\mathbf{Y}_N = \mathbf{y};\phi) = \\
&= \sum_{\mathbf{y}^*\in\{0,1\}^{N\times N}:\mathbf{S}_{N-1,N}\mathbf{y}^*=\mathbf{y}} \Pr(\mathbf{Y}_N = \mathbf{y}^*;\phi) \\
&= \sum_{\mathbf{y}^*\in\{0,1\}^{N\times N}:\mathbf{S}_{N-1,N}\mathbf{y}^*=\mathbf{y}} \int_{[0,1]^N} \prod_{1\leq i\neq j\leq N} \phi(u_i,u_j)^{y^*_{ij}}(1-\phi(u_i,u_j))^{1-y^*_{ij}}\, du_1\cdots du_N \\
&= \int_{[0,1]^N} \sum_{\mathbf{y}^*\in\{0,1\}^{N\times N}:\mathbf{S}_{N-1,N}\mathbf{y}^*=\mathbf{y}} \prod_{1\leq i\neq j\leq N} \phi(u_i,u_j)^{y^*_{ij}}(1-\phi(u_i,u_j))^{1-y^*_{ij}}\, du_1\cdots du_N \\
&= \int_{[0,1]^N} \prod_{1\leq i\neq j\leq N-1} \phi(u_i,u_j)^{y_{ij}}(1-\phi(u_i,u_j))^{1-y_{ij}} \times \\
&\qquad \times \prod_{i=1}^{N-1} (\phi(u_i,u_N) + (1-\phi(u_i,u_N)))(\phi(u_N,u_i) + (1-\phi(u_N,u_i)))du_1\cdots du_N \\
&= \int_{[0,1]^{N-1}} \prod_{1\leq i\neq j\leq N-1} \phi(u_i,u_j)^{y_{ij}}(1-\phi(u_i,u_j))^{1-y_{ij}}\, du_1\cdots du_{N-1} \\
&= \Pr(\mathbf{Y}_{N-1} = \mathbf{y};\phi),
\end{aligned}
\tag{6.17}
$$

for $\Pr(\mathbf{Y}_{N-1} = \mathbf{y};\phi)$ as in (6.14). Iterating this calculation for $N-2, N-3, \ldots$ shows that $\mathbf{S}_{n,N}\mathbf{Y}_N =_{\mathscr{D}} \mathbf{Y}_n$ for $\mathbf{Y}_n$ obeying a ϕ-process on $\{0,1\}^{n\times n}$, for all $1 \leq n \leq N$.

Theorem 6.2 *For any graphon* $\phi : [0,1] \times [0,1] \to [0,1]$, *let* $\mathbf{Y}_m$ *and* $\mathbf{Y}_n$ *follow the graphon distribution with parameter* ϕ *on* $\{0,1\}^{m\times m}$ *and* $\{0,1\}^{n\times n}$, *respectively, as in* (6.14). *Then* $\mathbf{Y}_m$ *and* $\mathbf{Y}_n$ *are consistent under selection.*

By Theorem 6.2, the family of graphon processes is consistent under selection in the sense that if $\mathbf{Y}_n$ obeys the ϕ-process, then so does $\mathbf{S}_{m,n}\mathbf{Y}_n$ for all $1 \leq m \leq n$. It follows that the subclass of models $(\{\mathcal{M}_n^\Psi\}_{n\geq 1}, \{\mathbf{S}_{m,n}\}_{n\geq m\geq 1})$ is coherent (Definition 5.2) for any subset of graphons $\Psi \subseteq \Phi$, where

$$\mathcal{M}_n^\Psi = \{\Pr(\mathbf{Y}_n = \cdot; \phi) : \phi \in \Psi\}, \quad n \geq 1.$$

Exercise 6.5 *Let $\mathbf{Y}_N$ be distributed according to a ϕ-process on $\{0,1\}^{N\times N}$, as in* (6.14). *Prove that $\mathbf{Y}_N$ is exchangeable.*

6.4.2 Aldous–Hoover theorem

Exercises 6.4 and 6.5 establish that graphons are exchangeable and projective in the sense of (4.4) from Chapter 4. In this section we see that graphons are canonical among all vertex exchangeable generative models on $\{0,1\}^{\mathbb{N}\times\mathbb{N}}$, in the sense that every such distribution can be expressed in terms of a graphon model.

From any function $f : [0,1]^4 \to \{0,1\}$, we construct a random array $\mathbf{Y}^* = (Y_{ij}^*)_{i,j\geq 1}$ by taking U_0, $(U_i)_{i\geq 1}$, and $(U_{ij})_{i,j\geq 1}$ to be i.i.d. Uniform$[0,1]$ random variables and putting

$$Y_{ij}^* = f(U_0, U_i, U_j, U_{ij}), \quad j \neq i \geq 1, \tag{6.18}$$

and $Y_{ii}^* = 0$ for all $i \geq 1$. It should clear from the construction that $\mathbf{Y}^*$ is exchangeable: since for any permutation $\sigma : \mathbb{N} \to \mathbb{N}$ the relabeled uniform random variables U_0, $(U_{\sigma(i)})_{i\geq 1}$, and $(U_{\sigma(i)\sigma(j)})_{i,j\geq 1}$ remain i.i.d. and Uniform$[0,1]$, relabeling the indices of $\mathbf{Y}^*$ in (6.18) does not affect its distribution. The Aldous–Hoover theorem states the converse, namely that any countable exchangeable random array admits such a construction.

Theorem 6.3 (Aldous–Hoover theorem [8, 92]) *Let $\mathbf{Y} = (Y_{ij})_{i,j\geq 1}$ be an exchangeable $\{0,1\}$-valued array. Then there exists a measurable function $f : [0,1]^4 \to \{0,1\}$ such that $\mathbf{Y} =_{\mathcal{D}} \mathbf{Y}^*$, for $\mathbf{Y}^*$ as constructed in* (6.18).[3]

Translated into the language of graphons, the Aldous–Hoover theorem says that to every infinite exchangeable random array $\mathbf{Y}$ there is a probability distribution φ on the space of graphons Φ such that $\mathbf{Y}$ can be constructed from a graphon process directed by ϕ chosen randomly according to φ. To see this connection explicitly, first suppose that $\mathbf{Y} = (Y_{ij})_{i,j\geq 1}$ is constructed as in (6.18) for some function f that is constant in its first argument, i.e., $f(a,\cdot,\cdot,\cdot) = f(b,\cdot,\cdot,\cdot)$ for all $a,b \in [0,1]$. Writing $f(-,\cdot,\cdot,\cdot)$ to indicate this function with an arbitrary first argument, we define a graphon $\phi_f : [0,1]\times[0,1] \to [0,1]$ by

$$\phi_f(u,v) = \int_0^1 f(-,u,v,w)\, dw, \quad u,v \in [0,1]. \tag{6.19}$$

As written, the defining integral for $\phi_f(u,v)$ in (6.19) accumulates the mass associated to the event that $f(-,u,v,w) = 1$, and thus $\phi_f(u,v)$ equals the conditional

[3]The version of this theorem for undirected graphs, i.e., symmetric arrays $\mathbf{Y} = (Y_{ij})_{i,j\geq 1}$, has the additional constraint that f is symmetric in its middle two arguments, i.e., $f(\cdot,u,v,\cdot) = f(\cdot,v,u,\cdot)$ for all $0 \leq u,v \leq 1$. See [9, Chapter 14].

probability of the event '$Y_{ij} = 1$' given $U_i = u$ and $U_j = v$. Conversely, from any graphon $\phi : [0,1]^2 \to [0,1]$, we define $f_\phi : [0,1]^4 \to \{0,1\}$ by

$$f_\phi(-,u,v,w) = \begin{cases} 1, & 0 \le w \le \phi(u,v), \\ 0, & \text{otherwise}, \end{cases} \tag{6.20}$$

to recover (6.18).

Exercise 6.6 *Verify that* $\mathbf{Y}^*$ *constructed from the* ϕ_f*-process, for* ϕ_f *defined in* (6.19), *is equal in distribution to* $\mathbf{Y}^*$ *constructed from* f *as in* (6.18).

If f is not constant in its first argument, then each $f(a,\cdot,\cdot,\cdot)$ determines a function indexed by $a \in [0,1]$,

$$\phi_{f,a}(u,v) = \int_0^1 f(a,u,v,w)\,dw, \quad u,v \in [0,1]. \tag{6.21}$$

In this case, the random argument U_0 in (6.18) is accounted for by choosing ϕ randomly from $\{\phi_{f,a} : a \in [0,1]\} \subseteq \Phi$ by taking $U_0 \sim \text{Uniform}[0,1]$ and putting $\phi = \phi_{f,a}$ on the event '$U_0 = a$'. We define φ as the probability distribution on Φ induced by this protocol.

The outcome of the Aldous–Hoover theorem can thus be translated as

every exchangeable random array in $\{0,1\}^{\mathbb{N}\times\mathbb{N}}$ *is distributed as a mixture of graphon processes.*

With this interpretation, the distribution of every vertex exchangeable family $(\mathbf{Y}_n)_{n\ge 1}$ that is consistent under selection can be expressed for each $n \ge 1$ by

$$\Pr(\mathbf{Y}_n = \mathbf{y}; \varphi) = \int_\Phi \left(\int_{[0,1]^n} \prod_{1\le i,j\le n} \phi(u_i,u_j)^{y_{ij}} (1-\phi(u_i,u_j))^{1-y_{ij}} du_1 \cdots du_n \right) \varphi(d\phi) \tag{6.22}$$

for some probability distribution φ on Φ. (Note that this distribution φ is not unique. See Section 6.6.2 below.)

6.4.3 Graphons and vertex exchangeability

The back and forth between graphons ϕ and the function f in the Aldous–Hoover theorem points to an important relationship between graphons and vertex exchangeable models. In particular, graphon models characterize the subclass of vertex exchangeable arrays $\mathbf{Y}$ that are *dissociated*, meaning that $\mathbf{Y}|_S$ and $\mathbf{Y}|_T$ are independent for all $S,T \subseteq \mathbb{N}$ such that $S \cap T = \emptyset$. In words, a dissociated random graph is one for which any two nonoverlapping subgraphs are independent. Consulting the Aldous–Hoover representation in (6.18) and the subsequent connection between the Aldous–Hoover theorem and graphons reveals the following relationship between graphons and dissociated random graphs.

Under what conditions is the random array $\mathbf{Y}^*$ constructed in (6.18) dissociated? To answer this, take any $S,T \subseteq \mathbb{N}$ with $S \cap T = \emptyset$ and consider what is necessary for the arrays

$$(f(U_0,U_i,U_j,U_{ij}))_{i,j\in S} \quad \text{and} \quad (f(U_0,U_{i'},U_{j'},U_{i'j'}))_{i',j'\in T}$$

to be independent. By assumption, the sets S and T indexing the independent uniform random variables in either subgraph are disjoint, and so the only common source of randomness between $(f(U_0,U_i,U_j,U_{ij}))_{i,j\in S}$ and $(f(U_0,U_{i'},U_{j'},U_{i'j'}))_{i',j'\in T}$ comes from the first argument U_0. It follows that these arrays can be independent if and only if f does not depend on its first argument, and we have already seen in (6.20) that such functions correspond precisely to graphons.

Dissociated exchangeable arrays make up the subclass of infinite exchangeable models which are *ergodic* with respect to the action of relabeling on $\{0,1\}^{\mathbb{N}\times\mathbb{N}}$. In statistical inference, ergodic measures correspond to the class of submodels whose finite sample statistics converge to a deterministic limit as the sample size grows. In Section 6.6.1, we further discuss the significance of ergodicity in network analysis and draw a connection between graphons and more familiar statistical applications involving sequences of random variables.

6.4.4 *Subsampling description*

The Aldous–Hoover theorem makes explicit how every infinite exchangeable model can be interpreted as a generative model; see the discussion surrounding (6.16) above. We have also seen how the graphon description translates easily into a coherent sampling model under selection sampling via (4.4). We now observe a more direct interpretation of vertex exchangeable models, and hence also graphon models, in terms of the distribution of subgraphs induced by simple random vertex sampling.

For $n\geq m\geq 1$, let $\psi:[m]\to[n]$ be any one-to-one function and recall the definition of the ψ-induced selection function $\mathbf{S}^{\psi}_{m,n}$ from (3.17). (The action $\mathbf{S}^{\psi}_{m,n}$ first selects an $m\times m$ submatrix by removing all rows and columns not indexed by $\psi(1),\ldots,\psi(m)$ and then relabels the vertices $\psi(1),\ldots,\psi(m)$ by $1,\ldots,m$, respectively, i.e.,

$$\mathbf{S}^{\psi}_{m,n}\mathbf{y}=\mathbf{y}^{\psi}=(y_{\psi(i)\psi(j)})_{1\leq i,j\leq m}.)$$

For $n\geq m\geq 1$, let $\mathbf{y}\in\{0,1\}^{n\times n}$ and $\mathbf{x}\in\{0,1\}^{m\times m}$. The statistic $\mathrm{ind}(\mathbf{x}:\mathbf{y})$ counts the number of *induced copies of* $\mathbf{x}$ *in* $\mathbf{y}$, defined formally as

$$\mathrm{ind}(\mathbf{x}:\mathbf{y})=\sum_{\text{injections }\psi:[m]\to[n]}\mathbf{1}(\mathbf{y}^{\psi}=\mathbf{x}).$$

In other words, $\mathrm{ind}(\mathbf{x}:\mathbf{y})$ is the number of different ways that one could observe $\mathbf{x}$ by performing ψ-selection on $\mathbf{y}$, for some $\psi:[m]\to[n]$. From this, the *density of* $\mathbf{x}$ *in* $\mathbf{y}$ is given by

$$\delta(\mathbf{x}:\mathbf{y})=\frac{1}{n^{\downarrow m}}\mathrm{ind}(\mathbf{x}:\mathbf{y}),\tag{6.23}$$

where $n^{\downarrow m}=n(n-1)\cdots(n-m+1)$ is the total number of injections $\psi:[m]\to[n]$. (To count all injections $[m]\to[n]$, note that a unique value $\psi(j)=i_j$ must be assigned for each $j=1,\ldots,m$. There are n options when assigning $\psi(1)$; once $\psi(1)$ is assigned, there are $n-1$ options left to assign $\psi(2)$; and so on until we assign $\psi(m)$ from one of the remaining $n-m+1$ elements, for a total of $n^{\downarrow m}$ possible choices.)

For a straightforward example, let $n = 3$ and $m = 2$ with

$$\mathbf{y} = \begin{pmatrix} 0 & 1 & 1 \\ 1 & 0 & 0 \\ 0 & 1 & 0 \end{pmatrix} \quad \text{and} \quad \mathbf{x} = \begin{pmatrix} 0 & 1 \\ 0 & 0 \end{pmatrix}. \tag{6.24}$$

There are $3^{\downarrow 2} = 6$ total injections $\psi : [2] \to [3]$, with $(\psi(1), \psi(2))$ given respectively by

$$(1,2), \quad (1,3), \quad (2,1), \quad (2,3), \quad (3,1), \quad (3,2).$$

Of these injections only $(1,3)$ and $(3,2)$ give $\mathbf{y}^{\psi} = \mathbf{x}$ for $\mathbf{y}$ and $\mathbf{x}$ as in (6.24), and thus $\delta(\mathbf{x} : \mathbf{y}) = 2/6$ in this case.

Since every injection $\psi : [m] \to [n]$ extracts a unique array $(y_{\psi(i),\psi(j)})_{1 \le i,j \le m} \in \{0,1\}^{m \times m}$, the densities $\delta(\mathbf{x} : \mathbf{y})$ for fixed $\mathbf{y}$ always satisfy

(i) $0 \le \delta(\mathbf{x} : \mathbf{y}) \le 1$ for all $\mathbf{x} \in \{0,1\}^{m \times m}$ and

(ii) $\sum_{\mathbf{x} \in \{0,1\}^{m \times m}} \delta(\mathbf{x} : \mathbf{y}) = 1$,

and therefore determine a probability distribution on $\{0,1\}^{m \times m}$, denoted

$$\Pr(\mathbf{Y}_m = \mathbf{x}; \mathbf{y}) = \delta(\mathbf{x} : \mathbf{y}), \quad \mathbf{x} \in \{0,1\}^{m \times m}. \tag{6.25}$$

The distribution in (6.25) can be interpreted as that of a random graph $\mathbf{Y}_m$ sampled uniformly from among all size m subgraphs of $\mathbf{y}$. To make this precise, we let $\Sigma_{m,n}$ be a uniform random sampling map with distribution

$$\Pr(\Sigma_{m,n} = \mathbf{S}^{\psi}_{m,n}) = 1/n^{\downarrow m}, \quad \psi : [m] \to [n], \tag{6.26}$$

and let $\mathbf{Y}_n = \mathbf{y}$ be fixed (i.e., $\Pr(\mathbf{Y}_n = \mathbf{y}) = 1$). Then

$$\Pr(\Sigma_{m,n} \mathbf{Y}_n = \mathbf{x}) = \delta(\mathbf{x} : \mathbf{y}), \quad \mathbf{x} \in \{0,1\}^{m \times m},$$

for $\delta(\mathbf{x} : \mathbf{y})$ as in (6.23). The distribution in (6.25) is therefore the proper distribution one would assign to $\mathbf{Y}_m$ obtained by simple random sampling from a fixed, known graph $\mathbf{y}$.

Notice that to every permutation $\sigma : [m] \to [m]$ and every injection $\psi : [m] \to [n]$ for which $\mathbf{y}^{\psi} = \mathbf{x}$ the composite $\psi \circ \sigma : [m] \to [n]$ is an injection such that $\mathbf{y}^{\psi \circ \sigma} = \mathbf{x}^{\sigma}$. It follows that $\mathrm{ind}(\mathbf{x} : \mathbf{y}) = \mathrm{ind}(\mathbf{x}^{\sigma} : \mathbf{y})$ for all $\mathbf{x} \in \{0,1\}^{m \times m}$ and all permutations $\sigma : [m] \to [m]$, and in particular $\delta(\mathbf{x} : \mathbf{y}) = \delta(\mathbf{x}^{\sigma} : \mathbf{y})$. Thus, the distribution defined in (6.25) is exchangeable by virtue of the simple random vertex sampling mechanism in (6.26).

Now, let $\mathbf{Y} = (Y_{ij})_{i,j \ge 1}$ be a random $\{0,1\}$-valued array generated by a graphon process directed by $\phi : [0,1] \times [0,1] \to [0,1]$. From the calculation in (6.17), the restriction $\mathbf{Y}_n = \mathbf{Y}|_{[n]} = (Y_{ij})_{1 \le i,j \le n}$ is also distributed according to the graphon distribution (6.14) with parameter ϕ. For $n \ge m \ge 1$ and $\mathbf{x} \in \{0,1\}^{m \times m}$, the (random) number of copies of $\mathbf{x}$ in $\mathbf{Y}_n$ is given by

$$\mathrm{ind}(\mathbf{x} : \mathbf{Y}_n) = \sum_{\text{injections } \psi:[m] \to [n]} \mathbf{1}(\mathbf{Y}_n^{\psi} = \mathbf{x}). \tag{6.27}$$

By linearity of the expectation operator E,[4] we have

$$\begin{aligned} E(\mathrm{ind}(\mathbf{x}:\mathbf{Y}_n)) &= E\left(\sum_{\text{injections } \psi:[m]\to[n]} \mathbf{1}(\mathbf{Y}_n^{\psi}=\mathbf{x})\right) \\ &= \sum_{\text{injections } \psi:[m]\to[n]} E(\mathbf{1}(\mathbf{Y}_n^{\psi}=\mathbf{x})) \\ &= \sum_{\text{injections } \psi:[m]\to[n]} \Pr(\mathbf{Y}_n^{\psi}=\mathbf{x};\phi) \\ &= \sum_{\text{injections } \psi:[m]\to[n]} \Pr(\mathbf{Y}_n|_{[m]}=\mathbf{x};\phi) \\ &= n^{\downarrow m}\Pr(\mathbf{Y}_m=\mathbf{x};\phi), \end{aligned}$$

where the second-to-last equality follows by exchangeability and consistency under selection of graphon models (Exercises 6.4 and 6.5). It follows that $E(\delta(\mathbf{x}:\mathbf{Y}_n)) = \Pr(\mathbf{Y}_m=\mathbf{x};\phi)$ for every $n\geq 1$ and $\mathbf{x}\in\{0,1\}^{m\times m}$, allowing us to define

$$\delta(\mathbf{x}:\phi) = \int_{[0,1]^m} \prod_{1\leq i\neq j\leq m} \phi(u_i,u_j)^{x_{ij}}(1-\phi(u_i,u_j))^{1-x_{ij}}\,du_1\cdots du_m, \quad \mathbf{x}\in\{0,1\}^{m\times m}, \tag{6.28}$$

as the expected density of $\mathbf{x}$ in a random graph distributed according to the ϕ-process. Thus, $\delta(\mathbf{x}:\phi)$ gives the distribution of a random subgraph obtained by sampling m vertices uniformly at random without replacement from a (finite) population graph distributed according to the ϕ-process.

Research Problem 6.4 *Let $\mathcal{M}_n^{\Phi}$ be the class of all graphon distributions on $\{0,1\}^{n\times n}$. Is it true that $(\{\mathcal{M}_n\}_{n\geq 1},\{\Sigma_{m,n}\}_{n\geq m\geq 1})$ is coherent only if the (random) sampling mechanism $\Sigma_{m,n}$ does not depend on $\mathbf{Y}_n$, for all $n\geq m\geq 1$? Articulate why this outcome is significant for the field of network analysis, particularly as it relates to the viability of graphon models for statistical applications.*

The exposition above shows how the homomorphism density in (6.23) corresponds to uniform random vertex sampling from a fixed $\mathbf{y}$. The expression in (6.28) extends this interpretation to mixtures over $\mathbf{y}$ generated from the ϕ-process. This can be further generalized by defining a probability distribution μ on the sampling operation $\Sigma_{m,n}$ and setting

$$\delta(\mathbf{x}:\mathbf{y};\mu) = \sum_{\text{injections } \psi:[m]\to[n]} \mathbf{1}(\mathbf{y}^{\psi}=\mathbf{x})\mu(\psi).$$

Putting these two together (with μ possibly depending on $\mathbf{y}$) results in a general family of distributions for $\Sigma_{m,n}\mathbf{Y}_n = \mathbf{Y}_m$ on $\{0,1\}^{m\times m}$, with $\mathbf{Y}_n \sim F$ on $\{0,1\}^{n\times n}$

[4]The expectation of a random variable Z with probability density f_Z on $(-\infty,\infty)$ is defined as $E(Z) = \int_{-\infty}^{\infty} z f_Z(z)\,dz$. In general, for a random variable Z distributed according to a probability measure μ, the expectation is given by the Lebesgue integral $E(Z) = \int_{-\infty}^{\infty} z\,\mu(dz)$. For any measurable set $A\subseteq\mathbb{R}$, the expectation of the indicator at A equals the probability that Z lies in A, i.e., $E(\mathbf{1}(Z\in A)) = \Pr(Z\in A)$.

and $\Sigma_{m,n} \sim \mu$:

$$\Pr(\mathbf{Y}_m = \mathbf{x}; F, \mu) = \sum_{\mathbf{y} \in \{0,1\}^{n \times n}} \sum_{\text{injections } \psi:[m] \to [n]} \mathbf{1}(\mathbf{y}^{\psi} = \mathbf{x}) F(\mathbf{y}) \mu(\psi \mid \mathbf{y}),$$

for each $\mathbf{x} \in \{0,1\}^{m \times m}$. It is left as an open problem to study the implications of this general setup.

6.5 Viability of graphon models

Having established the basic structure of graphon models, we can now revisit the implications of vertex exchangeability from Section 6.2, focusing especially on the viability of graphon models for statistical applications.

Statistical implications of the Aldous–Hoover theorem

The representation of exchangeable arrays as a mixture of dissociated arrays, as in (6.22), lays bare some limitations of graphon models for statistical analysis. Notice first that the Erdős–Rényi–Gilbert distribution with parameter $\theta \in (0,1)$, as defined in (2.10), corresponds to a graphon process directed by the constant function $\phi_\theta(-,-) \equiv \theta$. Going hand-in-hand with its simple structure, Erdős–Rényi–Gilbert random graphs do not exhibit the heterogeneous features that are commonly observed in real-world networks, such as sparsity, clustering, and heavy-tailed degree distributions.

The representation in (6.22) also makes clear why more general vertex exchangeable population models offer little or no improvement over Erdős–Rényi–Gilbert for modeling heterogeneous networks. Compare the Erdős–Rényi–Gilbert distribution (2.10) with parameter θ, for which

$$\Pr(\mathbf{Y}_n = \mathbf{y}; \theta) = \prod_{1 \le i \ne j \le n} \theta^{y_{ij}} (1-\theta)^{1-y_{ij}}, \quad \mathbf{y} \in \{0,1\}^{n \times n},$$

with the generic representation of the finite-dimensional distributions of countable vertex exchangeable models in (6.22), having

$$\Pr(\mathbf{Y}_n = \mathbf{y}; \varphi) =$$
$$= \int_{\Phi} \left(\int_{[0,1]^n} \prod_{1 \le i \ne j \le n} \phi(u_i,u_j)^{y_{ij}} (1 - \phi(u_i,u_j))^{1-y_{ij}} \, du_1 \cdots du_n \right) \varphi(d\phi).$$

Notice first in the inside integral that each vertex is independently assigned a latent feature U_i distributed uniformly in $[0,1]$. Given the latent features $U_1, \ldots, U_n$, $\mathbf{Y}_n$ behaves as a generalized Erdős–Rényi–Gilbert-type network with each edge (i,j) present conditionally independently with probability $\phi(U_i,U_j)$. Thus, although ϕ is able to encode some local heterogeneities in the network structure—for example, $\phi(u,v) = |u - v|$ implies that a vertex i with $U_i \approx 0$ will have degree approximately $n/2$ for large n whereas i with $U_i \approx 1/2$ will have degree approximately $n/4$ for large n—any such heterogeneities result from the random assignment of the latent vertex effects $U_1, U_2, \ldots$, and not any inherent, observable features of the population. Thus, ϕ-processes, and coherent vertex exchangeable models in general, are in effect glorified Erdős–Rényi–Gilbert models. The formulation, as in (6.13), which at first appears rather general turns out to be quite bland and not very useful for many statistical applications.

6.5.1 *Implication 1: Dense structure*

Let $\mathbf{Y} = (Y_{ij})_{i,j\geq 1}$ be an exchangeable random array taking values in $\{0,1\}^{\mathbb{N}\times\mathbb{N}}$. By the Aldous–Hoover theorem (Theorem 6.3), there exists a probability distribution φ on Φ such that $\mathbf{Y}$ obeys a ϕ-process for ϕ drawn randomly from φ. Define the *edge density* of $\mathbf{Y}$ by

$$\varepsilon(\mathbf{Y}) = \lim_{n\to\infty} \frac{1}{n(n-1)} \sum_{1\leq i\neq j\leq n} \mathbf{1}(Y_{ij} = 1). \tag{6.29}$$

For fixed $n \geq 1$, we compute the expectation

$$E\left(\frac{1}{n(n-1)} \sum_{1\leq i\neq j\leq n} \mathbf{1}(Y_{ij} = 1)\right) = \frac{1}{n(n-1)} \sum_{1\leq i\neq j\leq n} E(\mathbf{1}(Y_{ij} = 1))$$

by calculating

$$E(\mathbf{1}(Y_{ij} = 1)) = \Pr(Y_{ij} = 1) = \int_{[0,1]\times[0,1]} \phi(u,v)\, du\, dv$$

for every $1 \leq i \neq j \leq n$. Writing

$$\varepsilon(\phi) = \int_{[0,1]\times[0,1]} \phi(u,v)\, du\, dv$$

and noting that the limit (6.29) exists with probability 1 and is deterministic (equal to $\varepsilon(\phi)$) for any $\mathbf{Y}$ generated by the ϕ-process,[5] we see that in general $\varepsilon(\mathbf{Y})$ exists with probability 1 and is random with distribution

$$\Pr(\varepsilon(\mathbf{Y}) \in \cdot; \varphi) = \int_{\Phi} \mathbf{1}(\varepsilon(\phi) \in \cdot)\, \varphi(d\phi)$$

for any countable exchangeable random array $\mathbf{Y}$. Therefore, to study the edge density of any vertex exchangeable array $\mathbf{Y}$, it is sufficient to study the edge density of arrays distributed according to the ϕ-process, for fixed $\phi \in \Phi$.

The limit in (6.29) is either strictly positive or equal to 0. Any $\mathbf{Y}$ with $\varepsilon(\mathbf{Y}) > 0$ is called a 'dense graph sequence', or simply a 'dense graph'. The theory of dense graph sequences has been studied in depth in the work of Lovász and coauthors; see [115, 116] for more details and references. For our interest in network modeling, the sparse case (i.e., $\varepsilon(\mathbf{Y}) = 0$) is of greater interest.

[5]The fact that this limit is deterministic is related to the ergodicity property of graphons. See Section 6.4.3 and [9, Chapter 14].

'Sparse' case

In many applications, the observed relational array $\mathbf{Y}_n$ for a sample $[n] \subseteq \mathbb{N}$ has empirical density (6.29) that is judged to be 'small' relative to n.[6] When modeling a network with unbounded population size, the assessment that the edge density is 'small' is most often translated into an assumption that the edge density of the hypothetical population network $\mathbf{Y}$ satisfies $\varepsilon(\mathbf{Y}) = 0$ in the limit as $n \to \infty$. Under the ϕ-process construction, however, the limiting density satisfies $\varepsilon(\mathbf{Y}) = \varepsilon(\phi) = 0$ only if $\phi(u,v) \equiv 0$ for almost all $u, v \in [0,1]$, and any $\mathbf{Y}$ generated from such a ϕ will be empty, i.e., $Y_{ij} = 0$ for all $i \neq j$, with probability 1.

Exercise 6.7 *Prove the above assertion in one line: For any $\mathbf{Y}$ constructed from the ϕ-process, the limiting density $\varepsilon(\mathbf{Y}) = \varepsilon(\phi) = 0$ only if $\phi(u,v) \equiv 0$ for almost all $u, v \in [0,1]$.*

This observation yields the first major implication of the Aldous–Hoover theorem:

> *A countable vertex exchangeable random graph is dense or empty with probability 1.*

6.5.2 Implication 2: Representative sampling

In the networks literature, the incompatibility between vertex exchangeability and sparsity is often cited in arguments against the use of graphon models. But in light of the modeling paradigm of Chapter 5 and an intuitive understanding of how most real-world networks are observed, the inability of vertex exchangeable models to replicate sparsity may instead be seen as a byproduct of the inherent infeasibility of graphons for modeling real-world networks. Indeed, the inability of vertex exchangeable models to replicate common empirical properties signals that something is awry with graphons, but pointing to the failure of vertex exchangeable models to reproduce a single empirical property misses a much greater flaw in applying these models to modern networks problems.

Since neither the sampling behavior nor the distributional properties of graphon models reflect the behaviors found in many modern network datasets, it should not be surprising that the empirical properties of graphons are also misaligned with those of observed networks. As the edge density is just one of many network statistics, it should not be the lone factor for assessing the goodness of fit of a network model. A model's usefulness should instead be assessed within the context of the analysis, and the network statistics used for such an assessment ought to be chosen with the context in mind. Chapter 3 highlights different ways in which real-world networks are observed and how rare it is for the observed vertices to be representative of the population of all vertices, as is implicitly assumed in the graphon framework. (The illustration in Section 3.4 all but rules out this possibility when the population network

[6] In practice, this density will either equal 0, which is certainly 'small', or will be strictly positive. The justification for 'smallness' in the latter case is usually based on (i) a qualitative assessment about the structure of the observed network relative to what would otherwise be expected in a network with the given edge density and/or (ii) an assumption about how the network is growing. See Sections 1.7.1, 3.4, and 4.2 for more on this point.

is sparse.) Thus, by considering the implicit sampling context of graphon models, e.g., Section 6.4.4, we can immediately dismiss graphons as viable statistical models for most applications of interest.

6.5.3 *The emergence of graphons*

In statistical analysis, graphon models arise most naturally from the line of thinking presented above: posit a natural invariance principle, i.e., vertex exchangeability, and derive graphons as the characteristic set of (ergodic) models in that class. In the combinatorics literature, where the term 'graphon' was first introduced, graphons emerge from a vastly different line of thought. In this case, a sequence of finite graphs is taken as primitive, and the graphon is derived as a summary of all limiting properties of that sequence.

For $n \geq 1$, let $\mathbf{y}_n$ be a $\{0,1\}$-valued array of arbitrary finite size $V(\mathbf{y}_n)$ (i.e., $V(\mathbf{y}_n)$ is the number of vertices in $\mathbf{y}_n$). The sequence $(\mathbf{y}_n)_{n\geq 1}$ is said to *converge* if the limit $\delta(\mathbf{x} : (\mathbf{y}_n)_{n\geq 1}) = \lim_{n\to\infty} \delta(\mathbf{x} : \mathbf{y}_n)$ exists for all $\mathbf{x} \in \{0,1\}^{m\times m}$, for all $m \in \mathbb{N}$, where

$$\delta(\mathbf{x} : \mathbf{y}_n) = \lim_{n\to\infty} \frac{1}{V(\mathbf{y}_n)^{\downarrow m}} \sum_{\text{injections } \psi:[m]\to[V(\mathbf{y}_n)]} \mathbf{1}(\mathbf{y}_n^{\psi} = \mathbf{x}). \qquad (6.30)$$

Note that (6.30) is consistent with the definition of the homomorphism density in (6.23), with the main difference being that the components of $(\mathbf{y}_n)_{n\geq 1}$ need not be related to one another (i.e., we do not require $\mathbf{y}_m = \mathbf{y}_n|_{[m]}$ for all $m \leq n$) and the number of vertices $V(\mathbf{y}_n)$ is arbitrary (i.e., we do not require $V(\mathbf{y}_n) = n$ for each $n \geq 1$). The limit (6.30), if it exists, is the limit of the densities of $\mathbf{x}$ in the sequence $(\mathbf{y}_n)_{n\geq 1}$, which cannot in general be interpreted as the limiting density of $\mathbf{x}$ in any particular infinite graph unless $(\mathbf{y}_n)_{n\geq 1}$ satisfies an additional compatibility condition for all large n.[7]

When the limit $\delta(\mathbf{x} : (\mathbf{y}_n)_{n\geq 1})$ exists for every $\mathbf{x} \in \bigcup_{m\geq 1}\{0,1\}^{m\times m}$, the sequence $(\mathbf{y}_n)_{n\geq 1}$ determines a probability distribution γ_m on $\{0,1\}^{m\times m}$, for every $m \geq 1$, by

$$\gamma_m(\mathbf{x}) = \delta(\mathbf{x} : (\mathbf{y}_n)_{n\geq 1}), \quad \mathbf{x} \in \{0,1\}^{m\times m}. \qquad (6.31)$$

Every such γ_m is exchangeable: for any permutation $\sigma : [m] \to [m]$,

$$\begin{aligned}
\gamma_m(\mathbf{x}^{\sigma}) &= \delta(\mathbf{x}^{\sigma}, (\mathbf{y}_n)_{n\geq 1}) \\
&= \lim_{n\to\infty} \frac{1}{V(\mathbf{y}_n)^{\downarrow m}} \sum_{\text{injections } \psi:[m]\to[V(\mathbf{y}_n)]} \mathbf{1}(\mathbf{y}_n^{\psi} = \mathbf{x}^{\sigma}) \\
&= \lim_{n\to\infty} \frac{1}{V(\mathbf{y}_n)^{\downarrow m}} \sum_{\text{injections } \psi:[m]\to[V(\mathbf{y}_n)]} \mathbf{1}(\mathbf{y}_n^{\psi\circ\sigma} = \mathbf{x}) \\
&= \lim_{n\to\infty} \frac{1}{V(\mathbf{y}_n)^{\downarrow m}} \sum_{\text{injections } \psi:[m]\to[V(\mathbf{y}_n)]} \mathbf{1}(\mathbf{y}_n^{\psi} = \mathbf{x}) \\
&= \gamma_m(\mathbf{x}),
\end{aligned}$$

[7]Unless $(\mathbf{y}_n)_{n\geq 1}$ is compatible for all large n, i.e., there exists $N \geq 1$ such that $\mathbf{y}_m|_{[k]} = \mathbf{y}_k$ for all $m \geq k \geq N$, we cannot associate the limiting density $\delta(\mathbf{x} : (\mathbf{y}_n)_{n\geq 1})$ of the sequence $(\mathbf{y}_n)_{n\geq 1}$ to any particular element of $\{0,1\}^{\mathbb{N}\times\mathbb{N}}$.

since averaging of all injections $\psi : [m] \to [V(\mathbf{y}_n)]$ is the same as averaging over all injections $\psi \circ \sigma$. And moreover, the collection of measures $(\gamma_n)_{n\geq 1}$ is consistent under selection sampling,

$$\gamma_m(\mathbf{x}) = \gamma_n(\{\mathbf{x}^* \in \{0,1\}^{n\times n} : \mathbf{x}^*|_{[m]} = \mathbf{x}\}), \quad \mathbf{x} \in \{0,1\}^{m\times m}, \tag{6.32}$$

for all $n \geq m \geq 1$.

Exercise 6.8 *Prove* (6.32) *for* γ_m *and* γ_n *defined as in* (6.31).

In words, condition (6.32) says that a random graph $\mathbf{Y}_m$ drawn from γ_m has the same distribution as the restriction $\mathbf{Y}_n|_{[m]}$ of $\mathbf{Y}_n$ drawn from γ_n for all $n \geq m \geq 1$. By Carathéodory's theorem from measure theory (the details of which we do not cover here), $(\gamma_m)_{m\geq 1}$ determines a unique exchangeable probability distribution γ on $\{0,1\}^{\mathbb{N}\times\mathbb{N}}$ such that

$$\gamma(\{\mathbf{x}^* \in \{0,1\}^{\mathbb{N}\times\mathbb{N}} : \mathbf{x}^*|_{[m]} = \mathbf{x}\}) = \gamma_m(\mathbf{x}) \quad \text{for all } \mathbf{x} \in \{0,1\}^{m\times m}, \quad m \in \mathbb{N}.^8 \tag{6.33}$$

By the Aldous–Hoover theorem (Theorem 6.3), any such γ corresponds to a probability measure φ on the space Φ of all functions $\phi : [0,1] \times [0,1] \to [0,1]$, and γ is the distribution of an exchangeable random array $\mathbf{Y}$ constructed from a ϕ-process for $\phi \sim \varphi$. But since the limiting densities are deterministic, φ must be concentrated on the subset of $\phi' \in \Phi$ for which $\delta(\mathbf{x} : \phi') = \delta(\mathbf{x} : (\mathbf{y}_n)_{n\geq 1})$ for all finite arrays $\mathbf{x}$. Any such ϕ' corresponds to the same probability measure γ on $\{0,1\}^{\mathbb{N}\times\mathbb{N}}$, implying that the limit of $(\mathbf{y}_n)_{n\geq 1}$ is only determined up to equivalence of the distribution determined by its limiting densities. There is thus no unique graphon ϕ corresponding to the sequence $(\mathbf{y}_n)_{n\geq 1}$ but there is a unique probability distribution γ, as defined through (6.33) and Carathéodory's theorem. In this sense, we can rightly call γ in (6.33) the *graph limit* of $(\mathbf{y}_n)_{n\geq 1}$, but we should not conflate γ with any particular graphon associated to this limit. Having said that, I caution the reader that many authors do use the term graphon and graph limit interchangeably.

Ergodicity

I conclude this section by connecting the combinatorial interpretation of graphons, as limits of graph sequences, to the statistical interpretation of graphons, as the class of ergodic measures for vertex exchangeable random graphs. Let ϕ be any graphon, let $\mathbf{x} \in \{0,1\}^{m\times m}$ be any finite graph, and let $\mathbf{Y}_n$, $n \geq 1$, be constructed from the ϕ-process as in (6.13). Two aspects of the above discussion figure into the coming remark:

(i) By consistency under selection of the ϕ-process, $(\mathbf{Y}_n)_{n\geq 1}$ can be treated as the finite restrictions $\mathbf{Y}_n = \mathbf{Y}|_{[n]}$ of some infinite exchangeable random array $\mathbf{Y}$ distributed according to a ϕ-process on $\{0,1\}^{\mathbb{N}\times\mathbb{N}}$.

[8] Readers unfamiliar with measure-theoretic probability will likely find that I have moved through this argument too quickly. Indeed I have, but the technical details are too much to discuss here. Readers interested to learn more are encouraged to consult a textbook on measure-theoretic probability. It is otherwise safe to continue reading without this technical detail.

(ii) Statistical intuition (e.g., the strong law of large numbers) suggests that the densities $\delta(\mathbf{x} : \mathbf{Y}_n)$ should satisfy

$$\lim_{n\to\infty} \delta(\mathbf{x} : \mathbf{Y}_n) = \delta(\mathbf{x} : \phi) \quad \text{with probability 1,} \tag{6.34}$$

for $\delta(\mathbf{x} : \phi)$ as defined in (6.28).

Under (i), the distribution of each $\mathbf{Y}_n$ is given by (6.14), so that

$$\Pr(\mathbf{Y}_n = \mathbf{x}; \phi) = \delta(\mathbf{x} : \phi), \quad \mathbf{x} \in \{0,1\}^{n\times n}.$$

Under (ii), the limiting densities $\delta(\mathbf{x} : \mathbf{Y}_n)$ should converge to $\delta(\mathbf{x} : \phi)$ as the sample size grows large.

If all terms in the sum (6.27) were independent, then (6.34) would be an immediate consequence of the strong law of large numbers (SLLN), as intuition suggests. In this case, however, there may be dependence among observations Y_{ij} and $Y_{i'j'}$ for which $\{i,j\} \cap \{i',j'\} \neq \emptyset$. (If $\{i,j\}$ and $\{i',j'\}$ overlap in an index, say, $r = i = i'$, then the Aldous–Hoover construction in (6.18) allows for the possible dependence between Y_{rj} and $Y_{rj'}$ as a result of their common dependence on U_r.) Nevertheless, the effect of such overlap on the distribution of $\mathbf{Y}_n$ is negligible as n increases, and the limit (6.34) does in fact hold for ϕ-processes. (Aldous [9] gives an explicit calculation of this fact.) It follows that the subgraph statistics (6.23) are deterministic (i.e., ergodic) for any $\mathbf{Y}$ generated from a graphon process.

6.6 Potential benefits of graphon models

The foregoing discussion speaks to the need for network modeling to move beyond the limitations of graphon models and vertex exchangeability. I take up this charge throughout Chapters 7–10. But before moving on, I identify some potentially redeeming qualities of graphon models. I first highlight a connection between the theory of vertex exchangeable random graphs (through graphons and the Aldous–Hoover theorem) and exchangeable random sequences (through de Finetti's theorem), which in turn suggests how graphons might be useful for identifying departures from exchangeability by detecting differences between the empirical graphon and its expected 'shape'. Admittedly, the content of Section 6.6.2 is speculative, but I mention it here as a topic worthy of some consideration.

6.6.1 Connection to de Finetti's theorem

For this section only, we divert away from our main discussion of networks, graphs, and $\{0,1\}$-valued arrays and instead discuss exchangeable sequences $\mathbf{X} = (X_1, X_2, \ldots)$ of $\{0,1\}$-valued random variables. By comparison to exchangeable sequences, which have a long history in probability and statistics, the theory of exchangeable arrays is lesser known and not as well understood. There are, however, several analogous concepts linking exchangeable arrays and the Aldous–Hoover theorem on the one hand to exchangeable sequences and de Finetti's theorem on the other. It is through these similarities that some more technical aspects of graphon models can perhaps be better understood.

Call a sequence $\mathbf{X} = (X_i)_{i \in V}$ *exchangeable* if $\mathbf{X}^\sigma = (X_{\sigma(i)})_{i \in V} =_{\mathscr{D}} \mathbf{X}$ for all permutations $\sigma : V \to V$. As for arrays, this definition applies for V both finite and countably infinite, but we restrict to the countably infinite case here. In this 1-dimensional setting, the role of the Aldous–Hoover theorem is played by its predecessor, de Finetti's theorem, according to which every infinite, exchangeable $\{0,1\}$-valued sequence $X_1, X_2, \ldots$ corresponds to a unique probability measure μ on $[0,1]$ such that $X_1, X_2, \ldots$ is distributed as a conditionally i.i.d. sequence of Bernoulli random variables with random success probability $P \sim \mu$. This formulation is sometimes also expressed as

$$P \sim \mu$$
$$X_1, X_2, \ldots \mid P \sim_{\text{i.i.d.}} \text{Bernoulli}(P).$$

To parallel the above presentation of ϕ-processes and graphons as closely as possible, let $\phi : [0,1] \to [0,1]$ be a function on the unit interval. (Note here that ϕ is a function of one argument, instead of two as in the 2-dimensional setting of Section 6.4.1.) To generate a sequence of random variables $X_1, X_2, \ldots$ in $\{0,1\}$, first draw $U_1, U_2, \ldots$ i.i.d. Uniform$[0,1]$ and, given $U_1, U_2, \ldots$, assign the value of each X_i, $i = 1, 2, \ldots$, conditionally independently according to

$$\Pr(X_i = 1 \mid U_1, U_2, \ldots; \phi) = \phi(U_i) \quad \text{and} \quad \Pr(X_i = 0 \mid U_1, U_2, \ldots; \phi) = 1 - \phi(U_i). \tag{6.35}$$

Call any sequence $(X_1, X_2, \ldots)$ generated in this way a *1-dimensional ϕ-process.*

Since $U_1, U_2, \ldots$ are i.i.d., it follows immediately that $(X_1, X_2, \ldots)$ is *exchangeable*. In fact, $(X_1, X_2, \ldots)$ are i.i.d. and, thus, must follow the Bernoulli distribution with success probability

$$\Pr(X_i = 1; \phi) = \int_{[0,1]} \phi(u)\, du.$$

The characterization of exchangeable $\{0,1\}$-valued sequences can thus be stated in parallel form to Theorem 6.3.

Theorem 6.4 (de Finetti's theorem [60]) *Let $\mathbf{X} = (X_1, X_2, \ldots)$ be an infinite, exchangeable sequence in $\{0,1\}$. Then there exists a function $f : [0,1]^2 \to \{0,1\}$ such that $\mathbf{X} =_{\mathscr{D}} \mathbf{X}^*$, with $\mathbf{X}^* = (X_1^*, X_2^*, \ldots)$ given by*

$$X_i^* = f(U_0, U_i), \quad i \geq 1,$$

for $U_0, U_1, \ldots$ i.i.d. Uniform$[0,1]$.

In parallel to our interpretation of the Aldous–Hoover theorem in Section 6.4.1, we translate Theorem 6.4 as

> *every infinite, exchangeable $\{0,1\}$-valued sequence is distributed as a mixture of i.i.d. sequences.*

Remark 6.3 *Theorem 6.4 immediately implies that the outcome of a* 1*-dimensional* ϕ*-process is equivalent in distribution to an i.i.d. Bernoulli sequence with success probability* $\int_0^1 \phi(u)du$*. But since the distribution of the sequence depends on* ϕ *only through the integral* $\int_0^1 \phi(u)du$*,* ϕ *and* $\phi \circ T$ *both determine the same distribution for any* $T : [0,1] \to [0,1]$ *that preserves Lebesgue measure, as defined in* (6.38) *below. In particular, any* $X_1, X_2, \ldots$ *from the 1-dimensional* ϕ*-process can also be described as a draw from a* 1*-dimensional* ϕ_p*-process, with* ϕ_p *defined as the constant function*

$$\phi_p : [0,1] \to [0,1]$$
$$\phi_p(u) \equiv p = \int_0^1 \phi(u)du. \tag{6.36}$$

Induced subsequence densities

The connection between vertex exchangeable random graphs, the Aldous–Hoover theorem, and graphons is paralleled by a connection between exchangeable random sequences, de Finetti's theorem, and limits associated to finite induced subsequences. When speaking of $\{0,1\}$-valued sequences, the analogs to the induced subgraph densities $\delta(\mathbf{x} : \mathbf{y})$ in (6.23) are induced subsequence densities defined as follows. In parallel to (6.23), let $\mathbf{x} \in \{0,1\}^n$, $\mathbf{w} \in \{0,1\}^m$, and define

$$\delta(\mathbf{w} : \mathbf{x}) = \lim_{n\to\infty} \frac{1}{n^{\downarrow m}} \sum_{\text{injections } \psi:[m]\to[n]} \mathbf{1}(\mathbf{x}^{\psi} = \mathbf{w}), \tag{6.37}$$

where $\mathbf{x}^{\psi} = (x_{\psi(i)})_{1\le i\le m}$ is the subsequence of $\mathbf{x}$ induced by ψ. By the strong law of large numbers, the limiting fraction of 1s in an i.i.d. sequence $\mathbf{X}$, i.e.,

$$P(\mathbf{X}) = \lim_{n\to\infty} n^{-1} \sum_{i=1}^{n} X_i,$$

exists with probability 1, and in fact determines the limiting density $\delta(\mathbf{w} : \mathbf{X})$ of any $\mathbf{w} \in \{0,1\}^m$ in $\mathbf{x}$ by

$$\lim_{n\to\infty} \delta(\mathbf{w} : \mathbf{X}|_{[n]}) = \Pr(\mathbf{X}^{\psi} = \mathbf{w} \mid P(\mathbf{X}) = p) = p^{\sum_{j=1}^m w_j}(1-p)^{m-\sum_{j=1}^m w_j}$$

with probability 1. Thus, $P(\mathbf{X}) = \delta((1) : \mathbf{X})$ (i.e., the frequency of 1s in $\mathbf{X}$) alone determines the distribution of an i.i.d. sequence of Bernoulli random variables with success probability $P(\mathbf{X})$.

Conversely, taking the analog to the 'graph convergence' perspective from the combinatorics literature (Section 6.5.3), let $(\mathbf{w}_n)_{n\ge1}$ be an infinite sequence of finite-length $\{0,1\}$-valued sequences, so that each $\mathbf{w}_n$ is a sequence $(w_n(1), \ldots, w_n(\ell_n))$ of finite length $\ell_n \ge 1$. For each $n \ge 1$, let $|\mathbf{w}_n| = \ell_n$ denote the length of $\mathbf{w}_n$ and write $p_n = |\mathbf{w}_n|^{-1} \sum_{i=1}^{|\mathbf{w}_n|} w_n(i)$ to denote the empirical frequency of 1s in $\mathbf{w}_n$. We say that $(\mathbf{w}_n)_{n\ge1}$ *converges* if $\lim_{n\to\infty} p_n = p$ exists. In this case, $p \in [0,1]$ corresponds to the limit of an i.i.d. sequence of Bernoulli(p) random variables, which can be represented trivially by the constant function ϕ_p in (6.36) or, alternatively, by any function $\phi : [0,1] \to [0,1]$ which integrates to p over $[0,1]$, cf. Remark 6.3.

6.6.2 Graphon estimation

Let $\mathbf{X} = (X_1, X_2, \ldots)$ be a random $\{0,1\}$-valued sequence, from which we observe an initial segment $\mathbf{X}_n = (X_1, \ldots, X_n)$. Assuming $\mathbf{X}$ has been generated from a 1-dimensional ϕ-process, as in (6.35), we know that the distribution of $\mathbf{X}$ is determined solely by the integral $\int_0^1 \phi(u)du$. We have already noted in Remark 6.3 that the graphon ϕ is not identifiable for $\mathbf{X}$: for any Lebesgue measure-preserving transformation $T : [0,1] \to [0,1]$, i.e., $T : [0,1] \to [0,1]$ satisfying

$$\int_{T^{-1}(a,b)} dx = b - a \tag{6.38}$$

for all $0 \leq a \leq b \leq 1$, and any $\phi : [0,1] \to [0,1]$, the function $\phi'(u) = \phi(T(u))$ determines the same distribution as ϕ. Therefore, even if given the complete realization of the infinite sequence $\mathbf{X}$ from the 1-dimensional ϕ-process, and thus also the exact limiting density $P(\mathbf{X})$, we cannot recover the function ϕ. We can, at best, determine the equivalence class of functions ϕ' that determine the same distribution on $\{0,1\}^{\mathbb{N}}$. We might then wonder what additional information about $\mathbf{X}$, beyond the integral $\int_0^1 \phi(u)du$, is encoded by the function ϕ?

If $\mathbf{X}$ is only suspected, but not known, to be i.i.d., then perhaps the additional structure encoded by ϕ could be valuable in assessing departures of $\mathbf{X}$ from exchangeability. With this in mind, define the *empirical graphon induced by* $\mathbf{X}_n$ as

$$\hat{\phi}_n(u) = X_i, \quad \text{for } u \in [(i-1)/n, i/n).^9 \tag{6.39}$$

This function is piecewise constant, taking values 0 and 1 over a partition of $[0,1]$ into equally spaced intervals of length $1/n$. Under the assumption that $\mathbf{X}$ is i.i.d., the conditional distribution of $\mathbf{X}_n$, given the empirical frequency $P_n = n^{-1}\sum_{i=1}^n X_i$, is i.i.d. Bernoulli with success probability P_n, for which each possible outcome $\mathbf{w}_n = (w_i)_{1 \leq i \leq n}$ with nP_n 1s and $n(1-P_n)$ 0s is equally probable given P_n. It follows that every empirical graphon $\hat{\phi}_n$ that is compatible with P_n is equally probable given P_n.

For an extreme example, let n be a large even integer and suppose we observe

$$\mathbf{X}_n = (\underbrace{1,1,\ldots,1}_{n/2 \text{ times}}, \underbrace{0,0,\ldots,0}_{n/2 \text{ times}}), \tag{6.40}$$

which under the hypothesis that $\mathbf{X}_n$ is i.i.d. is just as likely as any of the other $\binom{n}{n/2}$ outcomes with empirical frequency $P_n = 1/2$. But if n is large and $\mathbf{X}_n$ were truly behaving as i.i.d. Bernoulli, then we would expect the mass of 1s and 0s in $\mathbf{X}_n$ to be relatively evenly dispersed over $[0,1]$. To address this, we can specify a kernel $K : [0,1] \to [0,\infty)$ and instead estimate the *K-smoothed graphon* by

$$\hat{\phi}_n^K(u) = \int_0^1 K(\|u - u'\|)\hat{\phi}_n(u')du',$$

[9] We can define $\hat{\phi}_n(1)$ arbitrarily, since this is a set of measure 0. Putting $\hat{\phi}_n(1) = X_n$ will suffice.

where $\hat{\phi}_n$ is as in (6.39) and $\|\cdot\|$ is a norm on $[0,1]$. If K is sufficiently well-behaved and $\mathbf{X}_n$ follows the 1-dimensional ϕ-process, then we expect

$$\hat{\phi}_n^K(u) \to \int_0^1 \phi(u')\,du' \quad \text{as } n \to \infty \text{ for all } u \in [0,1] \text{ with probability 1.}$$

But for an observation $\mathbf{X}_n$ as in (6.40), we see that

$$\hat{\phi}_n^K(0) = \int_0^{1/2} K(u')\,du' \quad \text{and}$$

$$\hat{\phi}_n^K(1) = \int_0^{1/2} K(1-u')\,du' = \int_{1/2}^1 K(u')\,du',$$

so that unless $K(\cdot)$ is a constant function there in general exists an $\varepsilon > 0$ and $0 \leq u \neq v \leq 1$ such that

$$|\hat{\phi}_n^K(u) - \hat{\phi}_n^K(v)| \geq \varepsilon,$$

preventing $\hat{\phi}_n^K$ from converging to a constant function. One can then consider a hypothesis test for exchangeability by comparing the empirical smoothed graphons $\hat{\phi}_n^K$ and $\tilde{\phi}_n^K$, for $\tilde{\phi}_n$ defined as the empirical graphon in (6.39) for a sequence $X_1',\ldots,X_n'$ from the 1-dimensional $\hat{\phi}_n$-process.

The connection between de Finetti's theorem and the Aldous–Hoover theorem suggests how graphon estimation could be useful for detecting departures from vertex exchangeability in networks that are observed sequentially. Generalizing from the above discussion for sequences, consider what information a graphon $\phi : [0,1] \times [0,1] \to [0,1]$ conveys about a sequence $(\mathbf{y}_n)_{n\geq 1}$ of $\{0,1\}$-valued arrays as $n \to \infty$. For the sake of illustration, suppose that $|V(\mathbf{y}_n)| = n$ for each $n = 1,2,\ldots$ and define the *empirical graphon* $\hat{\phi}_n : [0,1] \times [0,1] \to [0,1]$ to be

$$\hat{\phi}_n(u,v) = \mathbf{y}_n(\lfloor nx \rfloor, \lfloor ny \rfloor), \quad 0 \leq u,v \leq 1, \tag{6.41}$$

where here I write $\mathbf{y}_n(i,j)$ to denote the (i,j) entry of $\mathbf{y}_n$ and $\lfloor z \rfloor$ to denote the largest integer smaller than z, for $z \in (-\infty,\infty)$. (In words, $\hat{\phi}_n$ "squeezes" the adjacency matrix of $\mathbf{y}_n$ into $[0,1]$ by splitting $[0,1]$ into n equally spaced regions $[0,1/n),[1/n,2/n),\ldots,[(n-1)/n,1]$ corresponding to the vertices $1,2,\ldots,n$, respectively. The value of $\hat{\phi}_n$ is constant and equals $\mathbf{y}_n(i,j)$ on the region $[(i-1)/n,i/n) \times [(j-1)/n,j/n)$.)

Following the discussion for sequences above, we specify a kernel $K : [0,1] \to [0,\infty)$ and define the *K-smoothed empirical graphon* by

$$\hat{\phi}_n^K(u,v) = \int_0^1 \int_0^1 K(d((u,v),(u',v')))\hat{\phi}_n(u',v')\,du'\,dv',$$

where $d : [0,1] \times [0,1] \to \mathbb{R}$ is any distance metric, e.g.,

$$d((u,v),(u',v')) = |u-u'| + |v-v'|.$$

The same rationale applies as for testing exchangeability in sequences. The details of this approach have not yet been worked out, and it is unclear whether it amounts to anything fruitful. I leave the details as an open research problem.

Research Problem 6.5 *Can graphons be useful for testing departures from exchangeability? In particular, if $\mathbf{Y}_n = (Y_{ij})_{1\leq i,j\leq n}$ is a realization from a probability distribution on graphs, can we test for exchangeability by comparing the empirical graphon $\hat{\phi}_n$ based on $\mathbf{Y}_n$ to the empirical graphon $\tilde{\phi}_n$ obtained from a realization of a $\hat{\phi}_n$-process? If we detect a departure from exchangeability, can we characterize the nature of this departure? See [158] for some related work on this question.*

6.7 Further reading

By now the exchangeability literature is extensive, admitting a number of refinements and generalizations which have been developed in the eighty-plus years since de Finetti [60]. Some recent extensions relevant to network analysis will be discussed over the next several chapters. For a broader discussion of exchangeability, the reader will find Aldous's lecture notes [9] and Kallenberg's book on probabilistic symmetry [98] the most thorough references. A more recent survey of exchangeability from a Bayesian nonparametrics perspective is given in [128]. Still other extensions to so-called Markov exchangeability can be found in the earlier work of Diaconis and Freedman [61].

Before moving on from the Aldous–Hoover theorem and its numerous implications, I conclude by commenting that the above discussion of exchangeability for 1-dimensional sequences (Theorem 6.4) and 2-dimensional arrays (Theorem 6.3) extends in a natural way to exchangeable d-dimensional arrays $\mathbf{Y} = (Y_{i_1\cdots i_d})_{i_1,\dots,i_d\geq 1}$ for any $1 \leq d < \infty$. Such models are the natural extension of vertex exchangeability to network data taking the form of a hypergraph. For example, if $\mathbf{Y}$ represents relations formed by email interactions, then each relation can in general involve more than two vertices, e.g., an email exchanged among a group of people $i_1,\dots,i_d$ is expressed as a single relation $Y_{i_1\cdots i_d} = 1$ in the associated array. In this setting, the definition of exchangeability in (6.2) generalizes to

$$\mathbf{Y}^\sigma = (\mathbf{Y}_{\sigma(i_1)\cdots\sigma(i_d)})_{i_1,\dots,i_d\geq 1} =_{\mathscr{D}} \mathbf{Y} \quad \text{for all permutations } \sigma : \mathbb{N} \to \mathbb{N}.$$

The Aldous–Hoover–Kallenberg theorem gives the analog to Theorem 6.3 for such exchangeable higher-order arrays. There is an associated construction from i.i.d. Uniform$[0,1]$ random variables, akin to that in (6.13) in the 2-dimensional case. The reader is referred to [98] for details.

6.8 Solutions to exercises

6.8.1 *Exercise 6.1*

Let $\mathbf{Y}_N$ be an exchangeable random array in $\{0,1\}^{N\times N}$ and $g : \{0,1\}^{N\times N} \to \mathscr{X}$ be any statistic. Then to prove

$$\mathbf{Y}_N =_{\mathscr{D}} \mathbf{Y}_N^\sigma \quad \Longrightarrow \quad g(\mathbf{Y}_N) =_{\mathscr{D}} g(\mathbf{Y}_N^\sigma)$$

for all permutations $\sigma : [N] \to [N]$, we must show

$$\Pr(g(\mathbf{Y}_N) \in A) = \Pr(g(\mathbf{Y}_N^\sigma) \in A) \quad \text{for all measurable } A \subseteq \mathscr{X}.$$

We see this immediately by

$$\Pr(g(\mathbf{Y}_N) \in A) = \Pr(\mathbf{Y}_N \in g^{-1}(A)) = \Pr(\mathbf{Y}_N^{\sigma} \in g^{-1}(A)) = \Pr(g(\mathbf{Y}_N^{\sigma}) \in A),$$

where the first and third equalities follow by definition of the induced distribution of $g(\mathbf{Y}_N)$ and $g(\mathbf{Y}_N^{\sigma})$, respectively, and the second equality follows by exchangeability of $\mathbf{Y}_N$.

6.8.2 Exercise 6.2

By assumption that $\Sigma_{n,N}$ is independent of $\mathbf{Y}_N$, the distribution of $\Sigma_{n,N}\mathbf{Y}_N$ is given by

$$\Pr(\Sigma_{n,N}\mathbf{Y}_N = \mathbf{y}) = \sum_{\psi:[n]\to[N]} \Pr(\Sigma_{n,N} = \mathbf{S}_{n,N}^{\psi}) \Pr(\mathbf{S}_{n,N}^{\psi}\mathbf{Y}_N = \mathbf{y})$$

as in (6.7). By exchangeability of $\mathbf{Y}_N$ and a corollary to Exercise 6.1, $\mathbf{S}_{n,N}^{\psi}\mathbf{Y}_N =_{\mathscr{D}} \mathbf{S}_{n,N}\mathbf{Y}_N$ for all $\psi : [n] \to [N]$, giving

$$\Pr(\Sigma_{n,N}\mathbf{Y}_N = \mathbf{y}) = \Pr(\mathbf{S}_{n,N}\mathbf{Y}_N = \mathbf{y}) \sum_{\psi:[n]\to[N]} \Pr(\Sigma_{n,N} = \mathbf{S}_{n,N}^{\psi}) = \Pr(\mathbf{S}_{n,N}\mathbf{Y}_N = \mathbf{y}),$$

since the probabilities of events '$\Sigma_{n,N} = \mathbf{S}_{n,N}^{\psi}$' sum to 1.

6.8.3 Exercise 6.3

Let $\mathbf{Y}_N$ be an exchangeable random graph on $\{0,1\}^{N\times N}$. By the law of total probability, we can express the distribution of $\mathbf{Y}_N$ by conditioning on its 'shape' $\langle \mathbf{Y}_N \rangle_{\cong}$:

$$\Pr(\mathbf{Y}_N = \mathbf{y}) = \sum_{\mathbf{u}\in\mathscr{U}_N} \Pr(\mathbf{Y}_N = \mathbf{y} \mid \langle \mathbf{Y}_N \rangle_{\cong} = \mathbf{u}) \Pr(\langle \mathbf{Y}_N \rangle_{\cong} = \mathbf{u}).$$

For every $\mathbf{y} \in \{0,1\}^{N\times N}$ and $\mathbf{u} \in \mathscr{U}_N$, we have

$$\Pr(\mathbf{Y}_N = \mathbf{y} \mid \langle \mathbf{Y}_N \rangle_{\cong} = \mathbf{u}) = \begin{cases} 1/|\mathbf{u}|, & \langle \mathbf{y} \rangle_{\cong} = \mathbf{u}, \\ 0, & \text{otherwise.} \end{cases} \tag{6.42}$$

And for any $\mathbf{y}$ and $\mathbf{y}'$ with $\langle \mathbf{y} \rangle_{\cong} = \langle \mathbf{y}' \rangle_{\cong}$, we have

$$\Pr(\mathbf{Y}_N = \mathbf{y} \mid \langle \mathbf{Y}_N \rangle_{\cong} = \mathbf{u}) = \Pr(\mathbf{Y}_N = \mathbf{y}' \mid \langle \mathbf{Y}_N \rangle_{\cong} = \mathbf{u}) = \begin{cases} 1/|\mathbf{u}|, & \langle \mathbf{y} \rangle_{\cong} = \langle \mathbf{y}' \rangle_{\cong} = \mathbf{u}, \\ 0, & \text{otherwise,} \end{cases}$$

where $|\mathbf{u}|$ is the cardinality of $\mathbf{u}$. Finally, since (6.42) is non-zero only for $\mathbf{u} = \langle \mathbf{y} \rangle_{\cong}$, we can express

$$\Pr(\mathbf{Y}_N = \mathbf{y}) = \Pr(\langle \mathbf{Y}_N \rangle_{\cong} = \langle \mathbf{y} \rangle_{\cong}) / |\langle \mathbf{y} \rangle_{\cong}|.$$

Defining $\mathbf{p}$ on $\mathscr{U}_N$ by

$$\mathbf{p}(\mathbf{u}) = \Pr(\langle \mathbf{Y}_N \rangle_{\cong} = \mathbf{u}) = \sum_{\mathbf{y}\in\{0,1\}^{N\times N}} \Pr(\mathbf{Y}_N = \mathbf{y}) \mathbf{1}(\langle \mathbf{y} \rangle_{\cong} = \mathbf{u})$$

completes the proof of Theorem 6.1.

6.8.4 Exercise 6.4

Fix any $\phi : [0,1] \times [0,1] \to [0,1]$ and let $\mathbf{Y}_m$ have the distribution in (6.14),

$$\Pr(\mathbf{Y}_m = \mathbf{y}; \phi) = \int_{[0,1]^m} \prod_{1 \le i \ne j \le m} \phi(u_i, u_j)^{y_{ij}} (1 - \phi(u_i, u_j))^{1-y_{ij}} du_1 \cdots du_m.$$

By definition $\Pi^{\phi}_{m,n} \mathbf{Y}_m$ is a random array in $\{0,1\}^{n \times n}$ obtained by a ϕ-process with $U_1, \ldots, U_n$ drawn conditionally i.i.d. Uniform$[0,1]$ given $\mathbf{Y}_m$. By the law of total probability, we have

$$\Pr(\Pi^{\phi}_{m,n} \mathbf{Y}_m = \mathbf{y}; \phi) = \Pr(\mathbf{Y}_m = \mathbf{y}|_{[m]}; \phi) \Pr(\Pi^{\phi}_{m,n} \mathbf{Y}_m = \mathbf{y} \mid \mathbf{Y}_m = \mathbf{y}|_{[m]}; \phi).$$

Given '$\mathbf{Y}_m = \mathbf{y}$', the conditional density of $U_1, \ldots, U_n$ is

$$\begin{aligned} & h(u_1, \ldots, u_n \mid \mathbf{Y}_m = \mathbf{y}; \phi) = \\ & = \frac{\prod_{1 \le i \ne j \le m} \phi(u_i, u_j)^{y_{ij}} (1 - \phi(u_i, u_j))^{1-y_{ij}}}{\int_{[0,1]^m} \prod_{1 \le i \ne j \le m} \phi(u_i, u_j)^{y_{ij}} (1 - \phi(u_i, u_j))^{1-y_{ij}}\, du_1 \cdots du_m}. \end{aligned}$$

By construction of $\Pi^{\phi}_{m,n} \mathbf{Y}_m$, we have

$$\begin{aligned} & \Pr(\Pi^{\phi}_{m,n} \mathbf{Y}_m = \mathbf{y}; \phi) = \\ & = \Pr(\mathbf{Y}_m = \mathbf{y}|_{[m]}; \phi) \Pr(\Pi^{\phi}_{m,n} \mathbf{Y}_m = \mathbf{y} \mid \mathbf{Y}_m = \mathbf{y}|_{[m]}; \phi) \\ & = \int_{[0,1]^m} \prod_{1 \le i \ne j \le m} \phi(u'_i, u'_j)^{y_{ij}} (1 - \phi(u'_i, u'_j))^{1-y_{ij}}\, du'_1 \cdots du'_m \times \\ & \quad \times \Pr(\Pi^{\phi}_{m,n} \mathbf{Y}_m = \mathbf{y} \mid \mathbf{Y}_m = \mathbf{y}|_{[m]}; \phi) \\ & = \left(\int_{[0,1]^m} \prod_{1 \le i \ne j \le m} \phi(u'_i, u'_j)^{y_{ij}} (1 - \phi(u'_i, u'_j))^{1-y_{ij}}\, du'_1 \cdots du'_m \right) \times \\ & \quad \times \int_{[0,1]^n} \left[\left(\prod_{m+1 \le i \le n} \prod_{1 \le j \le n} \phi(u_i, u_j)^{y_{ij}} (1 - \phi(u_i, u_j))^{1-y_{ij}} \right) \times \right. \\ & \quad \times \left(\prod_{1 \le i \le n} \prod_{m+1 \le j \le n} \phi(u_i, u_j)^{y_{ij}} (1 - \phi(u_i, u_j))^{1-y_{ij}} \right) \times \\ & \quad \left. \times\ h(u_1, \ldots, u_n \mid \mathbf{Y}_m = \mathbf{y}|_{[m]}; \phi)\, du_1 \cdots du_n \right] \\ & = \left(\int_{[0,1]^m} \prod_{1 \le i \ne j \le m} \phi(u'_i, u'_j)^{y_{ij}} (1 - \phi(u'_i, u'_j))^{1-y_{ij}} du'_1 \cdots du'_m \right) \times \\ & \quad \times \int_{[0,1]^n} \left[\left(\prod_{m+1 \le i \le n} \prod_{1 \le j \le n} \phi(u_i, u_j)^{y_{ij}} (1 - \phi(u_i, u_j))^{1-y_{ij}} \right) \times \right. \\ & \quad \times \left(\prod_{1 \le i \le n} \prod_{m+1 \le j \le n} \phi(u_i, u_j)^{y_{ij}} (1 - \phi(u_i, u_j))^{1-y_{ij}} \right) \times \\ & \quad \left. \times \frac{\prod_{1 \le i \ne j \le m} \phi(u_i, u_j)^{y_{ij}} (1 - \phi(u_i, u_j))^{1-y_{ij}}\, du_1 \cdots du_n}{\int_{[0,1]^m} \prod_{1 \le i \ne j \le m} \phi(u_i, u_j)^{y_{ij}} (1 - \phi(u_i, u_j))^{1-y_{ij}}\, du_1 \cdots du_m} \right] \end{aligned}$$

$$= \int_{[0,1]^n} \prod_{1\le i\ne j\le n} \phi(u_i,u_j)^{y_{ij}}(1-\phi(u_i,u_j))^{1-y_{ij}}\, du_1\cdots du_n$$
$$= \Pr(\mathbf{Y}_n = \mathbf{y};\phi),$$

as claimed.

6.8.5 Exercise 6.5

Recall the definition of relabeling $\mathbf{y} \mapsto \mathbf{y}^\sigma$ in (6.1) and let $\mathbf{Y}_N$ be distributed as in (6.14) for some $\phi : [0,1]\times[0,1] \to [0,1]$. From the expression in (6.14), we see that

$$\begin{aligned}
&\Pr(\mathbf{Y}_N = \mathbf{y}^\sigma;\phi) = \\
&= \int_{[0,1]^N} \prod_{1\le i\ne j\le N} \phi(u_i,u_j)^{y_{\sigma(i)\sigma(j)}}(1-\phi(u_i,u_j))^{1-y_{\sigma(i)\sigma(j)}}\, du_1\cdots du_N \\
&= \int_{[0,1]^N} \prod_{1\le i\ne j\le N} \phi(u_{\sigma^{-1}(i)},u_{\sigma^{-1}(j)})^{y_{ij}}(1-\phi(u_{\sigma^{-1}(i)},u_{\sigma^{-1}(j)}))^{1-y_{ij}} \times \\
&\qquad \times du_{\sigma^{-1}(1)}\cdots du_{\sigma^{-1}(N)} \\
&= \int_{[0,1]^N} \prod_{1\le i\ne j\le N} \phi(u_i,u_j)^{y_{ij}}(1-\phi(u_i,u_j))^{1-y_{ij}}\, du_1\cdots du_N \\
&= \Pr(\mathbf{Y}_N = \mathbf{y};\phi),
\end{aligned}$$

for any permutation $\sigma : [N] \to [N]$, and therefore every graphon distribution on $\{0,1\}^{N\times N}$ is exchangeable, for every $N \ge 1$.

6.8.6 Exercise 6.6

Let $\mathbf{Y}^*$ be constructed as in (6.18) for f satisfying $f(a,\cdot,\cdot,\cdot) = f(b,\cdot,\cdot,\cdot)$ for all $0 \le a,b \le 1$. Then compute the finite-dimensional distributions of $\mathbf{Y}^*$ by

$$\begin{aligned}
&\Pr(\mathbf{Y}^*|_{[n]} = \mathbf{y}; f) = \\
&= \int_{[0,1]^n} \prod_{1\le i\ne j\le n} \left(\int_0^1 f(-,u_i,u_j,u_{ij})\, du_{ij}\right)^{y_{ij}} \times \\
&\qquad \times \left(\int_0^1 (1-f(-,u_i,u_j,u_{ij}))\, du_{ij}\right)^{1-y_{ij}} du_1\cdots du_n \\
&= \int_{[0,1]^n} \prod_{1\le i\ne j\le n} \left(\int_0^1 f(-,u_i,u_j,u_{ij})\, du_{ij}\right)^{y_{ij}} \times \\
&\qquad \times \left(1-\int_0^1 f(-,u_i,u_j,u_{ij})\, du_{ij}\right)^{1-y_{ij}} du_1\cdots du_n \\
&= \int_{[0,1]^n} \prod_{1\le i\ne j\le n} \phi_f(u_i,u_j)^{y_{ij}}(1-\phi_f(u_i,u_j))^{1-y_{ij}}\, du_1\cdots du_n,
\end{aligned}$$

for ϕ_f as defined in (6.19). The last line coincides with the finite-dimensional distributions of a ϕ_f-process, as defined in (6.14), thus proving the equivalence between (6.18) and the ϕ_f-process, as claimed.

6.8.7 Exercise 6.7

Ergodicity of the ϕ-process implies that

$$\varepsilon(\mathbf{Y}) = \int_{[0,1]\times[0,1]} \phi(u,v)\,du\,dv \geq \delta \int_{\{(u,v):\phi(u,v)\geq\delta\}} du\,dv = \delta\,\Pr(\phi(U,V) \geq \delta)$$

for every $\delta > 0$, for U and V i.i.d. Uniform$[0,1]$; thus, $\varepsilon(\mathbf{Y}) = 0$ if and only if $\Pr(\phi(U,V) \geq \delta) = 0$ for all $\delta > 0$, and $\Pr(\phi(U,V) = 0) = 1$, as claimed.

6.8.8 Exercise 6.8

Let γ_m and γ_n be as defined in (6.31). Without loss of generality, assume $V(\mathbf{y}_k) = k$ for every $k \geq 1$, so that we can express γ_m and γ_n by

$$\gamma_m(\mathbf{x}) = \delta(\mathbf{x} : (\mathbf{y}_k)_{k\geq1}) = \lim_{k\to\infty} \frac{1}{k^{\downarrow m}} \sum_{\psi:[m]\to[k]} \mathbf{1}(\mathbf{y}_k^{\psi} = \mathbf{x}), \quad \mathbf{x} \in \{0,1\}^{m\times m}$$

$$\gamma_n(\mathbf{x}^*) = \delta(\mathbf{x}^* : (\mathbf{y}_k)_{k\geq1}) = \lim_{k\to\infty} \frac{1}{k^{\downarrow n}} \sum_{\psi:[n]\to[k]} \mathbf{1}(\mathbf{y}_k^{\psi} = \mathbf{x}^*), \quad \mathbf{x}^* \in \{0,1\}^{n\times n}. \quad (6.43)$$

For every $\mathbf{x} \in \{0,1\}^{m\times m}$, we want to show

$$\gamma_m(\mathbf{x}) = \gamma_n(\{\mathbf{x}^* \in \{0,1\}^{n\times n} : \mathbf{x}^*|_{[m]} = \mathbf{x}\}) = \sum_{\{\mathbf{x}^*\in\{0,1\}^{n\times n}:\mathbf{x}^*|_{[m]}=\mathbf{x}\}} \gamma_n(\mathbf{x}^*).$$

By (6.43), we have

$$\begin{aligned}
&\gamma_n(\{\mathbf{x}^* \in \{0,1\}^{n\times n} : \mathbf{x}^*|_{[m]} = \mathbf{x}\}) = \\
&= \sum_{\{\mathbf{x}^*\in\{0,1\}^{n\times n}:\mathbf{x}^*|_{[m]}=\mathbf{x}\}} \lim_{k\to\infty} \frac{1}{k^{\downarrow n}} \sum_{\psi:[n]\to[k]} \mathbf{1}(\mathbf{y}_k^{\psi} = \mathbf{x}^*) \\
&= \sum_{\{\mathbf{x}^*\in\{0,1\}^{n\times n}:\mathbf{x}^*|_{[m]}=\mathbf{x}\}} \lim_{k\to\infty} \frac{1}{k^{\downarrow m}} \frac{1}{(k-m)^{\downarrow(n-m)}} \times \\
&\quad \times \sum_{\psi:[m]\to[k]} \sum_{\psi^*:[n]\to[k] \text{ s.t. } \psi^*|_{[m]}=\psi} \mathbf{1}(\mathbf{y}_k^{\psi^*} = \mathbf{x}^*) \\
&= \lim_{k\to\infty} \frac{1}{k^{\downarrow m}} \sum_{\psi:[m]\to[k]} \frac{1}{(k-m)^{\downarrow(n-m)}} \times \\
&\quad \times \left(\sum_{\psi^*:[n]\to[k] \text{ s.t. } \psi^*|_{[m]}=\psi} \; \sum_{\{\mathbf{x}^*\in\{0,1\}^{n\times n}:\mathbf{x}^*|_{[m]}=\mathbf{x}\}} \mathbf{1}(\mathbf{y}_k^{\psi^*} = \mathbf{x}^*) \right)
\end{aligned}$$

$$
\begin{aligned}
&= \lim_{k\to\infty} \frac{1}{k^{\downarrow m}} \sum_{\psi:[m]\to[k]} \frac{1}{(k-m)^{\downarrow(n-m)}} \times (k-m)^{\downarrow(n-m)} \mathbf{1}(\mathbf{y}_k^{\psi} = \mathbf{x}) \\
&= \lim_{k\to\infty} \frac{1}{k^{\downarrow m}} \sum_{\psi:[m]\to[k]} \mathbf{1}(\mathbf{y}_k^{\psi} = \mathbf{x}) \\
&= \gamma_m(\mathbf{x}),
\end{aligned}
$$

where beginning in the fourth line above I write $\psi^*|_{[m]}$ for the domain restriction of ψ^* to $[m]$ (i.e., $\psi^*|_{[m]} : [m] \to [k]$ defined by $\psi^*|_{[m]}(i) = \psi^*(i)$ for $1 \le i \le m$) and the identity

$$
\sum_{\psi^*:[n]\to[k] \text{ s.t. } \psi^*|_{[m]}=\psi} \; \sum_{\{\mathbf{x}^*\in\{0,1\}^{n\times n}:\mathbf{x}^*|_{[m]}=\mathbf{x}\}} \mathbf{1}(\mathbf{y}_k^{\psi^*} = \mathbf{x}^*) = (k-m)^{\downarrow(n-m)} \mathbf{1}(\mathbf{y}_k^{\psi} = \mathbf{x})
$$

follows since there are exactly $(k-m)^{\downarrow(n-m)}$ ways to extend $\psi : [m] \to [k]$ to $\psi^* : [n] \to [k]$, and for any such extension

$$
\sum_{\{\mathbf{x}^*\in\{0,1\}^{n\times n}:\mathbf{x}^*|_{[m]}=\mathbf{x}\}} \mathbf{1}(\mathbf{y}_k^{\psi^*} = \mathbf{x}^*) = \mathbf{1}(\mathbf{y}_k^{\psi} = \mathbf{x}).
$$

Chapter 7

Getting beyond graphons

As the preceding chapters indicate, modern network analysis is challenged by the apparent incompatibility between the basic model properties of

(i) vertex exchangeability and

(ii) coherence

and the empirical properties of

(iii) sparsity and

(iv) power law degree distribution.

Whereas (iii) and (iv) are believed to be widely observed in real-world networks, (i) and (ii) are relevant, and in some cases necessary, to elicit meaningful model-based statistical inferences from network data.

Viewed in this light, the preceding chapter paints a rather bleak picture. By the Aldous–Hoover theorem (Theorem 6.3), the only random graph models satisfying (i) and (ii) are mixtures of graphons. The connection between graphons and *dense* graph sequences points to the trivial 0-graphon, $\phi(-,-) \equiv 0$, as the lone model in the intersection of (i), (ii), and (iii) (see, e.g., Section 6.5). Adding (iv) to the mix reduces the number of models satisfying (i)–(iv) to *zero*. Plainly, if statistical thinking is to provide any insights into this highly relevant class of networks problems, then it must move beyond vertex exchangeability, or else give up on at least one of criteria (ii)–(iv). The move away from exchangeable models can easily be accomplished—e.g., the Barabási–Albert preferential attachment model [14] satisfies (ii)–(iv) but not (i) (see Section 4.2)—but the challenge remains to justify such a move on statistical grounds. Before describing some initial approaches to this issue in Sections 7.2 and 7.3, I first consider which of (i)–(iv), if any, are up for discussion and which should be held sacred in the quest of complex network analysis.

Remark 7.1 (Disclaimer) *I highlight items (i)–(iv), in particular vertex exchangeability, sparsity, and power law, to illustrate the difficulties facing modern network analysis, and the inadequacy of standard approaches for handling these difficulties. I do not wish to give the impression that sparsity and power law are the only, or even the most important, network attributes to study. Whereas sparsity/power law are especially apt in some applications, they are not in others. I discuss them here merely as a test case for the potential of statistical tools to model complex data structures.*

7.1 Exchangeability, coherence, sparsity, or power law: Something must go

Items (i)–(iv) above spell out four generally desired properties of statistical network models. It is known that there is no model satisfying all four, raising the conundrum: if all four properties are indeed essential, then the enterprise of statistical network modeling is a non-starter. But before declaring the mission hopeless, let's consider each property in turn, deferring exchangeability until last.

Coherence

In a widely discussed article from the early 2000s, McCullagh [120] examined the fundamental structure of statistical models using concepts from category theory. Central to McCullagh's treatment is *functoriality*, a concept from category theory which seems closely aligned with the condition of coherence from Chapter 5. Although its category-theoretic presentation is obscure to most statisticians, the rationale underlying McCullagh's framework and its connection to coherence can be easily detected in statements such as

> *The sense of the model and the meaning of the parameter [...] may not be affected by accidental or capricious choices such as sample size or experimental design.* [120, p. 1237]

Indeed, for network sampling models, condition (C) in Definition 5.2 parallels McCullagh's criterion: A model is coherent precisely when inferences from it are robust to artificial and arbitrary choices.

I mention McCullagh's viewpoint here not as validation of my own, but rather to highlight that coherence, though perhaps not articulated in the same way, already appears in other forms throughout the statistics literature. By and large, the idea of coherence is intuitively understood by practitioners of statistics, who know very well the importance of accounting for context in statistical inference. But understanding seems to be lacking among theoreticians, whose predilection for mathematics often overshadows the context for which their theory was initially intended. For our purposes of establishing the probabilistic foundations of statistical network analysis, coherence is the glue that binds theory and practice, and shall not be sacrificed at any cost.

Sparsity and power law degree distributions

If one or both of sparsity and/or power law degree distribution is known to be absent from a given application, then the missing property should not be incorporated into the model. But if these properties are known or strongly believed to be present, then they ought to be accounted for. The principle is basic: if data is known to exhibit a specific property, then any model for that data ought to replicate that property. After all, it is only in replicating real-world behaviors that a *model* lives up to its name. If a model does not replicate a property which is known to be present, then how can it be trusted to describe properties about which little is known?[1]

In the early pages of their book on theoretical statistics, Cox and Hinkley [42, p. 5] echo this principle, stating that a model should exhibit "consistency with known limiting behaviour." Though the principle extends beyond asymptotic analyses, Cox and Hinkley's focus on "known limiting behaviour" is especially apt in the present discussion of sparsity and power law, both of which are asymptotic properties. Since sparsity and power law figure prominently in many modern networks applications, both are non-negotiable for models aiming at a realistic account of networks with those properties. And since we are primarily interested here in laying down the foundations of network modeling for modern applications, we insist that models for sparse, power law networks exhibit such properties themselves.

But before moving on to discuss exchangeability, we should address those critics who disagree with Cox and Hinkley, and therefore do not believe that a model for a sparse, power law network must necessarily replicate those properties of the network which it purports to model. Box and Draper's (tired) cliché is often cited to defend this point of view:

> *All models are wrong but some are useful.* [26]

The rationale of the criticism goes: since all models are wrong, but some are useful, it is therefore possible that our model is wrong in how it handles sparsity and/or power law but nevertheless remains useful, especially if neither sparsity nor power law are of primary interest in the given application.

To such critics, I ask how a model can be assessed as 'useful' when it is faulty for describing known behaviors? In other words, if it is bad at modeling known properties, then how can it be trusted to be 'useful' for modeling properties for which less is known? One criterion for usefulness is coherence: a 'statistical model' is useful only insofar as it can be used to perform inference, and such inferences are possible only if the model is coherent (i.e., inferences from it 'make sense'); see Section 5.4 for further discussion on this point. An additional practical criterion is the replication of known properties, e.g.,

(P) If the data is known to exhibit a given property 'P' prior to any observation, then the model ought to contain candidate distributions under which 'P' occurs with strictly positive probability.

As an immediate application of this principle, recall from the opening discussion of this chapter that the collection of all graphon models (Section 6.4.1.1) does not contain a single candidate distribution that replicates the properties of sparsity and power law degree distribution. By principle (P), the class of graphon models does not make sense, and is therefore not useful, for modeling networks with those properties.

[1] Misspecified models in one metric can still perform well in other metrics. It is conceivable, for example, that specifying a non-sparse model for a network which is known to be sparse could perform well in fitting the degree of clustering, community structure, etc. But evidence of such robustness to model misspecification, either in the form of mathematical proof or empirical evidence, should be given in support of such a claim.

Exchangeability

Finally, exchangeability. Is it essential? Nothing in the framework of Chapter 5 says that it is. In fact, the major considerations of Chapters 3–6 indicate that exchangeability may be inappropriate for most networks applications. As the previous chapter makes clear, vertex exchangeability entails an implicit assumption of a homogeneous population and representative vertex sampling, neither of which is accurate for many networks of current interest. And yet, exchangeability is a prominent theme throughout the rest of this chapter and the next four.

Exchangeability is not necessary for coherent network modeling. But as a practical matter the symmetry induced by exchangeability, or any other suitable invariance principle, is often needed to make statistical inference tractable and/or interpretable. As mentioned earlier, exchangeability establishes the link between observed and unobserved parts of the network being studied, in effect articulating how the data is representative of the population and, thus, on what grounds inferences beyond the sample can be justified. So rather than contemplating whether exchangeability is essential to statistical modeling, we are better served by recalling the purpose of exchangeability for statistical inference, and whether the notion of exchangeability associated to graphon models serves this purpose. When dealing with networks that are known to be sparse and/or have power law degree distribution, vertex exchangeability clearly does not serve its intended purpose, and therefore must not be essential. But since the observation mechanism for most network data is such that the observed vertices are non-representative of the population of all vertices, the implicit context of vertex exchangeable models can be ruled out even without consideration of any known empirical properties.

If we dispense with vertex exchangeability, then what will replace it? The next several chapters present some possibilities, and establish that exchangeability is compatible with criteria (ii)–(iv), if thought about from the right perspective.

7.2 Sparse graphon models

One of the earliest attempts to get beyond the aforementioned limitations of graphons appeared in the work of Bickel and Chen [19]. Recall that a compatible family of finite graphs $(\mathbf{Y}_n)_{n\geq 1}$ (i.e., $\mathbf{Y}_n\,|_{[m]} = \mathbf{Y}_m$ for all $m \leq n$) can be regarded as the finite restrictions of a single process for an infinite graph $\mathbf{Y} = (Y_{ij})_{i,j\geq 1}$ by defining $\mathbf{Y}_n = \mathbf{Y}\,|_{[n]}$ for each $n \geq 1$. With this representation, the edge density of $(\mathbf{Y}_n)_{n\geq 1}$ from the graphon construction in (6.13) satisfies

$$\lim_{n\to\infty} \frac{1}{n(n-1)} \sum_{1\leq i\neq j\leq n} Y_{ij} = \int_0^1 \int_0^1 \phi(u,v)\, du\, dv \quad \text{with probability 1.} \tag{7.1}$$

This limit equals 0 and $(\mathbf{Y}_n)_{n\geq 1}$ is sparse if and only if each $\mathbf{Y}_n$ is empty (i.e., has no edges) with probability 1, and any such 'sparse' network corresponds to the vertex exchangeable model parameterized by the trivial 0-graphon; see Exercise 6.7.

With this observation, Bickel and Chen [19, p. 21069] proposed to decompose

$\phi(-,-)$ into the product of its expected edge density

$$\varepsilon(\phi) = \Pr(Y_{12} = 1; \phi) = \int_0^1 \int_0^1 \phi(u,v)\, du\, dv$$

and the conditional density $w(u,v) = \phi(u,v)/\varepsilon(\phi)$ of the latent variables (U_1, U_2) given that there is an edge between vertices 1 and 2. Bickel and Chen suggested that the decomposition

$$\phi(u,v) = \varepsilon(\phi) w(u,v), \quad 0 \leq u, v \leq 1,$$

be interpreted as a "decoupling" of the expected degree of each vertex in $\mathbf{Y}_n = (Y_{ij})_{1 \leq i,j \leq n}$,

$$E(\deg_{\mathbf{Y}_n}(i)) = E\left(\sum_{j \neq i} Y_{ij}\right) = (n-1)\varepsilon(\phi) \propto \varepsilon(\phi),$$

and the so-called "inhomogeneity structure" captured by the conditional density $w(u,v)$. The rationale underlying Bickel and Chen's approach is summarized in a single sentence,

"It is natural to let $\varepsilon(\phi)$ depend on n but $w(\cdot,\cdot)$ to be fixed" [19].[2]

In allowing $\varepsilon(\phi)$ to depend on n while holding $w(u,v)$ fixed, Bickel and Chen in effect propose to construct each finite graph $\mathbf{Y}_n$ according to a ϕ_n-process, for a family of graphons $(\phi_n)_{n \geq 1}$ defined by

$$\phi_n(u,v) = \rho_n^{-1} w(u,v), \quad 0 \leq u, v \leq 1, \tag{7.2}$$

where $\rho_n^{-1} = \varepsilon(\phi_n)$ is the expected edge density. The most direct way to accomplish this is by fixing a graphon $\phi : [0,1] \times [0,1] \to [0,1]$ and letting $(\rho_n)_{n \geq 1}$ be a sequence for which $\rho_n \to \infty$ and

$$\lim_{n \to \infty} \rho_n^{-1} \int_0^1 \int_0^1 \phi(u,v)\, du\, dv = 0. \tag{7.3}$$

Now, for $\mathbf{Y}_n = (Y_{ij}^{(n)})_{1 \leq i,j \leq n}$ distributed according to a ϕ_n-process, for each $n \geq 1$, the analog to (7.1) is given by

$$\begin{aligned}
\lim_{n \to \infty} \frac{1}{n(n-1)} \sum_{1 \leq i \neq j \leq n} Y_{ij}^{(n)} &= \lim_{n \to \infty} \int_0^1 \int_0^1 \phi_n(u,v)\, du\, dv \quad \text{with probability 1} \\
&= \lim_{n \to \infty} \rho_n^{-1} \int_0^1 \int_0^1 \phi(u,v)\, du\, dv \\
&= 0,
\end{aligned}$$

implying that the sequence of ϕ_n-processes $(\mathbf{Y}_n)_{n \geq 1}$ generated under these conditions is sparse. Notice that the incoherent model in Section 5.2 has this form with $\phi(-,-) \equiv \theta$ and $\rho_n = n$.

[2] I write '$\varepsilon(\phi)$' in place of Bickel and Chen's notation 'ρ' for consistency with the rest of this text.

Exercise 7.1 *Take a moment to think about what exactly this model describes, how the 'decoupling' in (7.2) can be seen as 'natural', and how the resulting sparsity property might be interpreted in light of a real application.*

In the paradigm of Chapter 5, the Bickel–Chen model takes

$$\mathscr{M}_n = \{\rho_n^{-1}\phi \mid \phi : [0,1] \times [0,1] \to [0,1],\ \rho_n \to \infty \text{ and satisfies (7.3)}\}, \quad (7.4)$$

for each $n \geq 1$. If each $\mathbf{Y}_n$ is modeled by $\mathscr{M}_n$, it is apparent that the sequence $(\mathbf{Y}_n)_{n\geq 1}$ will be sparse with probability 1, in the sense of Lovász–Szegedy convergence of graph sequences (Section 6.5.3). But for the purpose of inference this model conveys no relationship among observations of different sizes. In particular, the Bickel–Chen model does not specify a context within which the family $\{\mathscr{M}_n\}_{n\geq 1}$ is to be interpreted. Furthermore, the models $\mathscr{M}_n$ for different n cannot be interpreted coherently under the standard selection sampling context. So while the statement about the limit of edge densities for $(\mathbf{Y}_n)_{n\geq 1}$ is valid mathematically, the construction of each $\mathbf{Y}_n$ from separate ϕ_n-processes, without any logical connection between them as n varies, raises questions as to what is actually being modeled and, furthermore, how inferences from such a model are to be interpreted. Thus, overall the Bickel–Chen and related sparse graphon approaches achieve sparsity but at the expense of other properties necessary to make the resulting models useful for statistical inference.

It is also notable that while Bickel and Chen's approach does, in some limited sense, provide a remedy for the fact that ϕ-processes produce dense graphs, the homogeneity inherent to ϕ-processes leaves many more 'complex' features of network data, such as heavy-tailed degree distributions, beyond its reach. The same coherence issues and homogeneity properties are present in the general family of L_p graphon models, e.g., [25], which came about after Bickel and Chen's initial proposal.

As this discussion indicates, the importance of consistency under subsampling, even the special case of consistency under selection, has been downplayed in the mathematical statistics literature on networks, and still coherence seems to be poorly understood. This trend has begun to change more recently, as several authors [52, 54, 128, 138] have emphasized the need for compatibility between sample and population in statistical network models. This renewed emphasis has led to a couple of new ideas. One of these is Caron and Fox's proposal to model networks using completely random measures and exchangeable point processes. The other is Crane and Dempsey's edge exchangeable framework for modeling interaction networks. Both approaches get beyond previous limitations of graphons and vertex exchangeability by representing networks as something other than a graph with labeled vertices, in the spirit of Section 1.2. I discuss the Caron–Fox approach in the next section and the Crane–Dempsey framework in Chapters 9 and 10.

7.3 Completely random measures and graphex models

Caron and Fox [31, 32] propose to expand the graphon framework in a way that overcomes the limitations highlighted in Sections 6.5 and 7.1. Instead of modeling network data as a random graph whose distribution is parameterized by a *graphon*

$\phi : [0,1] \times [0,1] \to [0,1]$ (as in Section 6.4.2), they propose to represent a network as an exchangeable point process in the plane $[0,\infty) \times [0,\infty)$ whose distribution is parameterized by a so-called *graphex* [147]. Because it draws on ideas from the theory of point processes and completely random measures, Caron and Fox's approach takes us outside the usual realm of network modeling. As such a departure from convention is needed in order to surmount the limitations of the standard paradigm, it is worth describing their construction in some detail.

Before beginning, I note that the authors have promoted their model's ability to replicate sparsity and power law degree distribution as its main selling point [31, 32]. But while accounting for sparsity and power law is a positive feature, especially in light of the concerns voiced throughout Section 7.1, the value of the Caron–Fox approach to the foundations of network analysis will ultimately be determined by whether and how it allows us to 'see' things that we couldn't see before; refer to Sections 1.2–1.3 for more on this point. With this in mind, I focus here on the new perspective provided by this approach and refer the reader to [31, 32] and related work for more details on empirical properties of the model.

Among the main questions to ponder while digesting the upcoming model construction are:

1. Is the representation of networks as point processes a natural way to think about network data or is it an artificial mathematical abstraction? Under what circumstances is it the former? Under what circumstances is it the latter?
2. What implications (good or bad) does this representation have for practical applications?
3. How are the generating dynamics of this family of processes related to the way in which real networks form? In particular, does the interpretation in terms of '*p*-sampling' (Section 7.3.7) align with the context in which real-world networks are observed?

These questions point to one drawback of the Caron–Fox model: there is currently no clearcut motivating example for the exchangeable point process representation of network data. To date, the primary focus of study, e.g., [23, 24, 31, 32, 33, 147, 148], has been on theory, without much attention to any realistic context in which the theory could be applied. To better motivate this model class, I first describe a practical scenario for which it could be natural. See [47] for further discussion of the above questions about the Caron–Fox model.

7.3.1 *Scenario: Formation of Facebook friendships*

Consider a social media platform, e.g., Facebook, on which users are related to one another through their 'friendships' (i.e., two users who agree to be 'friends' are able to access each others' posts on the site, post content on each others' profile, etc.). In addition to observing whether two users are friends, suppose we also observe the time t at which each user joined the social media site. One way to summarize this data is as a pair $(\mathbf{x},\mathbf{y})$, where $\mathbf{x} = (x_i)_{1\leq i\leq n}$ records the times at which each of the observed vertices (labeled $1,\ldots,n$) joined the site and $\mathbf{y} = (y_{ij})_{1\leq i,j\leq n}$ records the presence

($y_{ij} = 1$) or absence ($y_{ij} = 0$) of a friendship between vertices labeled i and j. The data can be alternatively represented as a subset $\mathbf{Y} \subseteq [0,\infty) \times [0,\infty)$, where $(t,t') \in \mathbf{Y}$ indicates a friendship between two users, one who opened an account at time t and another at time t'. The following model class is geared toward data represented in this latter form.

7.3.2 *Network representation*

The discussion of vertex exchangeable models throughout Chapter 6 and Section 7.2 reveals a few limitations of the networks-as-graphs mindset (Section 1.2). A first step toward overcoming these limitations is to represent the data in such a way that vertex exchangeability is no longer the predominant or default invariance principle for network analysis. Caron and Fox achieve this by representing a network as a point pattern in $\mathbb{R}_+^2 = [0,\infty) \times [0,\infty)$ instead of as a $\{0,1\}$-valued array.

A *point pattern* in $\mathbb{R}_+^2$ is a (symmetric) subset $\mathbf{y} \subseteq [0,\infty) \times [0,\infty)$.[3] Sometimes it is convenient to represent $\mathbf{y}$ equivalently as a measure which assigns a positive real number to every (measurable) subset of $[0,\infty) \times [0,\infty)$ by

$$\mathbf{y}(A) = |\{a \in A : a \in \mathbf{y}\}|, \quad A \subseteq [0,\infty) \times [0,\infty), \tag{7.5}$$

where $|S|$ denotes the cardinality of $S \subseteq [0,\infty) \times [0,\infty)$. Defined in this way, $\mathbf{y}(A)$ counts the number of points in $\mathbf{y}$ that are also in A. Since the point pattern $\mathbf{y} \subseteq [0,\infty) \times [0,\infty)$ and the associated measure (7.5) are equivalent, we use the same notation $\mathbf{y}$ to refer to both, often without explicit mention of which interpretation we prefer.

Exercise 7.2 *Show that the representation of* $\mathbf{y}$ *as a subset of* $[0,\infty) \times [0,\infty)$ *and as a measure* $\mathbf{y}(\cdot)$ *in* (7.5) *are equivalent.*

Note a key distinction between the point process representation and the more conventional representation of a network as a $\{0,1\}$-valued array $(y_{ij})_{i,j\geq 1}$. In the traditional representation, every index $i = 1,2,\ldots$ corresponds to a vertex. As a result, the network representation $(y_{ij})_{i,j\geq 1}$ records the presence/absence of every possible edge in the network through y_{ij}. In the point process representation, however, $[0,\infty)$ is a set of *potential* vertex labels, in the sense that if

$$\{(t,s) \in \mathbf{y} : 0 \leq s < \infty\} = \emptyset$$

for some $t \geq 0$, i.e., there is no edge in $\mathbf{y}$ involving a vertex labeled by t, then there might as well be no vertex corresponding to t. In other words, the set $\mathbf{y}$ contains only those edges that are present, without any record of absent edges or vertices. In the scenario of Section 7.3.1, for example, a Facebook account that is created at time t but which never interacts with another account via 'friendship' is not treated

---

[3] Symmetry, i.e., $(x,x') \in \mathbf{y}$ if and only if $(x',x) \in \mathbf{y}$, is not required, but we assume it here to most closely mirror the presentation in [32, 147] and elsewhere. The asymmetric case, for representing directed networks, is analogous and can be treated without issue. The setup also allows for multigraphs and weighted graphs, by associating each point $(x,x') \in \mathbf{y}$ to a positive weight. I defer to [31, Section 3] for a discussion of these alternatives.

as part of the 'Facebook network' in the point process representation. Similar to the forthcoming discussion of edge exchangeable models (Chapter 9), this aspect of the point process representation seems to better align with the concept of 'network', which naturally invokes the notion of interaction/connectivity in a way that is not captured by more traditional graphical representations.

7.3.3 Interpretation of vertex labels

For representing network data, the point pattern $\mathbf{y} \subseteq [0,\infty) \times [0,\infty)$ is interpreted as a set of edges with vertices corresponding to points in $[0,\infty)$ and the occurrence of $(t,t') \in \mathbf{y}$ interpreted to mean that there is an edge between the vertices labeled t and t'. Caron and Fox [31, Section 3.5] suggest to interpret the vertex labels as the "time at which a *potential node* enters the network and has the opportunity to link with other existing nodes." In concrete scenarios, e.g., Section 7.3.1, it seems difficult to articulate the concept of a 'potential node', and so we instead interpret the label $t > 0$ assigned to a vertex as the time at which that vertex first enters the system.

When adopting this interpretation, the reader must be careful not to interpret the vertex labels as the 'times' at which each vertex first appears in the observed *network*. Instead, the label t assigned to a vertex is better understood as the time at which that vertex is 'born' into the *system*, and thus the first time at which the vertex *could* appear in the network. A vertex does not appear in the network until it interacts with another vertex. Note well the distinction between the 'network', i.e., interactions/relationships among vertices, and the 'system', i.e., the platform/setting in which the network is formed. In the scenario of Section 7.3.1, for example, the 'Facebook network' is comprised of 'friendships' between users on the social media platform called 'Facebook'. Each vertex label is the time at which a Facebook account was created, regardless of whether the user initiated any friendships at that time. This distinction is important when interpreting the forthcoming sampling operation associated to the Caron–Fox model.

Following the discussion about statistical units in Section 3.7, 'time' is the implicit unit of observation for networks represented as point processes. In this case, a sample of size t is an observation of the process for t units of time, and we can think of $\mathbf{Y} \subseteq [0,\infty) \times [0,\infty)$ as the result of a temporally evolving sequence $(\mathbf{Y}_t)_{t\geq 0}$ with each $\mathbf{Y}_t$ determined by edges between vertices that arrive before time t,

$$\mathbf{Y}_t = \mathbf{Y} \cap [0,t]^2, \quad t \geq 0.$$

The restriction $\mathbf{Y} \cap [s,t)^2$, for $s \leq t$, is the set of edges between only those vertices which arrived between times s and t.

Although many real-world networks evolve over time, it is rare in practice to observe the complete temporal evolution of the network, as we have assumed in Section 7.3.1. More often, the temporal structure in $\mathbf{Y}$ is observed only as a $\{0,1\}$-valued array $\mathbf{X} = (X_{ij})_{i,j\geq 1}$ with

$$X_{ij} = \begin{cases} 1, & (\theta_i, \theta_j) \in \mathbf{Y}, \\ 0, & \text{otherwise}, \end{cases} \tag{7.6}$$

for $\theta_1, \theta_2, \ldots \in [0,\infty)$ denoting the arrival times of vertices in the point process. In particular, even if the time t and the relative ordering of vertex arrivals are known, it is rare to also observe the times $0 < \theta_1 < \cdots < \theta_{n_t} < t$ at which observed vertices first appeared in the system. So while the representation as a point process $\mathbf{Y}$ gives the clearest interpretation of the forthcoming model, with 'time' as the natural unit of observation and exchangeability interpreted as invariance with respect to observations over a given length of time, this interpretation becomes muddled when passing to the coarser structure $\mathbf{X}$ in (7.6).

Another intermediate possibility occurs if the time t is observed, but the arrival times of specific vertices are not. In this case, the data could be represented by the equivalence class of all point processes $\mathbf{Y}' \subseteq [0,t] \times [0,t]$ that are compatible with the observed edge pattern. In particular, for any $\mathbf{y} \subseteq [0,t] \times [0,t]$, let $\mathbf{X}(\mathbf{y})$ denote its induced $\{0,1\}$-valued array as in (7.6). For a time $t > 0$ and a pattern of edges represented as a $\{0,1\}$-valued array $\mathbf{x}$, define its equivalence class of point processes by

$$\tilde{\mathbf{x}} = \{\mathbf{y} \subseteq [0,t] \times [0,t] \mid \mathbf{X}(\mathbf{y}) = \mathbf{x}\}. \tag{7.7}$$

For any point process $\mathbf{Y}_t \subseteq [0,t] \times [0,t]$, the distribution of its associated equivalence class $\tilde{\mathbf{X}}_t$, for $\mathbf{X}_t$ as in (7.6) and $\tilde{\mathbf{X}}_t$ as in (7.7), is induced by the distribution of $\mathbf{Y}_t$, as in

$$\Pr(\tilde{\mathbf{X}}_t = \tilde{\mathbf{x}}) = \Pr(\mathbf{Y}_t \in \{\mathbf{y} \subseteq [0,t] \times [0,t] \mid \mathbf{X}(\mathbf{y}) = \mathbf{x}\}).$$

I do not consider this intermediate case any further below. See [147, 148] for further discussion about how to make sense of the real-valued sample size $t > 0$ for the $\{0,1\}$-valued representation $\mathbf{X}_t$ associated to $\mathbf{Y}_t = \mathbf{Y} \cap [0,t]^2$ via (7.6). The distinction between the point process $\mathbf{Y}$ and the associated 'graphical representation' $\mathbf{X}$ is especially important when interpreting exchangeability of the point process $\mathbf{Y}$ in the context of $\mathbf{X}$.

7.3.4 Exchangeable point process models

Following the terminology of Kallenberg [98, Chapter 9], a point process $\mathbf{Y} \subseteq [0,\infty) \times [0,\infty)$ is called *exchangeable* if its distribution is invariant under the joint action on $[0,\infty) \times [0,\infty)$ by Lebesgue measure-preserving transformations of $[0,\infty)$. To be precise, a function $T : [0,\infty) \to [0,\infty)$ is called a *Lebesgue measure-preserving transformation of* $[0,\infty)$ if

$$\int_A dx = \int_{T^{-1}(A)} dx \tag{7.8}$$

for all Borel measurable subsets $A \subseteq [0,\infty)$. For any such $T : [0,\infty) \to [0,\infty)$, we write $T \circ \mathbf{Y} \equiv \mathbf{Y}^T$ for the transformation of $\mathbf{Y}$ under T given by

$$(T(x), T(x')) \in \mathbf{Y}^T \quad \text{if and only if} \quad (x, x') \in \mathbf{Y}. \tag{7.9}$$

(In words, $\mathbf{Y}^T$ is the point process obtained from $\mathbf{Y}$ by transforming $[0,\infty)$ according to T.) Then $\mathbf{Y}$ is *exchangeable* if and only if

$$\mathbf{Y}^T =_{\mathscr{D}} \mathbf{Y} \text{ for all } T : [0,\infty) \to [0,\infty) \text{ satisfying (7.8)}. \tag{7.10}$$

With the interpretation of the vertex labels as 'arrival times' from Section 7.3.3, the 'exchangeability' condition in (7.10) means that every observation of the network over a time period of length $t \geq 0$ is representative of every other observation of the network over a period of length t; in particular, $\mathbf{Y}_t$ is representative of any other observation $\mathbf{Y} \cap (S \times S)$, for $S \subseteq [0, \infty)$ with Lebesgue measure t. With this understanding of exchangeability, it is important to realize that the 'graph structure' associated to an exchangeable point process $\mathbf{Y}$ through (7.6) is *not* exchangeable in the sense of Chapter 6. In particular, let $\mathbf{Y} \subseteq [0, \infty) \times [0, \infty)$ be an exchangeable point process with vertices labeled $0 < \theta_1 < \theta_2 < \cdots$. Then $\mathbf{X}$ defined in (7.6) does *not* satisfy

$$\mathbf{X}^\sigma =_{\mathscr{D}} \mathbf{X} \quad \text{for all permutations } \sigma : \mathbb{N} \to \mathbb{N}.$$

Exercise 7.3 *Explain the intuition for why* $\mathbf{X}$ *is not exchangeable. (Hint: Think of the interpretation of vertex labels as arrival times.)*

7.3.5 Oxymoron: 'Sparse exchangeable graphs'

The definition of exchangeability (for point processes) in (7.10) should not be conflated with the conventional definition of exchangeability (for graphs) in (6.2). And yet that is exactly what results from the use of 'exchangeable' to describe point processes, as in (7.10), and the oxymoronic term 'sparse exchangeable graphs' employed by some authors, e.g., in the titles of [23, 24]. Whereas exchangeability of a graph facilitates an intuitive interpretation as 'invariance under relabeling', exchangeability of a point process does not. In the point process representation, the set $[0, \infty)$ does not merely label distinct vertices; it also corresponds to the 'times' at which vertices arrive. In this way, the labeling set not only contains points but also possesses topological structure, and the 'exchangeability' condition in (7.10) is defined as an invariance with respect to only those transformations which preserve this structure (by preserving Lebesgue measure).

This distinction does not come across clearly in [32, Section 5.1], where exchangeability of $\mathbf{Y}$ is described as "invariance to the time of arrival of the nodes." (Recall that 'time' not only labels the vertices but also quantifies the duration over which the process has been observed.) Even more confusion results from the term "sparse exchangeable graphs," which conflates exchangeability of the point process in (7.10) with sparsity of the graph induced by the point process through (7.6). The trouble with this overloaded terminology is that graphs invoke their own notion of exchangeability, as in (6.2), which is distinct from exchangeability of the point process (7.10). It is both more accurate and more honest to instead regard the graphex model as the "class of random graphs arising from exchangeable random measures," as in the title of [147].

Instead of using 'exchangeability' to describe point processes satisfying (7.10), a better term would have been 'stationarity' with respect to time since, from condition (7.10), the process is not invariant with respect to rearrangements of arrival times (as the term 'exchangeability' suggests) but rather is 'stationary' with respect to observing the network over equal durations of time. The distinction should be clear from Exercise 7.3. I discuss this interpretation further in Section 7.3.7.

7.3.6 Graphex representation

Since the connection between the point process $\mathbf{Y}$ and $\{0,1\}$-valued array $\mathbf{X}$ in (7.6) is valid irrespective of exchangeability, we are left asking what statistical significance, if any, the condition (7.10) adds beyond the potential computational benefits mentioned in [32]? In particular, what additional assumptions does exchangeability (condition (7.10)) impose on the structure of $\mathbf{Y}$, and thus also on the array $\mathbf{X}$ it induces? On this latter point, Veitch and Roy [147] adapt a theorem of Kallenberg [97] to obtain a generic *graphex representation* for random graphs associated to exchangeable random measures via (7.6). Because this connection is a bit technical, and not essential to our broader discussion, I discuss it only briefly here, and leave the details to [147].

Following [147], we define a *graphex* as a triple (I,S,W) with $I \in [0,\infty)$, $S : [0,\infty) \to [0,\infty)$ an integrable function, and $W : [0,\infty)\times[0,\infty) \to [0,1]$ a symmetric measurable function satisfying some integrability conditions.[4] Let $\Theta \subseteq [0,\infty)^2$, $\Xi_i \subseteq [0,\infty)^2$ for each $i \geq 1$, and $R \subseteq [0,\infty)^3$ be independent unit rate Poisson point processes on their respective spaces. Writing $\Theta = \{(\theta_j,\vartheta_j)\}_{j\geq 1}$, $\Xi_i = \{(\sigma_{ij},\chi_{ij})\}_{j\geq 1}$ for each $i \geq 1$, and $R = \{(\rho_k,\rho'_k,\eta_k)\}_{k\geq 1}$, we construct a random measure $\mathbf{Y}^*(\cdot)$ on $[0,\infty)\times[0,\infty)$ by choosing a graphex (I,S,W) at random and, given (I,S,W), Θ, $\{\Xi_i\}_{i\geq 1}$, and R, putting

$$\begin{aligned}
\mathbf{Y}^*(\cdot) \quad = \quad & \sum_{i,j}\mathbf{1}(W(\vartheta_i,\vartheta_j) \leq \zeta_{\{i,j\}})\delta_{\theta_i,\theta_j}(\cdot) + && (7.11)\\
& + \sum_{j,k}\mathbf{1}(\chi_{jk} \leq S(\vartheta_j))(\delta_{\theta_j,\sigma_{jk}}(\cdot)+\delta_{\sigma_{jk},\theta_j}(\cdot)) + && (7.12)\\
& + \sum_{k}\mathbf{1}(\eta_k \leq I)(\delta_{\rho_k,\rho'_k}(\cdot)+\delta_{\rho'_k,\rho_k}(\cdot)), && (7.13)
\end{aligned}$$

where $\{\zeta_{\{i,j\}}\}$ is an independent family of Uniform$[0,1]$ random variables and $\delta_{x,x'}(\cdot)$ is the Dirac point mass at (x,x') for each $x,x' \in [0,\infty)$,

$$\delta_{x,x'}(A) = \begin{cases} 1, & (x,x') \in A, \\ 0, & \text{otherwise.} \end{cases}$$

The three components of the graphex representation (7.11), (7.12), and (7.13) decompose the structure of networks distributed according to these models according to three edge types:

1. The component in (7.11) is closely related to the graphon representation in (6.13). For random W and $(\theta_i,\vartheta_i),(\theta_j,\vartheta_j) \in \Theta$, the edge (θ_i,θ_j) appears in $\mathbf{Y}^*$ with probability
$$\Pr(W(\vartheta_i,\vartheta_j) \leq \zeta_{\{i,j\}}).$$
2. The component in (7.12) describes isolated 'stars' in $\mathbf{Y}^*$. For random S and $(\theta_j,\vartheta_j) \in \Theta$ for each $j \geq 1$, the Poisson point process $\{\sigma_{jk}\}_{k\geq 1} \subseteq [0,\infty)$ corresponds to a family of vertices for which each edge (θ_j,σ_{jk}), for $k \geq 1$, appears

[4] See [147] for a complete statement of the required technical conditions.

with probability

$$\Pr(\chi_{jk} \leq S(\vartheta_j)).$$

Since the Poisson processes Θ and Ξ_j are independent for all $j \geq 1$, the sets $\{\theta_j\}_{j\geq 1}$ and $\{\sigma_{jk}\}_{k\geq 1}$, $j \geq 1$, are non-overlapping with probability 1, and thus the vertices corresponding to $\{\theta_j\}$ are distinct from those in each $\{\sigma_{jk}\}$ with probability 1. As a result, each σ_{jk} can appear in at most one edge of $\mathbf{Y}^*$, namely (θ_j, σ_{jk}) (and (σ_{jk}, θ_j) by the enforced symmetry in (7.12)). Altogether, these potential edges (θ_j, σ_{jk}) and (σ_{jk}, θ_j), $k \geq 1$, result in a 'star' between θ_j and infinitely many (otherwise isolated) vertices.

3. Finally, component (7.13) describes isolated edges, i.e., edges between two vertices which are otherwise isolated in $\mathbf{Y}^*$. For random I, there is an edge between the pair (ρ_k, ρ'_k), $k \geq 1$, from the point process R with probability

$$\Pr(\eta_k \leq I).$$

Once again, because the point processes are independent, the points $\{\rho_k\} \cup \{\rho'_k\}$ are disjoint from both $\{\theta_j\}$ and $\{\sigma_{jk}\}$, $j \geq 1$, with probability 1. Therefore, each ρ_k and ρ'_k has exactly one opportunity to appear in $\mathbf{Y}^*$, through the edge (ρ_k, ρ'_k) (and (ρ'_k, ρ_k) by symmetry), forcing any such edge to be isolated whenever it appears.

This decomposition of edge types for graphex models mirrors a common structure of exchangeability. In Section 9.5, we observe similar behavior for edge exchangeable models.

7.3.7 Sampling context

The distinction between the point process $\mathbf{Y}$ and its associated array $\mathbf{X}$ in (7.6) is especially relevant for understanding the sampling context of Caron–Fox models. When treated as a point process, the role of ordinary vertex selection for graphs, as in (3.6), is played by *t-selection* $\mathbf{S}_t$, which I define here as the action in (7.9) induced by the inclusion map $i_t : [0,t] \to [0,\infty)$, $u \mapsto u$, i.e.,

$$\mathbf{S}_t \mathbf{Y} = i_t \circ \mathbf{Y} = \mathbf{Y} \cap [0,t]^2, \quad t \geq 0. \tag{7.14}$$

For $s \leq t$, the inclusion $i_{s,t} : [0,s] \to [0,t]$, $u \mapsto u$, induces the projection $\mathbf{S}_{s,t}$ from point processes on $[0,t]^2$ to point processes on $[0,s]^2$ by

$$\mathbf{S}_{s,t} \mathbf{Y}_t = i_{s,t} \circ \mathbf{Y}_t = \mathbf{Y}_t \cap [0,s]^2. \tag{7.15}$$

With each $\mathscr{M}_t$ defined as the set of all distributions of point processes $\mathbf{S}_t \mathbf{Y} \subseteq [0,t]^2$ obtained by t-selection from an exchangeable point process $\mathbf{Y}$ on $[0,\infty) \times [0,\infty)$, and with $\mathbf{S}_{s,t}$ defined as the selection map in (7.15), I define the *projective Caron–Fox model* by $(\{\mathscr{M}_t\}_{t\geq 0}, \{\mathbf{S}_{s,t}\}_{t\geq s\geq 0})$; see (7.16) for a more general definition of what I mean here by *Caron–Fox model*.[5] The generating mechanism determined

[5]Here I define the *projective Caron–Fox model* as $(\{\mathscr{M}_t\}_{t\geq 0}, \{\mathbf{S}_{s,t}\}_{t\geq s\geq 0})$, where each $\mathscr{M}_t$ consists of

by the exchangeable point process construction via the graphex characterization in (7.11)–(7.13) automatically makes the projective Caron–Fox model coherent in the sense of Definition 5.2.

To generalize the above setup, we define a *t-sampling map* for any $t \geq 0$ as a Lebesgue measure-preserving injection $T_t : [0,t] \to [0,\infty)$ and let $\mathscr{T}_t$ denote the set of all such t-sampling maps. By analogy to Section 3.9, recall the definition of ψ-selection $\mathbf{S}^{\psi}_{m,n} : \{0,1\}^{n\times n} \to \{0,1\}^{m\times m}$ as the sampling operation induced by an injection $\psi : [m] \to [n]$. With a network represented as a $\{0,1\}$-valued array of size n, ψ-selection corresponds to sampling a network by selecting m units from n according to ψ. In the present setting with a network represented as a point process in $[0,\infty) \times [0,\infty)$ and the sampling unit given by 'time', t-sampling corresponds to observing a network for a sample of size t, where the 'sample size' of t is ensured by the fact that T_t preserves the Lebesgue measure of $[0,t]$. Exchangeability (7.10) implies that all observations of $\mathbf{Y}$ over a time period of length t are representative of one another. In particular, what is *representative* for the network determined by $\mathbf{Y}$ is not the sample of 'vertices' in a given observed network but rather the pattern of edges that is observed among the vertices that appear within any time span of length t.

For the remainder of this chapter, I define *a Caron–Fox model* as a network model $(\{\mathscr{M}_t\}_{t\geq 0}, \{\Sigma_{s,t}\}_{t\geq s\geq 0})$ for which $\{\Sigma_{s,t}\}_{t\geq s\geq 0}$ is any (random) sampling context and each $\mathscr{M}_t$, $t \geq 0$, is the set of distributions induced by t-selection from an exchangeable point process on $[0,\infty) \times [0,\infty)$. To be precise, we define $\mathscr{T}_{s,t}$ as the set of all Lebesgue measure-preserving injections $[0,s] \to [0,t]$, and regard $\Sigma_{s,t}$ as the sampling operation induced by putting $\Sigma_{s,t} = T_{s,t} \circ \mathbf{Y}_t$ for $\mathbf{Y}_t \subseteq [0,t] \times [0,t]$ and $T_{s,t}$ chosen randomly from $\mathscr{T}_{s,t}$. The candidate distributions $\{\mathscr{M}_t\}_{t\geq 0}$ in any Caron–Fox model can be defined deductively by letting $\mathscr{M}$ consist of all distributions for exchangeable point processes on $[0,\infty) \times [0,\infty)$ and for each $t \geq 0$ defining $\mathscr{M}_t$ as the $\mathbf{S}_t$-induced model as in (5.6), i.e.,

$$\mathscr{M}_t = \{\mathbf{S}_t P : P \in \mathscr{M}\}, \quad t \geq 0, \tag{7.16}$$

where $\mathbf{S}_t P$ is the distribution of $\mathbf{S}_t \mathbf{Y}$ for $\mathbf{Y} \sim P$. The 'projective' version of this model studied in [31, 32] corresponds to the sampling context with each $\Sigma_{s,t}$ deterministic at $\mathbf{S}_{s,t}$. The next exercise and following two problems concern Caron–Fox models under more general sampling contexts.

Exercise 7.4 *By construction, it is immediate that the projective Caron–Fox model is coherent in the sense of Definition 5.2. Prove that a Caron–Fox model $(\{\mathscr{M}_t\}_{t\geq 0}, \{\Sigma_{s,t}\}_{t\geq s\geq 0})$ is coherent under any sampling context $\{\Sigma_{s,t}\}_{t\geq s\geq 0}$ such that $\Sigma_{s,t}$ is independent of $\mathbf{Y}_t$ for each $t \geq s \geq 0$.*

Exercise 7.4 suggests that the main results of [32] can be expanded to any sampling context $\{\Sigma_{s,t}\}_{t\geq s\geq 0}$ for which the random sampling operation $\Sigma_{s,t}$ is independent of the underlying point process for every $t \geq s \geq 0$. But just as we considered

all point processes on $[0,t] \times [0,t]$ obtained by restriction of an exchangeable point process on $[0,\infty) \times [0,\infty)$ via t-selection. Though the Caron–Fox model was not formally defined in this way by the original authors [32], I believe that my interpretation, with the assumed selection sampling context $\{\mathbf{S}_{s,t}\}_{t\geq s\geq 0}$, is consistent with the implicit context adopted in [32] and in follow-up work by other authors, e.g., [23, 147].

the possibility that a random sampling mechanism $\Sigma_{m,n}$ for $\{0,1\}$-valued arrays can depend on the structure of the network being sampled, e.g., by edge sampling, path sampling, or snowball sampling as in Sections 3.6 and 3.9, we can consider the possibility that a random t-sampling map is chosen in a way that depends on the point pattern $\mathbf{Y} \subseteq [0,\infty) \times [0,\infty)$.

Research Problem 7.1 *Relating back to Problem 6.4, is it true that a Caron–Fox model* $(\{\mathscr{M}_t\}_{t\geq 0}, \{\Sigma_{s,t}\}_{t\geq s\geq 0})$ *is coherent, in the sense of Definition 5.2, only if its sampling context is independent of the underlying point process?*

Research Problem 7.2 *Extend the above discussion of the Caron–Fox model to sampling contexts* $\{\Sigma_{s,t}\}_{t\geq s\geq 0}$ *for which each* $\Sigma_{s,t}$ *is allowed to depend on the underlying point process. For example, what can be said about the empirical properties of* $\Sigma_{s,t}\mathbf{Y}_t$, $s \leq t$, *when* $\mathbf{Y}_t = \mathbf{S}_t\mathbf{Y}$ *for an exchangeable point process* $\mathbf{Y}$ *and* $\Sigma_{s,t}$ *is some random sampling operation that depends on* $\mathbf{Y}_t$*? For this problem, it is best to start with a specific choice of* $\Sigma_{s,t}$*, though as of now it is not clear what a natural choice would be. For a given choice of* $\Sigma_{s,t}$*, how do the distributional properties of* $\Sigma_{s,t}\mathbf{Y}_t$ *compare to those of* $\mathbf{S}_{s,t}\mathbf{Y}_t$*?*

p-sampling

To see how exchangeability of $\mathbf{Y}$ and its associated sampling interpretation affects the model for the induced array $\mathbf{X}$ given in (7.6), we appeal to the construction of exchangeable random measures by independent unit rate Poisson point processes Θ, $\{\Xi_i\}_{i\geq 1}$, for $i \geq 1$, and R. (See the discussion surrounding (7.11), (7.12), and (7.13) and associated definition of Θ, Ξ, and R.) The superposition and thinning properties of Poisson processes permits the forthcoming p-sampling interpretation.

- The *superposition property* of Poisson processes says that independent Poisson point processes can be combined (i.e., 'superimposed' on one another) to form a new Poisson point process whose intensity measure is the sum of the previous two. More precisely, let Λ and Λ' be independent Poisson point processes with respective intensity measures λ and λ' on a common space $\mathscr{X}$. With Λ and Λ' represented as random measures as in (7.5), the process Λ^* defined by

$$\Lambda^*(A) = \Lambda(A) + \Lambda'(A), \quad A \subseteq \mathscr{X},$$

  is also a Poisson point process on $\mathscr{X}$ with intensity measure $\lambda + \lambda'$.
- The *thinning property* of Poisson processes says that a Poisson process Λ with intensity λ can be 'thinned out' according to an independent process of Bernoulli coin flips with success probability p, resulting in a Poisson process with intensity $p\lambda$. More precisely, let $\Lambda \subseteq \mathscr{X}$ be a Poisson point process with intensity λ and let $p \in [0,1]$. Given Λ, construct $\Lambda^p \subseteq \mathscr{X}$ by tossing a p-coin (i.e., a weighted coin with probability p of heads and $1-p$ of tails) independently for each $x \in \Lambda$. If the coin toss associated to x lands heads, then include x in Λ^p; otherwise, exclude x from Λ^p. The resulting point pattern $\Lambda^p \subseteq \Lambda$ is a subset of Λ (i.e., a 'thinned out version' of Λ according to the coin toss process). This thinned process is a Poisson process with intensity measure $p\lambda$.

In light of the above two properties and the construction of $\mathbf{Y}$ from independent Poisson processes Θ, $\{\Xi_i\}_{i\geq 1}$, and R, Veitch and Roy [148] derive the following 'p-sampling' interpretation for the array $\mathbf{X}$ associated to $\mathbf{Y}$. First, for any probability measure τ on $\mathscr{T}_t$ and $\mathbf{X}_t$ corresponding to the array (7.6) obtained from $\mathbf{Y}_t = \mathbf{Y} \cap [0,t]^2$, let $\mathbf{X}_t^\tau = (X_{ij}^\tau)$ denote the array given by

$$X_{ij}^\tau = \begin{cases} 1, & (\theta_i, \theta_j) \in T_t \circ \mathbf{Y}, \\ 0, & \text{otherwise}, \end{cases}$$

for $T_t \sim \tau$. By the exchangeability condition for $\mathbf{Y}$ in (7.10), $T_t \circ \mathbf{Y} =_{\mathscr{D}} \mathbf{Y}_t$ for any (fixed) measure preserving transformation $T_t : [0,t] \to [0,\infty)$. Thus, for $s \leq t$ we also have $T_{s,t} \circ \mathbf{Y}_t =_{\mathscr{D}} \mathbf{Y}_s$ for any (fixed) measure preserving $T_{s,t} : [0,s] \to [0,t]$. Given that there are n_t vertices in $\mathbf{X}_t$, their associated locations $\theta_1, \ldots, \theta_{n_t}$ in $\mathbf{Y}_t$ are conditionally uniformly distributed in the interval $[0,t]$ (by standard theory of Poisson processes); whence, by the graphex representation (7.11)–(7.13), there is conditional probability s/t that each vertex represented in $\mathbf{X}_t$ also appears in $\mathbf{X}_s$. This suggests the following notion of p-sampling for the induced process $(\mathbf{X}_t)_{t\geq 0}$ of edge sets.

Given the projective Caron–Fox model $(\{\mathscr{M}_t\}_{t\geq 0}, \{\mathbf{S}_{s,t}\}_{t\geq s\geq 0})$ for $\mathbf{Y} = (\mathbf{Y}_t)_{t\geq 0}$, each $\mathscr{M}_t$ induces a model $\mathscr{M}_t'$ for $\mathbf{X}_t$ through the relationship in (7.6). Since the vertices of $\mathbf{Y}$ are labeled by the atoms of independent unit rate Poisson processes on $[0,\infty)$, the selection sampling operations $\mathbf{S}_{s,t}$ in $(\{\mathscr{M}_t\}_{t\geq 0}, \{\mathbf{S}_{s,t}\}_{t\geq s\geq 0})$ translate for $\{\mathscr{M}_t'\}_{t\geq 0}$ to a sampling context $\Sigma_{s,t}'$ defined by thinning by a Bernoulli process with success probability s/t, for each $t \geq s \geq 0$. Specifically, let $\mathbf{X}_t$ be the $\{0,1\}$-valued array associated to $\mathbf{Y}_t$ through (7.6) and write $1, \ldots, n_t$ to denote the vertex labels of $\mathbf{X}_{t}$. Then for $t \geq s \geq 0$, define $\Sigma_{s,t}'$ by sampling a random subset $A \subseteq \{1, \ldots, n_t\}$ which includes each $i = 1, \ldots, n_t$ in A independently with probability s/t. Given $A = \{a_1 < \cdots < a_k\}$, define $\psi : [k] \to [n_t]$ by $\psi(i) = a_i$ and put $\Sigma_{s,t}' \mathbf{X}_t = \mathbf{S}_{k,n_t}^{\psi} \mathbf{X}_t = \mathbf{X}_t^{\psi}$ as defined in (3.17). By the thinning property of Poisson processes and the graphex construction of the projective Caron–Fox model, $\Sigma_{s,t}' \mathbf{X}_t =_{\mathscr{D}} \mathbf{X}_s$ for all $t \geq s \geq 0$, proving that the induced model $(\{\mathscr{M}_t'\}_{t\geq 0}, \{\Sigma_{s,t}'\}_{t\geq s\geq 0})$ is coherent in the p-sampling context. The above p-sampling interpretation was first observed and studied in [148], to which the reader is referred for further details. The interpretation in terms of p-sampling, however, raises concerns over the practical viability of graphex models, cf. the discussion in Section 6.2 and [47].

7.3.8 Further discussion

The construction of graphex models in [31, 147] suggests a number of extensions and modifications. I refer the reader to [23, 33, 129, 145, 148] for some ongoing work in this direction. As these references suggest, most of the present interest in graphex models is confined to a segment of the Bayesian nonparametrics literature and seems to be focused primarily on mathematical theory. The lack of a clearly articulated motivating example for the point process representation and implicit p-sampling context remains a major conceptual obstacle to the implementation of this model in applications. For some other pressing questions about this class of models, the reader is referred to [47]. In Section 8.7 I suggest an extension of the Caron–Fox model to account for 'relative invariance' with respect to a non-uniform base measure on $[0,\infty)$.

Research Problem 7.3 *So far the Caron–Fox model has been defined for point processes in* $[0,\infty)^2$. *But in light of upcoming discussions about multiway interactions, as in Chapter 10, it seems worthwhile to consider extensions of this approach for modeling networks in which more than two vertices can participate in any given edge. Directly analogous to the representation by point processes in* $[0,\infty)^2$ *in the preceding section, such hypergraphs could be represented by a point process in* $\bigcup_{d\geq 1}[0,\infty)^d$, *with a d-ary interaction represented as a point in* $[0,\infty)^d$. *Many aspects of this extension, including its associated graphex representation, are likely to be a straightforward generalization of the 2-dimensional case, but this extension has not yet been studied in any detail.*

7.4 Variants of invariance

Each of the next three chapters covers a different invariance principle for network analysis: relative exchangeability (Chapter 8), edge exchangeability (Chapter 9), and relational exchangeability (Chapter 10). I provide a brief synopsis of each concept here.

7.4.1 Relatively exchangeable models (Chapter 8)

Even before noticing the coherence issues of the sparse graphon proposal in Section 7.2, we notice that in many applications the population is known to be heterogeneous in a way that cannot be immediately accounted for by the graphon or graphex approach. The most straightforward instance of such heterogeneity occurs when the vertices cluster into disjoint communities. In the stochastic blockmodel (SBM) [89], for example, the vertices partition into nonoverlapping communities (or blocks) $B_1, B_2, \ldots$, so that two vertices v, v', one in block B_i and the other in block B_j, interact with block-dependent probability p_{ij}. The SBM is a special case of the more general class of relatively exchangeable network models, which emerges from work by Crane [48] and Crane and Towsner [59] but has not yet been integrated into statistical network analysis.

7.4.2 Edge exchangeable models (Chapter 9)

From the discussion of network sampling schemes in Chapter 3, it is apparent that many network datasets are constructed by sampling interactions among a population of individuals, e.g., by sampling phone calls from a database. In such a context, the edges are the statistical units and the network data is naturally represented by an edge-labeled graph instead of as a vertex-labeled graph or its corresponding adjacency array. *Edge exchangeable models* are suitable for modeling networks whose sampled edges are representative of a population of interactions. By contrast to the vertex exchangeable networks discussed above, the distribution of any edge exchangeable network is invariant with respect to relabeling of its edges. In addition to its straightforward interpretation for modeling interaction networks, edge exchangeability can also account for sparsity and power law degree distributions. See Chapter 9 and [54] for more details.

7.4.3 Relationally exchangeable models (Chapter 10)

Relational exchangeability extends edge exchangeability to networks constructed from a representative sample of arbitrary relations among individuals in a population. In the case of edge exchangeability, the relations are pairwise interactions, e.g., each phone call is an interaction between a caller-receiver pair. More generally, we can consider networks constructed by sampling academic articles or emails, in which case edge exchangeability gives way to hyperedge exchangeability. Or, if studying the topology of the Internet by sampling the paths traversed when transmitting a message between different servers, then the observed network might be exchangeable with respect to the sampling of these paths, invoking the concept of *path exchangeability*. In the language of Chapter 10, edge, hyperedge, and path exchangeability all arise as special cases of relational exchangeability.

7.5 Solutions to exercises

7.5.1 Exercise 7.1

Based on Chapters 3–5, I cannot think any application for which this 'decoupling' approach would be natural. As stated, the model defines a set of candidate distributions, as in (7.4), for every finite sample size $n \geq 1$. These finite sample models are not coherent with respect to the canonical selection sampling scheme, and the specification in [19] provides no alternative context in which to interpret this model. Consequently, it is hard to elicit any realistic scenario in which this model is useful for statistical inference (in the Boxian sense, Sections 1.1 and 5.4). The reader is encouraged, however, to suggest a possible justification.

7.5.2 Exercise 7.2

The equivalence between the representation of $\mathbf{y}$ as a subset of $[0,\infty) \times [0,\infty)$ and as a measure $\mathbf{y}(\cdot)$ in (7.5) can be seen as follows. Given a subset $\mathbf{y} \subseteq [0,\infty) \times [0,\infty)$, define a measure $\mathbf{y}(\cdot)$ just as in (7.5). Conversely, given a point measure $\mathbf{y}(\cdot)$ on $[0,\infty) \times [0,\infty)$, construct a subset $\mathbf{y}' \subseteq [0,\infty) \times [0,\infty)$ by

$$(x,x') \in \mathbf{y}' \quad \text{if and only if} \quad \mathbf{y}(\{(x,x')\}) = 1.$$

It is clear by construction that the measure $\mathbf{y}'(\cdot)$ defined from this subset, as in (7.5), coincides with the original measure $\mathbf{y}(\cdot)$ used to define the set $\mathbf{y}'$, thus proving that the two representations are equivalent.

7.5.3 Exercise 7.3

First, since $\mathbf{X}$ is constructed in (7.6) from a point process $\mathbf{Y}$ that is invariant with respect to measure-preserving transformations (not permutations), there is no *a priori* reason to expect that $\mathbf{X}$ is exchangeable in the sense of Chapter 6. Second, by the interpretation of $\mathbf{Y}$ as the point pattern formed by the edges of a network, the vertices that appear in any given sample from $\mathbf{Y}$, and thus the vertices that index the rows and columns of the induced adjacency array $\mathbf{X}$ in (7.6), consists only of those vertices which have had at least one interaction within the sampling timeframe. The vertices represented in $\mathbf{X}$ are, therefore, not exchangeable with the vertices that do not appear in $\mathbf{X}$, and thus we should not expect $\mathbf{X}$ to be exchangeable. Finally, by comparing the theory of vertex exchangeable models and dense graphs (Section 6.5.1) with the fact that the Caron–Fox model can replicate the properties of sparsity and power law degree distribution [32], it is immediate that $\mathbf{X}$ *cannot* be exchangeable (in general). From the opening discussion of Chapter 7 there are zero vertex exchangeable random graph models for sparse, power law networks.

7.5.4 Exercise 7.4

Let $\mathbf{Y} \subseteq [0,\infty) \times [0,\infty)$ be an exchangeable point process with distribution P and, for $t \geq s > 0$, let $\mathbf{Y}_t = \mathbf{S}_t \mathbf{Y}$ and let $\Sigma_{s,t}$ be a random sampling map that is independent of $\mathbf{Y}_t$. By definition, $\Sigma_{s,t}\mathbf{Y}_t = T_{s,t} \circ \mathbf{Y}_t$ for $T_{s,t}$ chosen randomly from the set $\mathscr{T}_{s,t}$ of all Lebesgue measure-preserving injections $[0,s] \to [0,t]$. Thus, for any $\mathbf{y} \subseteq [0,s]^2$ and $\mathbf{S}_t P \in \mathscr{M}_t$,

$$
\begin{aligned}
&\Pr(\Sigma_{s,t}\mathbf{Y}_t \in d\mathbf{y}) = \\
&\quad = \Pr(\Sigma_{s,t}(\mathbf{S}_t\mathbf{Y}) \in d\mathbf{y}) \\
&\quad = \int_{\{\mathbf{y}' \subseteq [0,t]\times[0,t]\}} \Pr(\Sigma_{s,t}\mathbf{y}' \in d\mathbf{y} \mid \mathbf{S}_t\mathbf{Y} \in d\mathbf{y}')(\mathbf{S}_t P)(d\mathbf{y}') \\
&\quad = \int_{\{\mathbf{y}' \subseteq [0,t]\times[0,t]\}} \Pr(\Sigma_{s,t}\mathbf{y}' \in d\mathbf{y})(\mathbf{S}_t P)(d\mathbf{y}') \\
&\quad = \int_{\{\mathbf{y}' \subseteq [0,t]\times[0,t]\}} \int_{\mathscr{T}_{s,t}} \mathbf{1}(T_{s,t} \circ \mathbf{y}' \in d\mathbf{y})\Sigma_{s,t}(dT_{s,t})(\mathbf{S}_t P)(d\mathbf{y}') \\
&\quad = \int_{\mathscr{T}_{s,t}} \left(\int_{\{\mathbf{y}' \subseteq [0,t]\times[0,t]\}} \mathbf{1}(T_{s,t} \circ \mathbf{y}' \in d\mathbf{y})(\mathbf{S}_t P)(d\mathbf{y}') \right) \Sigma_{s,t}(dT_{s,t}) \\
&\quad = \int_{\mathscr{T}_{s,t}} \Pr(T_{s,t} \circ \mathbf{Y}_t \in d\mathbf{y})\, \Sigma_{s,t}(dT_{s,t}) \\
&\quad = \int_{\mathscr{T}_{s,t}} \Pr(\mathbf{S}_{s,t}\mathbf{Y}_t \in d\mathbf{y})\, \Sigma_{s,t}(dT_{s,t}) \\
&\quad = \Pr(\mathbf{S}_{s,t}\mathbf{Y}_t \in d\mathbf{y}) \\
&\quad = (\mathbf{S}_s P)(d\mathbf{y}).
\end{aligned}
$$

It follows that the $\Sigma_{s,t}$-induced model satisfies $\Sigma_{s,t}\mathscr{M}_t = \mathbf{S}_{s,t}\mathscr{M}_t$ for all $t \geq s \geq 0$. Coherence follows by observing that $\mathbf{S}_s = \mathbf{S}_{s,t} \circ \mathbf{S}_t$ and

$$\mathbf{S}_{s,t}\mathscr{M}_t = \mathbf{S}_{s,t}(\mathbf{S}_t\mathscr{M}) = (\mathbf{S}_{s,t} \circ \mathbf{S}_t)\mathscr{M} = \mathbf{S}_s\mathscr{M} = \mathscr{M}_s$$

for all $t \geq s \geq 0$.

Chapter 8

Relatively exchangeable models

Few real-world networks are fully homogeneous in the way that vertex exchangeability implies. With this in mind, relative exchangeability refines vertex exchangeability by expressing the distributional symmetries of a network in terms of the symmetries of another (fixed) structure that is meant to capture the heterogeneity in the population. Here I discuss a generic class of relatively exchangeable network models which incorporates population heterogeneity into the graphon framework from Chapter 6. The stochastic blockmodel (Section 8.2) is a canonical subclass of relatively exchangeable models.

In the formal definition of relative exchangeability, $\mathbf{Y}$ is a random network (e.g., a $\{0,1\}$-valued array) and X is a fixed structure encoding the symmetries of $\mathbf{Y}$ (e.g., a classification of elements, another network, etc.). $\mathbf{Y}$ is *relatively exchangeable with respect to* X, or X*-exchangeable*, if

$$\mathbf{Y}^{\sigma} =_{\mathscr{D}} \mathbf{Y} \quad \text{for all permutations } \sigma : \mathbb{N} \to \mathbb{N} \text{ for which } X^{\sigma} = X,$$

where $\mathbf{Y}^{\sigma}$ and X^{σ} are understood as the 'relabeling' of $\mathbf{Y}$ and X, respectively, according to σ. This preliminary definition of relative exchangeability clarifies the sense in which the exchangeability of $\mathbf{Y}$ (i.e., distributional symmetry) can be interpreted as *relative to* the underlying population structure X (i.e., automorphisms/symmetries of the population). (Note well, the version of relative exchangeability specified in Definitions 8.1 and 8.2 is stronger than the one mentioned above.)

When $\mathbf{Y}$ (respectively, X) is a $\{0,1\}$-valued array, the relabeling $\mathbf{Y}^{\sigma}$ (resp., X^{σ}) is defined as in (6.1). Otherwise, I leave the general definition of relabeling undefined until it is needed later on. But while I have been imprecise in reference to X here, it is best to proceed with a vague interpretation of X as 'generic structure' on the population. As the chapter progresses, the various forms that X can take will become more clear, ultimately resulting in the general formulation of relative exchangeability given in Section 8.5. The full mathematical details of relative exchangeability are quite technical and far beyond the scope of this chapter. A more complete treatment can be found in [3, 58, 59].

8.1 Scenario: Heterogeneity in social networks

Consider a social network of high school students $\mathbf{Y}_N = (Y_{ij})_{1\le i,j\le N}$, with each Y_{ij} indicating the friendship status of students i and j,

$$Y_{ij} = \begin{cases} 1, & i \text{ regards } j \text{ as a friend,} \\ 0, & \text{otherwise.} \end{cases}$$

Along with $\mathbf{Y}_N$, suppose that each student is classified according to year $C : [N] \to \{1,2,3,4\}$, with

$$C(i) = \begin{cases} 1, & i \text{ is a freshman (first year),} \\ 2, & i \text{ is a sophomore (second year),} \\ 3, & i \text{ is a junior (third year),} \\ 4, & i \text{ is a senior (fourth year).} \end{cases}$$

From this network, suppose that an observation $\mathbf{Y}_n = (Y_{ij})_{1\le i,j\le n}$ is obtained by selection sampling, i.e., $\mathbf{Y}_n = \mathbf{S}_{n,N}\mathbf{Y}_N$ for $\mathbf{S}_{n,N} : \{0,1\}^{N\times N} \to \{0,1\}^{n\times n}$ as defined in (3.6). With $C|_{[n]}$ denoting the domain restriction of C to $[n]$, i.e., $C|_{[n]} : [n] \to \{1,2,3,4\}$ is a function $i \mapsto C(i)$ which classifies students $1,\dots,n$ according to their class year, the observed data for a sample of n students includes their class years $C|_{[n]} = (C(i))_{1\le i\le n}$ together with their friendship network $\mathbf{Y}_n = \mathbf{S}_{n,N}\mathbf{Y}_N$.

We might expect that inhomogeneities in friendship patterns can be explained (at least partially) by inhomogeneities in C: freshmen are more likely to be friends with other freshmen, sophomores with other sophomores, etc. When specifying a model for $\mathbf{Y}_N$ subject to the classification C, it makes sense to regard students of the same class (i.e., 2 freshmen, 2 sophomores, etc.) as indistinguishable and, therefore, as interchangeable (or 'exchangeable') but to distinguish between students in different classes (e.g., a freshman and a senior). For example, for three students i,j,k with $C(i) = C(j)$ and $C(i) \neq C(k)$, we generally expect the marginal distribution of Y_{ij} to differ from that of Y_{ik}: in the absence of further information, the probability that two students (i and j) in the same class are friends is likely to be higher than the probability that two students (i and k) in different classes are friends. We, therefore, do not expect the pair (Y_{ij}, Y_{ik}) to be exchangeable; that is, we do *not* assume $(Y_{ij}, Y_{ik}) =_{\mathscr{D}} (Y_{ik}, Y_{ij})$. On the other hand, if $C(i) = C(j) = C(k)$ and the labels i,j,k are assumed to have been arbitrarily assigned, then C does not distinguish between Y_{ij} and Y_{ik}. It seems reasonable to treat (Y_{ij}, Y_{ik}) as exchangeable since in this case the friendship status of students i and j is *a priori* indistinguishable from the friendship status of students i and k.

8.2 Stochastic blockmodels

The *stochastic blockmodel* (SBM) is a default model for networks $\mathbf{Y}_N$ whose heterogeneity is encoded by a classification factor C as in Section 8.1. For $K \ge 1$, let $w : [K]\times[K] \to [0,1]$ be a symmetric weight function and let $C : [N] \to [K]$ assign each individual to a class $1,\dots,K$. (For example, $K = 4$ in Section 8.1.) Under the *stochastic blockmodel with block structure C and weight function w*, each Y_{ij}, $1 \le i \neq j \le N$,

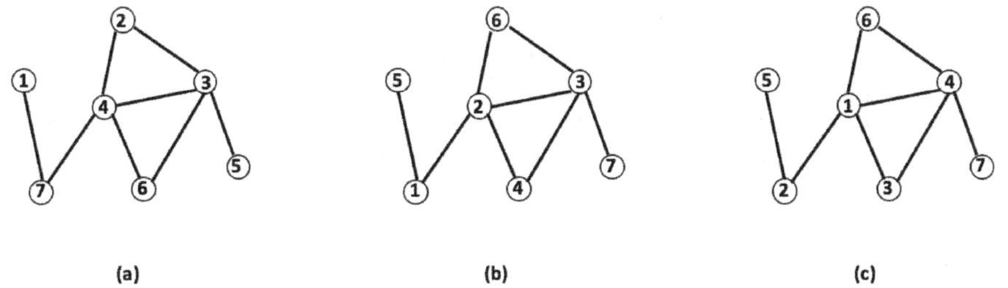

Figure 8.1 *Under the SBM with classification factor* $C : [7] \to \{1,2\}$ *such that* $C(i) = 1$ *for* $i = 2,4,6$ *(even) and* $C(i) = 2$ *for* $i = 1,3,5,7$ *(odd), the graphs in (a) and (b) are assigned the same probability, but the graph in (c) may be assigned a different probability. Between (a) and (b) there is a permutation of vertices which preserves the structure of C (i.e., permutes evens with evens and odds with odds). There is no such permutation for relating (a) to (c) or (b) to (c).*

is distributed independently according to

$$\Pr(Y_{ij} = 1; w, C) = w(C(i), C(j)) \quad \text{and} \quad \Pr(Y_{ij} = 0; w, C) = 1 - w(C(i), C(j)), \tag{8.1}$$

and altogether the distribution of $\mathbf{Y}_N = (Y_{ij})_{1 \le i,j \le N}$ is given by

$$\Pr(\mathbf{Y}_N = \mathbf{y}; w, C) = \prod_{1 \le i \ne j \le N} w(C(i), C(j))^{y_{ij}} (1 - w(C(i), C(j)))^{1-y_{ij}}, \ \mathbf{y} \in \{0,1\}^{N \times N}. \tag{8.2}$$

By (8.1), vertices interact according to a distribution that depends only on their class membership C; that is, the distribution of Y_{ij} for any pair (i, j) for which $(C(i), C(j)) = (c, c')$ is a Bernoulli random variable with success probability $w(c, c')$. If $K = 1$, then $w(C(i), C(j))$ is constant for all i and j and (8.2) coincides with the Erdős–Rényi–Gilbert distribution in (3.16). For $K \ge 2$, the distribution in (8.2) is not exchangeable, since in general any permutation $\sigma : [N] \to [N]$ for which $C \circ \sigma \ne C$ does not preserve the distribution in (8.2). (If two vertices i and j with $C(i) \ne C(j)$ are interchanged by σ, then the distribution of $\mathbf{Y}_N$ in (8.2) may change.) But (8.2) is invariant with respect to permutations that preserve C, that is, $\sigma : [N] \to [N]$ for which $C(\sigma(i)) = C(i)$ for all $i = 1, \ldots, N$. The family of distributions in (8.2) also satisfies $\Pr(\mathbf{Y}_N = \mathbf{y}; w, C) = \Pr(\mathbf{Y}_N = \mathbf{y}^\sigma; w, C^\sigma)$ for all permutations $\sigma : [N] \to [N]$; see Section 8.3.4 for related discussion. Thus, in the context of Section 8.1, the SBM would regard freshmen as interchangeable with other freshmen, sophomores with other sophomores, etc., but freshmen would not be interchangeable with seniors, for example. Figure 8.1 illustrates this property of the SBM.

Exercise 8.1 *Let* $\mathbf{y}_a$ *denote the adjacency array of the graph in Figure 8.1(a) and* $\mathbf{y}_b$ *denote the adjacency array of the graph in Figure 8.1(b). Give a permutation* $\sigma : [7] \to [7]$ *such that* $\mathbf{y}_a^\sigma = \mathbf{y}_b$ *and* $C \circ \sigma = C$*, for C as defined in the caption of Figure 8.1. Deduce that* $\mathbf{y}_a$ *and* $\mathbf{y}_b$ *have the same probability assignment under the SBM with block structure C and arbitrary weight function w, as defined in* (8.2).

Furthermore, since each Y_{ij} is drawn independently according to (8.1) and in a way that depends on C only through the pair $(C(i),C(j))$, the marginal distribution of any restriction $\mathbf{Y}_N|_S = (Y_{ij})_{i,j\in S}$, $S \subseteq [N]$, depends on C only through its domain restriction to S, i.e., $C|_S : S \to [K]$, $i \mapsto C(i)$. In this case, we call $\mathbf{Y}_N$ *relatively exchangeable with respect to C* if for all $S \subseteq [N]$ the marginal distribution of $\mathbf{Y}_N|_S$ is invariant with respect to all permutations $\sigma : S \to S$ for which $C|_S \circ \sigma = C|_S$, i.e.,

$$\mathbf{Y}_N|_S^\sigma =_{\mathscr{D}} \mathbf{Y}_N|_S \text{ for all permutations } \sigma : S \to S \text{ such that } (C(\sigma(i)))_{i\in S} = (C(i))_{i\in S}.^1$$

If a permutation σ leaves C invariant, then the distribution in (8.2) cannot distinguish between $\mathbf{Y}_N$ and $\mathbf{Y}_N^\sigma$. Under this condition, relative exchangeability is equivalent to

$$\Pr(\mathbf{Y}_N|_S = \mathbf{y}) = \Pr(\mathbf{Y}_N|_S = \mathbf{y}^\sigma), \quad \mathbf{y} \in \{0,1\}^{S\times S}, \tag{8.3}$$

for all permutations $\sigma : S \to S$ such that $C \circ \sigma = C$.

Theorem 8.1 *A random array $\mathbf{Y}_N$ distributed according to the stochastic blockmodel with block structure C and any weight function w, as defined in* (8.2)*, is relatively exchangeable with respect to C.*

Exercise 8.2 *Prove Theorem 8.1.*

Remark 8.1 (Dependence on sampling scheme) *I have so far suppressed the dependence of relative exchangeability* (8.3) *on the assumed sampling scheme, which is implicitly taken to be selection in the above case. But while most of this chapter follows the developments in [59], which takes selection sampling as the default, the behavior of relatively exchangeable models under different sampling schemes remains unexamined, and is left as an open class of problems. In Section 8.6 I discuss how the concept of relative exchangeability could possibly be extended to more general sampling contexts.*

8.2.1 Generalized blockmodels

The connection between the SBM in (8.1) and the Erdős–Rényi–Gilbert distribution together with the relationship between the Erdős–Rényi–Gilbert distribution and graphon models (see Section 6.4.1) suggest the following extension to a larger class of relatively exchangeable ϕ-processes. For $K \geq 1$, specify a function $\phi : [K]^2 \times [0,1] \times [0,1] \to [0,1]$, let $C : \mathbb{N} \to [K]$ be a classification of the vertices, and let $U_1, U_2, \ldots$ be i.i.d. Uniform$[0,1]$ random variables.[2] Given ϕ, C, and $U_1, U_2, \ldots$, construct $\mathbf{Y}^* = (Y^*_{ij})_{i,j\geq 1}$ by choosing each Y^*_{ij} conditionally independently according to

$$\begin{aligned} \Pr(Y^*_{ij} = 1 \mid U_1, U_2, \ldots; \phi, C) &= \phi(C|_{(i,j)}, U_i, U_j) \quad \text{and} \\ \Pr(Y^*_{ij} = 0 \mid U_1, U_2, \ldots; \phi, C) &= 1 - \phi(C|_{(i,j)}, U_i, U_j), \end{aligned} \tag{8.4}$$

[1]Note well the order of operations in the above expression $\mathbf{Y}_N|_S^\sigma$: the restriction to S always occurs before the relabeling by σ. $\mathbf{Y}_N|_S^\sigma$ could be written more pedantically as $(\mathbf{Y}_N|_S)^\sigma$.

[2]If $K = 1$, then the classification factor C and the role of the first argument $[K]^2$ in ϕ are moot, reducing ϕ to a graphon as in Chapter 6.

where $C|_{(i,j)} : \{i,j\} \to [K]$ is determined by the pair $(C(i),C(j))$. The finite-dimensional distributions of $\mathbf{Y}^*$ are thus given by

$$\Pr(\mathbf{Y}^*|_{[N]} = \mathbf{y}; \phi, C) = \tag{8.5}$$
$$= \int_{[0,1]^N} \prod_{1\leq i\neq j\leq N} \phi(C|_{(i,j)}, u_i, u_j)^{y_{ij}} (1-\phi(C|_{(i,j)}, u_i, u_j))^{1-y_{ij}}\, du_1 \cdots du_N,$$

for $\mathbf{y} \in \{0,1\}^{N\times N}$, for all $N \geq 1$.

Comparing (6.14) with (8.5) clarifies how the relatively exchangeable SBM refines vertex exchangeable graphon models. Because all entries of $\mathbf{Y}^* = (Y^*_{ij})_{i,j\geq 1}$ are conditionally independent given $U_1, U_2, \ldots$, the distribution of each restriction $\mathbf{Y}^*|_S$ depends only on $C|_S$ and is invariant with respect to permutations that fix $C|_S$. In particular, we immediately see that the restriction $\mathbf{Y}^*|_{[n]}$ of any $\mathbf{Y}^*$ distributed according to (8.5) also satisfies (8.5) with parameter ϕ and classification factor $C|_{[n]}$ on $\{0,1\}^{n\times n}$.

As in Chapter 6, the class of distributions in (8.5) enjoys special status among infinite random arrays $\mathbf{Y} = (Y_{ij})_{i,j\geq 1}$ that are relatively exchangeable with respect to $C : \mathbb{N} \to [K]$ satisfying a technical regularity condition.[3] In particular, if $\mathbf{Y}$ is relatively exchangeable with respect to C, then there is a probability measure φ on the space of functions $[K]^2 \times [0,1]^2 \to [0,1]$ such that the construction in (8.4) holds.

Theorem 8.2 (Crane–Towsner [59]) *Fix $K = 1,2,\ldots,\infty$ and $C : \mathbb{N} \to [K]$, and let $\mathbf{Y} = (Y_{ij})_{i,j\geq 1}$ be relatively exchangeable with respect to C. Assume further that each set $\{i \geq 1 : C(i) = j\}$, $j = 1,\ldots,K$, is infinite and has positive limiting frequency. Then there exists a probability measure φ on the space Φ_K of functions $\phi : [K]^2 \times [0,1] \times [0,1] \to [0,1]$ such that $\mathbf{Y} =_{\mathscr{D}} \mathbf{Y}^*$, for $\mathbf{Y}^*$ constructed as in* (8.4) *for ϕ distributed according to φ.*

To parallel the Aldous–Hoover theorem (Theorem 6.3) more closely, Theorem 8.2 can be restated in terms of a function $f : [K]^2 \times [0,1]^4 \to \{0,1\}$ by taking U_0, $(U_i)_{i\geq 1}$, and $(U_{ij})_{i,j\geq 1}$ to be i.i.d. Uniform$[0,1]$ random variables and putting

$$Y^*_{ij} = f(C|_{(i,j)}, U_0, U_i, U_j, U_{ij}), \quad i,j \geq 1. \tag{8.6}$$

The reader can confirm that (8.6) is equivalent to the description in Theorem 8.2 by drawing parallels with our previous discussion about the relationship between graphon models and the Aldous–Hoover representation in Section 6.4.2.

---

[3]The regularity condition requires that each set $\{i \geq 1 : C(i) = j\}$, $j = 1,\ldots,K$, is infinite with positive limiting frequency, i.e.,

$$\lim_{n\to\infty} \frac{\#\{1 \leq i \leq n : C(i) = j\}}{n} > 0 \quad \text{for all } j = 1,\ldots,K.$$

This condition is needed in order for the representation in Theorem 8.2 to hold, where $\#\{1 \leq i \leq n : C(i) = j\}$ denotes the cardinality (i.e., number of elements) of $\{1 \leq i \leq n : C(i) = j\}$. To see how the representation might fail, note that if one of the sets $S_j = \{i \geq 1 : C(i) = j\}$ were finite and nonempty, then the restriction $\mathbf{Y}|_{S_j}$ is marginally distributed as a finite exchangeable array, which need not have a ϕ-process representation; see Section 6.3 and also [43, 59] for a more detailed technical account.

8.2.2 Community detection and Bayesian versions of SBM

Whereas the original blockmodel of Holland, Laskey, and Leinhardt [89] takes C to be fixed, and perhaps known, the SBM more commonly appears in statistical applications for community detection, in which C is regarded as the unknown parameter of interest. In this context, the class of finite sample models $\{\mathscr{M}_n\}_{n\geq 1}$ is parameterized by classification factors $C:[n]\to[K]$ and weight functions w, with the selection sampling context $\{\mathbf{S}_{n,N}\}_{N\geq n\geq 1}$ taken for granted. In applied work, the classification of vertices according to C is often called 'community structure' and the sets $C^{-1}(\ell)=\{i:C(i)=\ell\}$ corresponding to each class label ℓ are called 'communities'.

Assume that $\mathbf{Y}_N=(Y_{ij})_{1\leq i,j\leq N}$ is relatively exchangeable with respect to some $C:[N]\to[K]$ and consider the problem of inferring the community structure $C|_{[n]}$ based on an observation $\mathbf{S}_{n,N}\mathbf{Y}_n$ obtained by selection sampling. A common approach, popularized in work by Snijders and Nowicki [141], is to formulate the problem in a Bayesian setup by assigning a prior distribution to C. The most direct way to define a Bayesian version of the SBM is to specify an exchangeable prior distribution on C and, given C, model $\mathbf{Y}$ as relatively exchangeable with respect to C.

As a special case, we can parameterize the model by $(\mathbf{t},\mathbf{a})$, where $\mathbf{t}=\{0=t_0<t_1<\cdots<t_{K-1}<t_K=1\}$ is a sequence that partitions $(0,1]$ into the K subintervals $(t_i,t_{i+1}]$, $i=0,1,\ldots,K-1$, and $\mathbf{a}=(a_{ij})_{1\leq i,j\leq K}$ is an array taking values in $[0,1]$.[4] From $(\mathbf{t},\mathbf{a})$, we define a probability distribution $\mathbf{p}=(p_1,\ldots,p_K)$ on the classes $1,\ldots,K$ by

$$p_i=t_i-t_{i-1},\quad i=1,\ldots,K,$$

and define $\phi_{\mathbf{a}}:[K]^2\times[0,1]\times[0,1]\to[0,1]$ by

$$\phi_{\mathbf{a}}((i,j),u,v)=a_{ij},\quad 1\leq i,j\leq K,\quad u,v\in[0,1]. \tag{8.7}$$

We then define the distribution of $\mathbf{Y}$ as that of an array generated by first drawing $C:\mathbb{N}\to[K]$ i.i.d. according to $\mathbf{p}$, i.e., $\Pr(C(i)=j;\mathbf{p})=p_j$ for each $j=1,\ldots,K$, and then, given C, taking $\mathbf{Y}$ to be relatively exchangeable according to the $\phi_{\mathbf{a}}$-process as in (8.4). We call $\mathbf{Y}$ generated this way a $(\mathbf{t},\mathbf{a})$*-graphon process*.

In the $(\mathbf{t},\mathbf{a})$-graphon process description, $\mathbf{p}$ acts as a prior distribution for an i.i.d. assignment of classes C. Given C, the model is relatively exchangeable and inference for C proceeds as usual, e.g., by Bayesian posterior inference. But in many treatments of the SBM, the combination of exchangeable prior $\mathbf{p}$ for C and relatively exchangeable conditional distribution for $\mathbf{Y}$ given C often leads to an assumption that $\mathbf{Y}$ is distributed according to a piecewise constant graphon model with $\phi:[0,1]\times[0,1]\to[0,1]$ defined as follows.

Given $\mathbf{t}$ and $\mathbf{a}$ as in the $(\mathbf{t},\mathbf{a})$-graphon process (8.7), let $\phi_{\mathbf{a},\mathbf{t}}:(0,1]\times(0,1]\to(0,1]$ be defined by

$$\phi_{\mathbf{a},\mathbf{t}}(u,v)=a_{ij}\quad\text{for all }(u,v)\in(t_{i-1},t_i]\times(t_{j-1},t_j],\quad 1\leq i,j\leq K. \tag{8.8}$$

[4] Notice here that I partition $(0,1]$, instead of $[0,1]$, into subintervals. Below, when I define the $(\mathbf{t},\mathbf{a})$-graphon process, I define the graphon as a function $(0,1]\times(0,1]\to[0,1]$, instead of $[0,1]\times[0,1]\to[0,1]$ as in Chapter 6. These choices are a matter of notational convenience. Because the event '$U_i=0$' has probability 0 for any uniform random variable on $[0,1]$, this choice has no effect on any of the claims made in this section.

Call any such ϕ a *blockwise constant graphon* with parameter $(\mathbf{t},\mathbf{a})$, or simply a $(\mathbf{t},\mathbf{a})$-*graphon*. To see the connection with (8.7), note that the occurrence of u_i in the subinterval $(t_{\ell-1},t_\ell]$ can be interpreted as the random block assignment $C(i)=\ell$ with probability $p_\ell = \mathbf{p}(\ell)$, since each of the events '$U_i \in (t_{\ell-1},t_\ell]$' occurs independently with probability $t_\ell - t_{\ell-1}$ as in the distribution $\mathbf{p}$. This way, the occurrence of u_i and u_j in the same subinterval is interpreted as $C(i)=C(j)$, and the description of $\mathbf{Y}$ as being relatively exchangeable with respect to C follows from its conditional construction according to the $\phi_{\mathbf{a}}$-process.

The relatively exchangeable conditional distributions in (8.7) and the fully exchangeable distributions induced by the $(\mathbf{t},\mathbf{a})$-graphon in (8.8) are a special case of the following more general class of Bayesian SBMs. First, let $C:\mathbb{N}\to[K]$ be an exchangeable $[K]$-valued sequence with full support. Then, given C, take $\mathbf{Y}$ to be relatively exchangeable with respect to C. This setup can be summarized by

$$\begin{aligned} C &\sim \mu && \text{(exchangeable)} \\ \mathbf{Y} \mid C &\sim \Pr(\mathbf{Y}\in\cdot\,;C,\varphi) && \text{(relatively exchangeable),} \end{aligned}$$

where the distribution of $\mathbf{Y}$ is determined by its finite-dimensional distributions

$$\Pr(\mathbf{Y}|_{[n]} = \cdot\,;C,\varphi) = \tag{8.9}$$
$$= \int_{\Phi_K}\left(\int_{[0,1]^n}\phi(C|_{(i,j)},u_i,u_j)^{y_{ij}}(1-\phi(C|_{(i,j)},u_i,u_j))^{1-y_{ij}}du_1\cdots du_n\right)\varphi(d\phi),$$

for φ a probability distribution on the space of functions $\phi : [K]\times[K]\times[0,1]^2 \to [0,1]$.

Note the difference between conditioning on C and averaging over it. Since we have modeled C as an exchangeable $[K]$-valued sequence, the marginal distribution of $\mathbf{Y}$, after integrating over the distribution of C, is exchangeable. (See Exercise 8.3.) But if interested in inferring the latent structure C and interpreting its meaning independently of the observed network structure $\mathbf{Y}$, then we should not average over C when specifying the model for $\mathbf{Y}$. Conditionally on C, $\mathbf{Y}$ is relatively exchangeable, while marginally $\mathbf{Y}$ is vertex exchangeable. Since the marginal distribution of $\mathbf{Y}$ is exchangeable, it falls under the class of vertex exchangeable models from Chapter 6, but the conditional distribution of $\mathbf{Y}$ given C is relatively exchangeable. If the distribution of $\mathbf{Y}$ is specified as relatively exchangeable with respect to C, then the posterior distribution of C given $\mathbf{Y}=\mathbf{y}$ can, in principle, be computed by Bayes rule.

So although the block structure inherent in $\phi_{\mathbf{a},\mathbf{t}}$ may induce block structure in any realization of a $\phi_{\mathbf{a},\mathbf{t}}$-process, the model specification in (8.8) (in terms of a vertex exchangeable graphon) alters the interpretation of this induced block structure vis-à-vis any perceived 'real' community structure in the observed network. By contrast to the SBM defined in (8.5), any community structure exhibited by $\mathbf{Y}$ from a $\phi_{\mathbf{a},\mathbf{t}}$-process cannot be interpreted as an inherent feature of the population. Since the community structure C is generated along with the network $\mathbf{Y}$, and therefore does not exist *prior to* constructing the data, any 'community-like' structure exhibited by $\mathbf{Y}$ is the result of random variation, not of any features of the population that can be explained independently and *a priori* to the observed network data.

Exercise 8.3 *Let $\mathbf{Y}_N$ be a random array in $\{0,1\}^{N\times N}$ whose distribution is defined by first choosing C according to an exchangeable prior distribution and, given C, taking $\mathbf{Y}_N$ to be relatively exchangeable with respect to C. Show that the marginal distribution of $\mathbf{Y}_N$, after integrating over the prior distribution of C, is exchangeable.*

Research Problem 8.1 *Does the above distinction between the blockwise constant graphon, which averages over community structures, and the Bayesian SBM, which conditions on an unknown C, have any practical or conceptual implications for statistical inference?*

8.2.3 Beyond SBMs and community detection

Community detection has been one of the most widely studied areas of network science for the past decade. The popularity of this subfield has inspired in-depth study of the SBM and its extensions. The reader looking for more information about community detection, stochastic blockmodels, and the like should have no trouble finding it elsewhere in the literature. But within the bigger picture of statistical network analysis, it is important to realize that networks exhibit far more intricate and interesting structure than can be captured by a simple classification of vertices into communities. And thus most of the research currently being conducted in this realm, including recent extensions to allow for vertices with multiple and possibly overlapping community memberships, is of an incremental nature, and is unlikely to have a noticeable impact on the future of network analysis. See, e.g., [6, 35] and references therein for extensions to mixed membership stochastic blockmodels and stochastic blockmodels with a growing number of classes. Refer to Chapter 1 for further discussion about how the probabilistic foundations laid down in this book fit into the bigger picture of network analysis.

The *degree-corrected stochastic blockmodel* (DC-SBM) [99] is an especially popular extension of the SBM which assigns to each vertex i an attribute $\theta_i \geq 0$. These attributes are intended to account for heterogeneous edge patterns found in observed networks. Given a classification $C : [N] \to [K]$, attributes $(\theta_i)_{i\geq 1}$, and a weight assignment $w : [K]^2 \to (0,1]$, each Y_{ij}, $1 \leq i \neq j \leq N$, is conditionally independent with distribution

$$\Pr(Y_{ij} = 0; w, C, \theta) = \exp\left\{\theta_i\theta_j \log(w(C(i), C(j)))\right\} \quad \text{and}$$
$$\Pr(Y_{ij} = 1; w, C, \theta) = 1 - \Pr(Y_{ij} = 0; w, C, \theta).$$

This could be generalized further, as in (8.4), by associating each vertex to i.i.d. Uniform$[0,1]$ random effects $U_1, U_2, \ldots$, specifying a generalized graphon $\phi : [K]^2 \times [0,1]^2 \to (0,1]$ and, given $U_1, U_2, \ldots$, assigning each edge conditionally independently with probability

$$\Pr(Y_{ij} = 0 \mid U_1, U_2, \ldots; \phi, C, \theta) = \exp\left\{\theta_i\theta_j \log(\phi(C|_{(i,j)}, U_i, U_j))\right\} \quad \text{and}$$
$$\Pr(Y_{ij} = 1 \mid U_1, U_2, \ldots; \phi, C, \theta) = 1 - \Pr(Y_{ij} = 0 \mid U_1, U_2, \ldots; \phi, C, \theta).$$

In the next section, I begin to expand upon the SBM in order to account for more complex heterogeneity in network data. To avoid ambiguity, throughout the rest of this chapter I treat C and any other heterogeneity structures as fixed, and work with the relatively exchangeable (i.e., 'frequentist') version of these models. The 'Bayesian' version of any of these models can be recovered by simply putting a prior on C.

8.3 Exchangeability relative to another network

8.3.1 Scenario: High school social network revisited

Instead of classifying students according to their year in school, as in Section 8.1, suppose that the population network $\mathbf{Y}_N$ is accompanied by a social network $G = (G_{ij})_{1\leq i,j\leq N}$ which records social media (e.g., Facebook) friendships among all N students in the school. We assume that $\mathbf{Y}_n$ is obtained from $\mathbf{Y}_N$ by selection sampling, so that $\mathbf{Y}_n = \mathbf{S}_{n,N}\mathbf{Y}_N = (Y_{ij})_{1\leq i,j\leq n}$, with

$$Y_{ij} = \begin{cases} 1, & i \text{ regards } j \text{ as a friend,} \\ 0, & \text{otherwise,} \end{cases}$$

is observed along with the social media friendships $G_n = \mathbf{S}_{n,N}G = (G_{ij})_{1\leq i,j\leq n}$ of the n sampled individuals,

$$G_{ij} = \begin{cases} 1, & i \text{ and } j \text{ are friends on Facebook,} \\ 0, & \text{otherwise.} \end{cases}$$

The goal is to understand whether and how the known relationships in G_n are informative about the structure of $\mathbf{Y}_n$. We might expect, for example, that Facebook friendship ($G_{ij} = 1$) corresponds to a higher probability that i and j are actually friends ($Y_{ij} = 1$ and $Y_{ji} = 1$). We should also expect, however, that some students i and j who are Facebook friends may not consider each other friends in the ordinary sense of the term; and it is possible that i and j are not Facebook friends even though both i and j regard each other as friends otherwise. The reason for any differences between G and $\mathbf{Y}_N$ could possibly be explained by the fact that interactions on social media reflect a different kind of relationship than what might ordinarily be considered 'friendship'.[5]

8.3.2 Exchangeability relative to a social network

In the above scenario, $\mathbf{S}_{n,N}\mathbf{Y}_N = (Y_{ij})_{1\leq i,j\leq n}$ is observed along with the Facebook friendship network $\mathbf{S}_{n,N}G = (G_{ij})_{1\leq i,j\leq n}$. There is no baseline information about students aside from their relationships through G. Based on $(G_n, \mathbf{Y}_n) = (\mathbf{S}_{n,N}G, \mathbf{S}_{n,N}\mathbf{Y}_N)$, we want to understand the extent to which the structure observed in $\mathbf{S}_{n,N}\mathbf{Y}_N$ can be explained by the relationships in G.

[5]The reader is encouraged to use his/her imagination when interpreting this example in the grander scheme of network analysis. One could imagine a number of applications in which a (more or less) known network G acts as a proxy for some other network of interest.

If only the Facebook relationships $G|_{[n]} = (G_{ij})_{1\leq i,j\leq n}$ for the sample $[n] \subseteq [N]$ are available, the data does not distinguish between observations $(Y_{ij})_{1\leq i,j\leq n}$ and $(Y_{\sigma(i)\sigma(j)})_{1\leq i,j\leq n}$ for any permutation $\sigma : [n] \to [n]$ that fixes $G|_{[n]}$. Thus, even though $\mathbf{Y}_n$ could be affected by unobserved relationships between sampled students and unsampled students, we cannot identify such relationships from the data and, therefore, cannot account for such information in the model. This presents an obvious constraint in that inferences based on $\mathbf{Y}_n$ only account for $G|_{[n]}$. For modeling $\mathbf{Y}_n$ based on $G|_{[n]}$, we assume that the unobserved interactions in G do not 'interfere' with the observed interactions in $\mathbf{Y}_n$.

8.3.3 *Lack of interference*

We borrow terminology from the experimental design literature and say that a model for $\mathbf{Y}_n$ parameterized by $G = (G_{ij})_{1\leq i,j\leq N}$ exhibits *non-interference with respect to selection*, or *lack of interference with respect to* $\mathbf{S}_{n,N}$, if the distribution of $\mathbf{Y}_n$ depends on G only through $\mathbf{S}_{n,N}G = (G_{ij})_{1\leq i,j\leq n}$. In words, the relationships of unsampled individuals $[N] \setminus [n]$ do not 'interfere' with the relationships of sampled individuals. For example, stochastic blockmodels exhibit non-interference since the distribution of each edge Y_{ij} in (8.4) depends only on the class status $C|_{(i,j)}$ of i and j. Note that this definition of non-interference tacitly assumes a selection sampling context. In Section 8.6, I briefly discuss the implications of non-interference and relative exchangeability in more general sampling contexts. Taken together, non-interference and invariance under relabeling combine to give relative exchangeability.

Definition 8.1 (Relative exchangeability, first version) *For* $N = 1,2,\ldots,\infty$, *let* $\mathbf{Y}_N = (Y_{ij})_{1\leq i,j\leq N}$ *be a random array and* $G = (G_{ij})_{1\leq i,j\leq N}$ *be a* $\{0,1\}$*-valued adjacency array for a population of individuals labeled in* $[N]$. *We say that* $\mathbf{Y}_N$ *is* relatively exchangeable with respect to G (in sampling context $\{\mathbf{S}_{n,N}\}_{N\geq n\geq 1}$) *if for every* $S \subseteq [N]$ *the marginal distribution of* $\mathbf{Y}_N|_S = (Y_{ij})_{i,j\in S}$ *is invariant under relabeling by any permutation* $\sigma : S \to S$ *that fixes* $G|_S = (G_{ij})_{i,j\in S}$. *In other words, if* $\sigma : S \to S$ *is such that* $G|_S^{\sigma} = (G_{\sigma(i)\sigma(j)})_{i,j\in S} = G|_S$, *then* $\mathbf{Y}_N|_S^{\sigma} = (Y_{\sigma(i)\sigma(j)})_{i,j\in S} =_{\mathscr{D}} \mathbf{Y}_N|_S$.

Remark 8.2 *Since for any given* $S \subseteq [N]$ *there can be many permutations* $\sigma : S \to S$ *that fix* $G|_S$ *but which do not extend to a symmetry of the population structure* G, *the definition of relative exchangeability in Definition 8.1 is stronger than the requirement that* $(Y_{ij})_{1\leq i,j\leq N}$ *is invariant with respect to the symmetries of the population structure* $(G_{ij})_{1\leq i,j\leq N}$. *See Figure 8.2 for illustration.*

The forthcoming representation theorem for relatively exchangeable networks with respect to G (Definition 8.1) holds as long as G is *ultrahomogeneous* and has the *n-disjoint amalgamation property* (n-DAP) for all $n \geq 1$. Ultrahomogeneity implies that every induced subgraph $G|_S$, for $S \subseteq \mathbb{N}$, is 'representative' of G, i.e., every permutation $\sigma : S \to S$ that fixes $G|_S$ extends to a permutation $\bar{\sigma} : \mathbb{N} \to \mathbb{N}$ that fixes all of G. Disjoint amalgamation is a further algebraic condition for which the reader is referred to [59, Section 2.2]. Both of these conditions stem from the requirement in Definition 8.1 that the distribution of $\mathbf{Y}|_S$ depends only on the symmetries of $G|_S$. In order for the distribution of $\mathbf{Y}|_S$ to be 'coherent' with the rest of $\mathbf{Y}$, the marginal

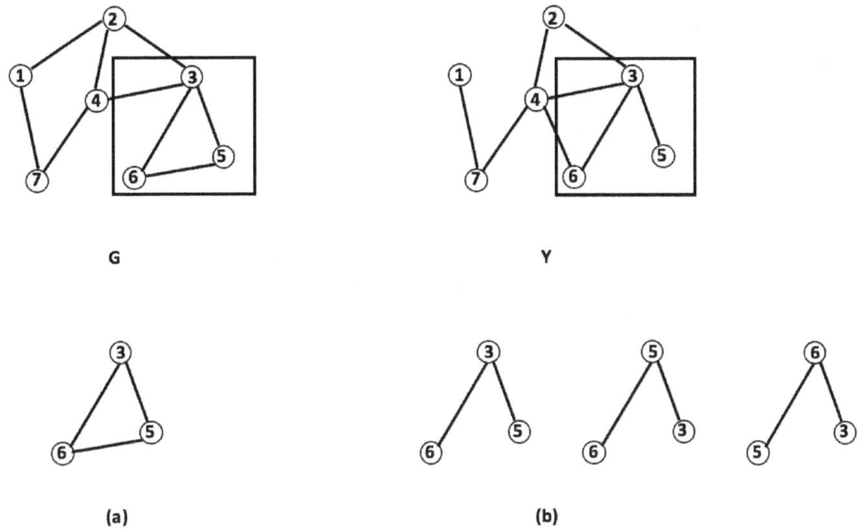

Figure 8.2 *Under the assumption that* $\mathbf{Y}$ *is relatively exchangeable with respect to G, Definition 8.1 implies that the marginal distribution of every subgraph of* $\mathbf{Y}$ *is invariant with respect to the symmetries of the corresponding subgraph in G. The figure highlights the subgraph induced by the restriction to vertices labeled 3, 5, and 6. Under relative exchangeability, the marginal distribution of* $\mathbf{Y}|_{\{3,5,6\}}$ *is exchangeable (assigning equal probability to all three graphs in the bottom of (b)) because the corresponding subgraph of G (shown at the bottom of part (a)) is fully symmetric.*

distributions of $\mathbf{Y}|_S$ and $\mathbf{Y}|_T$ must coincide for any $S,T \subseteq \mathbb{N}$ with $G|_S = G|_T$. In other words, the induced subgraphs $G|_S$ and $G|_T$ must be sufficient for determining the distributional symmetries of $\mathbf{Y}|_S$ and $\mathbf{Y}|_T$, respectively. I do not discuss these technical conditions any further here. The motivated reader is encouraged to consult [48, 59] for more details. The next theorem extends the Aldous–Hoover theorem (Section 6.4.2 and Theorem 6.3) to the relatively exchangeable setting.

Theorem 8.3 (Crane–Towsner [48, 59]) *Let* $G = (G_{ij})_{i,j\geq 1}$ *be ultrahomogeneous and have n-DAP for all* $n \geq 1$ *and suppose* $\mathbf{Y} = (Y_{ij})_{i,j\geq 1}$ *is relatively exchangeable with respect to G. Then there exists a probability measure* φ *on the space of functions* $\phi : \{0,1\}^{2\times 2} \times [0,1]^2 \to [0,1]$ *such that* $\mathbf{Y} =_{\mathcal{D}} \mathbf{Y}^*$*, for* $\mathbf{Y}^*$ *constructed by first taking* $\phi \sim \varphi$ *and* $U_1, U_2, \ldots$ *i.i.d. Uniform*$[0,1]$ *and, given* $\phi, U_1, U_2, \ldots$*, generating each* Y^*_{ij} *conditionally independently according to*

$$
\begin{aligned}
Pr(Y^*_{ij} = 1 \mid \phi, U_1, U_2, \ldots; G) &= \phi(G|_{\{i,j\}}, U_i, U_j) \\
Pr(Y^*_{ij} = 0 \mid \phi, U_1, U_2, \ldots; G) &= 1 - \phi(G|_{\{i,j\}}, U_i, U_j).
\end{aligned} \tag{8.10}
$$

According to (8.10), $\mathbf{Y}$ can be constructed so that each entry Y_{ij} depends on G

only through

$$G|_{\{i,j\}} = \begin{pmatrix} G_{ii} & G_{ij} \\ G_{ji} & G_{jj} \end{pmatrix}.$$

Without any restrictions on the entries of G, there are $2^4 = 16$ possibilities for each $G|_{\{i,j\}}$, and if G is undirected (i.e., symmetric) and irreflexive (i.e., no self-loops) then there are only four possibilities. So while the dependence on G allows for some heterogeneity in the distribution of $\mathbf{Y}$, the heterogeneity is limited by the local dependence of each Y_{ij} on $G|_{\{i,j\}}$. Since I am glossing over technical aspects of the above theorem, it is worth noting that the localized dependence property in (8.10) mirrors the requirements put on G by ultrahomogeneity and the disjoint amalgamation property. One could perhaps refine this theorem by relaxing these conditions, but proving such a result seems to require more sophisticated techniques than those in [59].

Research Problem 8.2 *According to [138], a family of ERGMs is consistent under selection if and only if its sufficient statistics have separable increments; see Section 2.3 for further discussion. By Theorem 8.3 and its more general version in Theorem 8.4 below, a relatively exchangeable model (with respect to an ultrahomogeneous population structure having an additional amalgamation property) depends on the underlying structure only through its local components. There seems to be a connection between separable increments (as defined in [138]), the non-interference property of relatively exchangeable models, and the dyad independence property of the p_1 model (Chapter 2), but the connection is not immediately clear.*

8.3.4 *Label equivariance*

Even when handling data which is not exchangeable, the model ought to be invariant with respect to relevant transformations of the data. For network data, the relevant transformations are most often permutations of the labels, e.g., the relabeling of vertices given by the action $\mathbf{Y}_n \mapsto \mathbf{Y}_n^\sigma$ above. In such cases, even though the data need not be exchangeable, inferences based on $\mathbf{Y}_n$ ought to be transferrable to inferences based on $\mathbf{Y}_n^\sigma$, and vice versa. This is the concept of label equivariance.

Let $\mathbf{Y}_n$ be modeled by $\mathscr{M}_n$ consisting of all distributions described by (8.10), with $G = (G_{ij})_{i,j\geq 1}$ and φ a probability distribution on the space of functions $\{0,1\}^{2\times 2} \times [0,1]^2 \to [0,1]$. Each candidate distribution in $\mathscr{M}_n$ is parameterized by a pair (φ, G), and every such distribution is relatively exchangeable with respect to G. For any (φ, G), relative exchangeability implies that the distribution of $\mathbf{Y}_n^\sigma$ is also parameterized by (φ, G) for any permutation $\sigma : [n] \to [n]$ for which $G^\sigma = G$. But if σ is left unrestricted, then $\mathbf{Y}_n^\sigma = (Y_{\sigma(i)\sigma(j)})_{1\leq i,j\leq n} =_{\mathscr{D}} \mathbf{Y}_n^*$ for $\mathbf{Y}_n^*$ distributed as in (8.10) with parameter (φ, G^σ). So while the distribution of $\mathbf{Y}_n$ is not invariant under relabeling by an arbitrary permutation, the *model* for both $\mathbf{Y}_n$ and its relabeling $\mathbf{Y}_n^\sigma$ are parameterized by all combinations of (φ, G) (because the distribution of $\mathbf{Y}_n^\sigma$ is also a candidate model in $\mathscr{M}_n$ for every permutation σ). Such a model is called *label equivariant*.

An important distinction to keep in mind is that exchangeability, as a property of a distribution, affects the interpretation of the generating and/or sampling scheme and the manner in which the data is observed (i.e., context). Label equivariance is a property of the set of distributions $\mathcal{M}_n$, and thus marks a robustness of the model to arbitrary labeling of the units. Most models used in practice are label equivariant, but there are exceptions. For a simple counterexample, suppose $\mathcal{M}_n = \{P\}$ consists of a single non-exchangeable distribution P. Since P is not exchangeable, there is a permutation $\sigma : [n] \to [n]$ such that $\mathbf{Y}_n^\sigma \neq_{\mathscr{D}} \mathbf{Y}_n$ when $\mathbf{Y}_n \sim P$. In this case, $\mathbf{Y}_n^\sigma$ is not distributed according to P and, therefore, $\mathbf{Y}_n^\sigma$ is not modeled by $\mathcal{M}_n$. The Barabási–Albert models parameterized over the full range $m \geq 1$ and $\delta > -m$ also fail to satisfy label equivariance. Refer to Section 4.2 for more discussion on the Barabási–Albert model.

8.4 Latent space models

The relatively exchangeable models discussed so far account for discrete structure in a population, either through the classification factor C in Section 8.1 or the social network G in Section 8.3. In many cases, however, network data is observed along with quantitative or categorical information, e.g., vertex-level covariates X_i, pairwise measurements such as the amount of time T_{ij} (in minutes) that i and j interacted during the previous week, and so on. Keeping with the scenario of Section 8.1, suppose that instead of the classification factor C, we observe covariate information $\mathbf{x} = (x_{ij})_{1\leq i,j\leq N}$, where each x_{ij} is the amount of time (in minutes) that students i and j spent interacting via text message, phone, or social media during the previous week. For modeling $\mathbf{Y}_N = (Y_{ij})_{1\leq i,j\leq N}$ based on $\mathbf{x}$, the *latent space model* (LSM) [88] specifies a parameter θ along with latent (random) positions $\mathbf{z} = (z_i)_{1\leq i\leq N}$ for each student, so that altogether $\mathbf{Y}_N$ is distributed as

$$\Pr(\mathbf{Y}_N = \mathbf{y} \mid \mathbf{z};\mathbf{x},\theta) = \prod_{1\leq i,j\leq N} p(y_{ij};x_{ij},z_i,z_j,\theta), \quad \mathbf{y} \in \{0,1\}^{N\times N}. \tag{8.11}$$

In practice, it is convenient to assume that $p(y_{ij};x_{ij},z_i,z_j,\theta)$ has the form

$$\log\left(\frac{\Pr(Y_{ij}=1 \mid z_i,z_j;x_{ij},\alpha,\beta)}{\Pr(Y_{ij}=0 \mid z_i,z_j;x_{ij},\alpha,\beta)}\right) = \alpha + \beta^T x_{ij} + f(z_i,z_j) \tag{8.12}$$

for $\theta = (\alpha,\beta)$ and some function $f : \mathbb{R}\times\mathbb{R} \to \mathbb{R}$, where β^T is the transpose of a vector-valued parameter β (in the event that x_{ij} is itself a vector). Often, f is chosen to be a distance, e.g., $f(z,z') = |z-z'|$, affording the interpretation of $f(z,z')$ as the distance between the latent positions z_i and z_j of individuals i and j. The assumed relationship (8.12) is sometimes also written as

$$\operatorname{logit}(\Pr(Y_{ij}=1 \mid z_i,z_j;x_{ij},\alpha,\beta)) = \alpha + \beta^T x_{ij} + f(z_i,z_j),$$

where

$$\operatorname{logit}(p) = \log\left(\frac{p}{1-p}\right)$$

is called the *log-odds*.

We can fit the LSM in (8.11) into the relatively exchangeable framework by letting $\mathscr{X}$ be the space in which the covariates $\mathbf{x}$ take values. (For the high school data, x_{ij} is the amount of time spent interacting, so that $\mathscr{X} = [0,\infty)$.) Following (8.10), we define $\phi_\theta : \mathscr{X} \times [0,1] \times [0,1] \to [0,1]$ for each $\theta = (\alpha, \beta)$ and let

$$p(y_{ij}; x_{ij}, z_i, z_j, \theta) = \begin{cases} \phi_\theta(x_{ij}, z_i, z_j), & y_{ij} = 1, \\ 1 - \phi_\theta(x_{ij}, z_i, z_j), & y_{ij} = 0. \end{cases} \tag{8.13}$$

Notice that the distribution determined by (8.13) is not exchangeable but does satisfy

$$\Pr(\mathbf{Y}_N = \mathbf{y}^\sigma; \mathbf{x}^\sigma, \mathbf{z}^\sigma, \theta) = \Pr(\mathbf{Y}_N = \mathbf{y}; \mathbf{x}, \mathbf{z}, \theta),$$

meaning that if σ is such that $\mathbf{x}^\sigma = \mathbf{x}$ and $\mathbf{z}^\sigma = \mathbf{z}$ then $\mathbf{Y}_N^\sigma =_{\mathscr{D}} \mathbf{Y}_N$. From this we easily see that LSMs are label equivariant, as defined in Section 8.3.4. For lack of interference, we observe that conditional independence in (8.11) implies that the covariate values and latent position of any vertex i only affect the distribution of entries Y_{ij} or Y_{ji} for some $j = 1, \ldots, N$, and thus Y_{ij} does not interfere with any $Y_{i'j'}$ not involving i.

Latent space models can account for vertex-specific covariate information in addition to the pairwise covariates $\mathbf{x}$ above. Supposing that there are elementwise measurements $\mathbf{x}' = (x'_i)_{1 \leq i \leq N}$, with each x'_i recording the academic performance (e.g., GPA) of student i, we augment the LSM by adding extra parameter vectors γ and γ' and a distance function g so that

$$\log\left(\frac{\Pr(Y_{ij} = 1 \mid z_i, z_j; x'_i, x'_j, x_{ij}, \alpha, \beta, \gamma, \gamma')}{\Pr(Y_{ij} = 0 \mid z_i, z_j; x'_i, x'_j, x_{ij}, \alpha, \beta, \gamma, \gamma')}\right) = \tag{8.14}$$
$$= \alpha + \gamma^T x'_i + \gamma'^T x'_j + \beta^T x_{ij} + g(x'_i, x'_j) + f(z_i, z_j).$$

By slight modification, the expression in (8.14) can still be written in the form of (8.13) by letting $\mathbf{x}$ record both pairwise and elementwise information. Note well, however, that the conditional independence structure of LSMs, and relatively exchangeable random graphs more generally, prevents the model from accounting for higher-order structure, such as relationships among three or more vertices. This observation ties in with Problem 8.2 above, again raising the question about the relationship between relative exchangeability, non-interference, and the separable increments property for ERGMs.

Here I have only introduced the basic idea behind latent space models in order to draw a clear connection to the more general concept of relative exchangeability. Readers interested in specific aspects and applications of LSMs are encouraged to consult [88] and more recent related work on random dot product graphs [11, 160].

8.5 Relatively exchangeable random graphs

Stochastic blockmodels, latent space models, and the models in Section 8.3 are all special kinds of relatively exchangeable network models. Because in each of these cases the structure of the population encodes only elementwise or pairwise information, the model can be described in a straightforward way, as in (8.2), (8.10), and (8.13). But is there an analog to these representations if the population structure records more than elementwise or pairwise information?

Consider the following mild extension of the SBM. Let $C : \mathbb{N} \to \mathbb{N}$ be a classification factor (with possibly infinitely many classes) and with the usual notation $C|_S : S \to \mathbb{N}$ denoting the domain restriction of C to $S \subset \mathbb{N}$. For each pair of labels $k, \ell \in \mathbb{N}$, define $\phi_{k,\ell} : [0,1] \times [0,1] \to [0,1]$ to be one of the graphon functions from Chapter 6. Given C and $(\phi_{k,\ell})_{k \neq \ell \geq 1}$, define the *generalized ϕ-process* with block structure C and parameter $(\phi_{k,\ell})_{k \neq \ell \geq 1}$ as the distribution of $\mathbf{Y} = (Y_{ij})_{i,j \geq 1}$ constructed as follows. Let $U_1, U_2, \ldots$ be i.i.d. Uniform$[0,1]$ and, given $U_1, U_2, \ldots$, construct $\mathbf{Y}$ according to

$$\begin{aligned} \Pr(Y_{ij} = 1 \mid U_1, U_2, \ldots; C, (\phi_{k,\ell})_{k \neq \ell \geq 1}) &= \phi_{C(i),C(j)}(U_i, U_j) \\ \Pr(Y_{ij} = 0 \mid U_1, U_2, \ldots; C, (\phi_{k,\ell})_{k \neq \ell \geq 1}) &= 1 - \phi_{C(i),C(j)}(U_i, U_j), \end{aligned} \tag{8.15}$$

conditionally independently for all $j \neq i \geq 1$. By this construction, $\mathbf{Y}$ is relatively exchangeable with respect to C. Subject to the regularity condition in Theorem 8.2, the distribution in (8.15) can be expressed in terms of a single function $\phi : (\mathbb{N} \times \mathbb{N}) \times [0,1] \times [0,1] \to [0,1]$ defined by

$$\phi((k,\ell), u, v) = \phi_{k,\ell}(u,v), \quad k, \ell \in \mathbb{N},\ u, v \in [0,1]. \tag{8.16}$$

With this, we see that relatively exchangeable models can be further extended by first choosing a collection $(\phi_{k,\ell})_{k \neq \ell \geq 1}$ at random and then proceeding with the construction as in (8.15). This observation is a precursor to Theorem 8.4 below.

8.5.1 *Relatively exchangeable ϕ-processes*

The construction in (8.16) generalizes to any relatively exchangeable random graph model by replacing $C : \mathbb{N} \to \mathbb{N}$ with an arbitrary relational structure X expressed in terms of a finite number of relations $X_1, \ldots, X_r$, where $X_j : \mathbb{N}^{a_j} \to \mathscr{X}_j$ assigns each tuple of length $a_j \geq 1$ a covariate value $X_j(a_j) \in \mathscr{X}_j$. We call $(a_1, \ldots, a_r)$ the *signature* of X. For example, the signature consisting only of $a_1 = 1$ and $\mathscr{X}_1 = [K]$ corresponds to the classification factor from Section 8.2; the signature consisting only of $a_1 = 2$ and $\mathscr{X}_1 = \{0,1\}$ corresponds to a binary relation, as in Section 8.3; their combination (a_1, a_2) with $a_1 = 1$, $\mathscr{X}_1 = [K]$, $a_2 = 2$, and $\mathscr{X}_2 = \{0,1\}$ gives a structure for which there is a binary relation on vertices in different communities, as would be the case if the scenarios of Sections 8.1 and 8.3.2 were combined to include a classification into class year ($a_1 = 1$) along with a social network of Facebook friendships ($a_2 = 2$); and the signature consisting of $a_1 = 1$ and $\mathscr{X} = \mathbb{R}^d$, for $d \geq 1$, corresponds to the latent space model with vertex-specific covariates $X(i) \in \mathbb{R}^d$ as in Section 8.4.

Any $X = (X_1, \ldots, X_r)$ can be relabeled and restricted in the usual way. For any injection $\psi : [n] \to \mathbb{N}$, the image of X under ψ-selection is given by $X^\psi = (X_1^\psi, \ldots, X_r^\psi)$, where

$$X_j^{\psi}(x_1,\ldots,x_{a_j}) = X_j(\psi(x_1),\ldots,\psi(x_{a_j})), \quad (x_1,\ldots,x_{a_j}) \in \mathbb{N}^{a_j}. \tag{8.17}$$

In particular, the insertion map $\psi\colon [n] \to \mathbb{N}$, $i \mapsto \psi(i) = i$, gives the domain restriction of X to $[n]$, written $X|_{[n]}$, and any permutation $\sigma : \mathbb{N} \to \mathbb{N}$ gives a relabeling of X by σ.

Definition 8.2 (Relative exchangeability [59]) *Let* $X = (X_1,\ldots,X_r)$ *have signature* $(a_1,\ldots,a_r)$. *A random binary array* $\mathbf{Y}$ *taking values in* $\{0,1\}^{\mathbb{N}\times\mathbb{N}}$ *is* relatively exchangeable with respect to X (in sampling context $\{\mathbf{S}_{m,n}\}_{n\geq m\geq 1}$) *if* $\mathbf{Y}\,|_S =_{\mathscr{D}} \mathbf{Y}\,|_S^{\sigma}$ *for all permutations* $\sigma : S \to S$ *such that* $X|_S^{\sigma} = X|_S$, *for all* $S \subset \mathbb{N}$.

For any $a \geq 1$ and $\mathscr{X}$, write $F(a,\mathscr{X})$ for the space of partial functions $[a] \to \mathscr{X}$. Define a *ϕ-process of signature* $(a_1,\ldots,a_r)$ by taking $\phi : F(a_1,\mathscr{X}_1)\times\cdots\times F(a_r,\mathscr{X}_r)\times[0,1]^2 \to [0,1]$ and constructing $\mathbf{Y}^* = (Y^*_{ij})_{i,j\geq 1}$ according to

$$\Pr(Y^*_{ij} = 1 \mid U_1,U_2,\ldots;X,\phi) = \phi(X_1|_{\{i,j\}},\ldots,X_r|_{\{i,j\}},U_i,U_j), \quad i,j\geq 1, \tag{8.18}$$

conditionally independently for all $i,j\geq 1$, for $(U_i)_{i\geq 1}$ i.i.d. Uniform$[0,1]$, where here we encode $X_k|_{\{i,j\}}$ as the partial function $\mathbb{N}^{a_j} \to \mathscr{X}_k$ with domain $\mathbb{N}^{a_k}\cap\{i,j\}^{a_k}$.[6]

Exercise 8.4 *Show that the distribution of* $\mathbf{Y}^*$ *defined in* (8.18) *is relatively exchangeable with respect to* X.

For a concrete example of such a relatively exchangeable model, combine the scenarios of Sections 8.1 and 8.3.1 and suppose that the friendship relation $\mathbf{Y} = (Y_{ij})_{i,j\geq 1}$, with $Y_{ij} = 1$ indicating that i considers j to be a friend, is observed along with the social media (e.g., Facebook) friendships G and the classification $C : \mathbb{N} \to [K]$ of students according to class year. The composite structure described by G and C can be interpreted as a single graph structure G with vertices 'colored' according to C, i.e., the classes $1,\ldots,K$ are interpreted as different colors. Relative exchangeability with respect to (G,C) requires invariance with respect to permutations that preserve (G,C), i.e., $\mathbf{Y}\,|_S^{\sigma} =_{\mathscr{D}} \mathbf{Y}\,|_S$ for all permutations $\sigma : S \to S$ such that both $G|_S^{\sigma} = G|_S$ and $C|_S \circ \sigma = C|_S$. In the notation for X presented above, the composite structure (G,C) is expressed as $X = (X_1,X_2)$ with $X_1 : \mathbb{N}\times\mathbb{N} \to \{0,1\}$ and $X_2 : \mathbb{N} \to [K]$. We arrive at the following analog to Theorem 8.3.

Theorem 8.4 (Crane–Towsner [59]) *Let G satisfy the same regularity condition as in Theorem 8.3,* $C : \mathbb{N} \to [K]$ *be such that* $\{i \geq 1 : C(i) = j\}$ *is infinite and has strictly positive limiting frequency for every* $j \in [K]$, *and* $\mathbf{Y}$ *be relatively exchangeable with respect to* (G,C). *Then there exists a probability distribution* φ *on the space of functions* $F(2,\{0,1\})\times F(1,[K])\times[0,1]^2 \to [0,1]$ *such that* $\mathbf{Y} =_{\mathscr{D}} \mathbf{Y}^*$, *for* $\mathbf{Y}^*$ *constructed from a randomly chosen* $\phi \sim \varphi$ *as in* (8.18).

[6] The description here in terms of partial functions is somewhat unfortunate, but should not obscure the main point. For readers interested in the basic idea of these models, without the full technical details, it is enough to interpret (8.18) as a model which constructs each edge conditionally independently in a way that depends only on the local restriction $X|_{\{i,j\}}$. Readers interested in the technical details are referred to [59].

Theorem 8.4 is but a special case of the more general Ackerman–Crane–Towsner theorem for relatively exchangeable structures [3, 59]. This general theorem requires several technical ideas from mathematical logic and is deferred to [59]. The reader not wishing to explore these technicalities can be reassured that the general representation obtained in [59] has the same form as (8.18). Even if its formal developments are somewhat technical, the core idea behind relative exchangeability is straightforward, reflecting the assumption that the structure of the population is sufficient for describing the heterogeneity in the data. By comparing (8.4), (8.10), and (8.18), we see that the probabilistic structure of relatively exchangeable networks $\mathbf{Y}$ (subject to regularity assumptions on the underlying structure) has the generic form

$$\begin{aligned} \Pr(Y_{ij}=1 \mid U_1,U_2,\ldots;\phi,X) &= \phi(X|_{(i,j)},U_i,U_j) \quad \text{and} \\ \Pr(Y_{ij}=0 \mid U_1,U_2,\ldots;\phi,X) &= 1-\phi(X|_{(i,j)},U_i,U_j) \end{aligned} \tag{8.19}$$

conditionally independently for all $i,j \geq 1$, regardless of what structure X represents.

8.6 Relative exchangeability under arbitrary sampling

So far we have taken selection as the default sampling context for relatively exchangeable models. In particular, Definitions 8.1 and 8.2 both state relative exchangeability as the invariance of each $\mathbf{Y}|_S$ with respect to the symmetries of $X|_S$. The discussion in Chapters 3–5 suggests that Definition 8.2 should be expanded to account for the wider range of sampling contexts that are relevant in network analysis.

For $N \geq n \geq 1$, let $\Sigma_{n,N}$ be a random sampling operation (possibly depending on $\mathbf{Y}$) with distribution

$$\Pr(\Sigma_{n,N}=\mathbf{S}^{\psi}_{n,N}), \quad \psi:[n]\to[N],$$

where $\mathbf{S}^{\psi}_{n,N}$ is the ψ-selection map defined in (3.17). This random sampling operation also induces an action $\Sigma_{n,N}$ on $X_N=(X_{N,1},\ldots,X_{N,r})$ with signature $(a_1,\ldots,a_r)$ by defining $\Sigma_{n,N}X_N=X^{\psi}_N$ on the event '$\Sigma_{n,N}=\mathbf{S}^{\psi}_{n,N}$', for X^{ψ}_N defined in (8.17). In the more general sampling context $\{\Sigma_{n,N}\}_{1\leq n\leq N}$, we refine the discussion of Section 8.3.3 to define non-interference with respect to $\{\Sigma_{n,N}\}_{1\leq n\leq N}$ as follows. Let $X_N=(X_1,\ldots,X_r)$ be a fixed relational structure with signature $(a_1,\ldots,a_r)$, as defined in Section 8.5. Under (deterministic) selection sampling, $\mathbf{S}_{n,N}X_N=X_N|_{[n]}$ is a fixed structure, but under the (possibly random) sampling by $\Sigma_{n,N}$, $\Sigma_{n,N}X_N$ corresponding to $\mathbf{S}^{\psi}_{n,N}X_N$ on the event '$\Sigma_{n,N}=\mathbf{S}^{\psi}_{n,N}$' is a (possibly random) structure. We say that $\mathbf{Y}_n$ parameterized by X_N exhibits *non-interference with respect to sampling from* $\Sigma_{n,N}$ if the distribution of $\mathbf{Y}_n$ with parameter X_N coincides with the conditional distribution of $\mathbf{Y}_n$ given $\Sigma_{n,N}X_N$, i.e.,

$$\Pr(\mathbf{Y}_n=\cdot;X_N)=\Pr(\mathbf{Y}_n=\cdot\mid\Sigma_{n,N}X_N).$$

Notice that the distribution on the left is parameterized by a fixed structure X_N whereas the distribution on the right is conditioned on a random structure $\Sigma_{n,N}X_N$. It remains to explore the implications of this definition and the following extension to relative exchangeability that it motivates.

Definition 8.3 *Given* $X_N = (X_{N,1},\dots,X_{N,r})$ *with signature* $(a_1,\dots,a_r)$ *and* $X_{N,j} : [N]^{a_j} \to \mathscr{X}_j$ *for each* $j = 1,\dots,r$, *we call* $\mathbf{Y}_N$ relatively exchangeable with respect to X_N in sampling context $\{\Sigma_{n,N}\}_{N\geq n\geq 1}$ *if*

$$(\Sigma_{n,N}\mathbf{Y}_N)^{\sigma} =_{\mathscr{D}} \Sigma_{n,N}\mathbf{Y}_N \tag{8.20}$$

for all permutations $\sigma : [n] \to [n]$ *such that* $(\Sigma_{n,N}X_N)^{\sigma} =_{\mathscr{D}} \Sigma_{n,N}X_N$, *for all* $1 \leq n \leq N$.

Note that in general Definition 8.3 is defined in terms of a distributional identity for $\Sigma_{n,N}X_N$. For example, suppose that $\Sigma_{n,N}$ is the random sampling map obtained by putting $\Sigma_{n,N} = \mathbf{S}^{\psi}_{n,N}$ (as defined in (3.17)) for ψ chosen uniformly at random among all injections $[n] \to [N]$. Then $\Sigma_{n,N}X_N$ is exchangeable, i.e., $(\Sigma_{n,N}X_N)^{\sigma} =_{\mathscr{D}} \Sigma_{n,N}X_N$ for all permutations $\sigma : [n] \to [n]$, and thus so is $\mathbf{Y}_N$. At the outset, it remains unclear whether or not Definition 8.3 is a substantive extension to the definition of relative exchangeability under selection sampling.

Research Problem 8.3 *Give an example of a relatively exchangeable network in the sense of Definition 8.3 which has non-trivial symmetries (i.e., there is a non-identity permutation* σ *for which* $(\Sigma_{n,N}X_N)^{\sigma} =_{\mathscr{D}} \Sigma_{n,N}X_N$*) and which is not fully exchangeable. If Definition 8.3 does not admit non-trivial models, propose another definition for relative exchangeability in arbitrary sampling contexts, and study the class of models it determines.*

Research Problem 8.4 *Assuming it admits non-trivial models (cf. Problem 8.3), study the more general version of relative exchangeability in Definition 8.3 by (i) extending the main theorems from [58, 59] to this setting, (ii) exploring its implications for statistical analysis, (iii) analyzing how to perform inference in this setting, and/or (iv) applying it on a real network data problem.*

For relatively exchangeable network models, it might also be natural for the random sampling scheme $\Sigma_{n,N}$ to be parameterized by the underlying population structure X_N. In particular, we might suppose that the distribution of $\Sigma_{n,N}$ is relatively exchangeable with respect to X_N, in the sense that

$$\Pr(\Sigma_{n,N} = \mathbf{S}^{\psi}_{n,N}; X_N) = \Pr(\Sigma_{n,N} = \mathbf{S}^{\psi'}_{n,N}; X_N)$$
$$\text{for all } \psi, \psi' : [n] \to [N] \text{ such that } \mathbf{S}^{\psi}_{n,N}X_N = \mathbf{S}^{\psi'}_{n,N}X_N.$$

Call any such $\Sigma_{n,N}$ an *X_N-exchangeable sampling scheme.* A few natural questions emerge, which I leave as a topic for future exploration.

Research Problem 8.5 *Let* X_N *be any structure. For* $1 \leq n \leq N$ *let* $\Sigma_{n,N}$ *and* $\mathbf{Y}_N$ *be independent and suppose both are* X_N*-exchangeable. Does the distribution of* $\Sigma_{n,N}\mathbf{Y}_N$ *satisfy any invariance properties? How do these invariance properties depend on the algebraic structure of* X_N*? Otherwise, let* $\mathbf{Y}_N$ *be generic and let* $\Sigma_{n,N}$ *be an* X_N*-exchangeable sampling scheme that is independent of* $\mathbf{Y}_N$. *What can be said of the distributional symmetries of* $\Sigma_{n,N}\mathbf{Y}_N$ *in this case?*

8.7 Relatively invariant graphex models

In view of the preceding discussion about relative exchangeability, I propose here the following refinement of the Caron–Fox model (Section 7.3). Under the Caron–Fox model, a network is represented as an 'exchangeable' point process $\mathbf{Y} \subset [0,\infty) \times [0,\infty)$, with exchangeability defined as the distributional invariance of $\mathbf{Y}$ with respect to Lebesgue measure-preserving transformations of $[0,\infty)$, as expressed in (7.10). To see how (7.10) fits in with the rest of this chapter, notice first that relative exchangeability (Definition 8.3) refines vertex exchangeability (Definition 6.1) by allowing the distributional symmetries of an array $\mathbf{Y} = (Y_{ij})_{i,j\geq 1}$ to be expressed in terms of the symmetries of a generic combinatorial structure X. Vertex exchangeability can then be recovered as a special case of relative exchangeability by taking X to be a fully symmetric structure, i.e., $X^\sigma = X$ for all permutations $\sigma : \mathbb{N} \to \mathbb{N}$.

Appealing to the same rationale, we extend the class of models in Section 7.3 by relaxing condition (7.10) so that the invariance of $\mathbf{Y}$ is instead defined with respect μ-preserving transformations, for an arbitrary measure μ on $[0,\infty)$. More precisely, we let μ be a σ-finite measure on $[0,\infty)$ and let $\mathbf{Y} \subseteq [0,\infty) \times [0,\infty)$ be a random point process. A (measurable) transformation $T : [0,\infty) \to [0,\infty)$ is called *μ-preserving* if

$$\mu(T^{-1}(A)) = \mu(A) \quad \text{for all measurable } A \subseteq [0,\infty).$$

With $T \circ \mathbf{Y}$ as defined in (7.9), we call $\mathbf{Y}$ *relatively invariant with respect to μ*, or simply *μ-invariant*, if

$$T \circ \mathbf{Y} =_{\mathscr{D}} \mathbf{Y} \quad \text{for all } \mu\text{-preserving transformations } T : [0,\infty) \to [0,\infty). \tag{8.21}$$

Note that (8.21) coincides with (7.10) when μ is taken to be Lebesgue measure on $[0,\infty)$, and thus we recover the 'exchangeable' point process models of Section 7.3 as a special case of (8.21). To my knowledge, the definition of μ-invariance for point processes $\mathbf{Y}$ is defined here for the first time. If such a definition or related work has appeared before, then I suspect it could be found in [98].

I present relatively invariant point process models as a wide open topic for future applied and theoretical research. Building off of the sampling interpretation of the Caron–Fox model (Section 7.3.7), the more general invariance in (8.21) should suggest a related sampling interpretation in terms of observing edge patterns over 'equivalent' durations of time, where 'equivalence' here is subject to 'warping time' according to μ. I conclude this section with a conjecture and open problem about relatively invariant point process models.

Conjecture 8.1 *Let μ be a σ-finite measure on $[0,\infty)$, perhaps satisfying an additional regularity condition, and let $\mathbf{Y} \subseteq [0,\infty) \times [0,\infty)$ be a μ-invariant point process representing a network. Then $\mathbf{Y}$ admits an analogous representation to the graphex characterization of exchangeable point process models in (7.11)–(7.13), but with the modification that the unit rate Poisson processes $\Theta \subseteq [0,\infty)^2$, $\Xi_i \subseteq [0,\infty)^2$, and $R \subseteq [0,\infty)^3$ are instead independent Poisson point processes with intensities $\mu \otimes dt$, $\mu \otimes dt$, and $\mu \otimes \mu \otimes dt$, where dt denotes Lebesgue measure on $[0,\infty)$ and $\mu \otimes dt$ denotes the product measure of μ and dt on $[0,\infty) \times [0,\infty)$.*

The conditions of this conjecture may need to be modified or strengthened in order for the representation to hold, but I leave it to the interested reader to work out those details.

Research Problem 8.6 *The above specification of relatively invariant point process models incorporates an additional measure-valued parameter μ into the model class from Section 7.3. The extra parameter seems to make the model more flexible, but it is unclear how this flexibility might manifest itself in practice. If μ is unknown, then it seems difficult to estimate it from data, but perhaps there is a reasonable parametric assumption for which some estimation theory is possible. I leave these and other related questions as future research directions.*

8.8 Final remarks and further reading

While the formal study of relatively exchangeable structures is new as of [3, 58, 59], the concept is implicit in the original stochastic blockmodel [89] and the latent space model [88]. In this chapter, I have emphasized the basic ideas underlying relative exchangeability, but as of now the available results involve obscure mathematical notions and difficult techniques which lie beyond the scope of a typical discussion on statistical network analysis. The core idea behind relative exchangeability, however, seems to be straightforward and natural for statistical applications involving complex structures, and there is ample room for future work on theory and applications of relative exchangeability. For example, recent work on community detection in the SBM for weighted graphs [157] is ripe for possible extensions to the more general relatively exchangeable models discussed here.

I conclude this chapter by noting that the form in (8.18) is not necessary for $\mathbf{Y}$ to be relatively exchangeable. There are relatively exchangeable structures $\mathbf{Y}$ which depend on X in a more general way than in (8.18). For example, instead of a classification factor $C : \mathbb{N} \to [K]$ suppose that the community structure is captured by an equivalence relation $b : \mathbb{N} \times \mathbb{N} \to \{0,1\}$ encoding

$$b(i,j) = \begin{cases} 1, & C(i) = C(j), \\ 0, & \text{otherwise.} \end{cases}$$

(The major difference between C and b is that b only records whether or not two vertices are in the same class. It does not keep track of which class label is associated to each vertex.) Any permutation $\sigma : \mathbb{N} \to \mathbb{N}$ acts on b by

$$b^{\sigma}(i,j) = b(\sigma^{-1}(i), \sigma^{-1}(j)),$$

meaning that

$$b^{\sigma}(i,j) = 1 \quad \text{if and only if} \quad b(\sigma^{-1}(i), \sigma^{-1}(j)) = 1.$$

Relative exchangeability with respect to b is defined in the analogous way to Definition 8.1: $\mathbf{Y}|_S^{\sigma} =_{\mathscr{D}} \mathbf{Y}|_S$ for all permutations $\sigma : S \to S$ for which $b|_S^{\sigma} = b|_S$, where $b|_S$ is the restriction of $b : \mathbb{N} \times \mathbb{N} \to \{0,1\}$ to $S \times S \to \{0,1\}$. But in this case, the marginal information $b(i,j)$ may not fully determine the distribution of $\mathbf{Y}_N|_S$. Take, for instance, the case $S = \{1,2,3,4\}$ with $C(1) = C(2) = 1$ and $C(3) = C(4) = 2$. Then $b(1,2) = b(2,1) = 1$, $b(3,4) = b(4,3) = 1$, and $b(i,j) = 0$ for all other combinations of i and j. But the marginal information $b(1,2) = b(3,4) = 1$ is not enough to determine that 1 and 2 occupy a different block than 3 and 4. (Note that $b(1,2) = b(3,4) = 1$ would also hold in the event that $C(1) = C(2) = C(3) = C(4)$.) Unlike the generic construction given in (8.19), in which the conditional distribution of each edge depends on X only through $X|_{\{i,j\}}$, the analogous construction of $\mathbf{Y}$ that is relatively exchangeable with respect to b must be built up sequentially in a way that takes the entire initial segment of b into account at each step. Specifically, $\mathbf{Y}_N$ can be constructed by first determining $\mathbf{Y}|_{[1]}$, then $\mathbf{Y}|_{[2]}$ conditional on $b|_{[2]}$ and $\mathbf{Y}|_{[1]}$, then $\mathbf{Y}|_{[3]}$ conditional on $b|_{[3]}$ and $\mathbf{Y}|_{[2]}$, and in general $\mathbf{Y}|_{[m]}$ conditional on $b|_{[m]}$ and $\mathbf{Y}|_{[m-1]}$. In this way, the construction 'keeps track' of structure in b which may not be reflected in the pairwise information used for the construction in (8.19). Harkening back to the conditions of ultrahomogeneity and disjoint amalgamation in Section 8.3.3, the construction in (8.19) fails in this case because b does not have the n-DAP property for all $n \geq 1$. For further details on the general case I defer to [58, 59].

8.9 Solutions to exercises

8.9.1 *Exercise 8.1*

The adjacency arrays corresponding to the graphs in Figures 8.1(a) and 8.1(b), respectively, are

$$\mathbf{y}_a = \begin{pmatrix} 0 & 0 & 0 & 0 & 0 & 0 & 1 \\ 0 & 0 & 1 & 1 & 0 & 0 & 0 \\ 0 & 1 & 0 & 1 & 1 & 1 & 0 \\ 0 & 1 & 1 & 0 & 0 & 1 & 1 \\ 0 & 0 & 1 & 0 & 0 & 0 & 0 \\ 0 & 0 & 1 & 1 & 0 & 0 & 0 \\ 1 & 0 & 0 & 1 & 0 & 0 & 0 \end{pmatrix} \quad \text{and} \quad \mathbf{y}_b = \begin{pmatrix} 0 & 1 & 0 & 0 & 1 & 0 & 0 \\ 1 & 0 & 1 & 1 & 0 & 1 & 0 \\ 0 & 1 & 0 & 1 & 0 & 1 & 1 \\ 0 & 1 & 1 & 0 & 0 & 0 & 0 \\ 1 & 0 & 0 & 0 & 0 & 0 & 0 \\ 0 & 1 & 1 & 0 & 0 & 0 & 0 \\ 0 & 0 & 1 & 0 & 0 & 0 & 0 \end{pmatrix},$$

and C is defined by

$$\begin{aligned} C : [7] &\to \{1,2\} \\ 1 &\mapsto 2 \\ 2 &\mapsto 1 \\ 3 &\mapsto 2 \\ 4 &\mapsto 1 \\ 5 &\mapsto 2 \\ 6 &\mapsto 1 \\ 7 &\mapsto 2. \end{aligned}$$

Define $\sigma : [7] \to [7]$ by

$$\begin{aligned} \sigma : [7] &\to [7] \\ 1 &\mapsto 7 \\ 2 &\mapsto 4 \\ 3 &\mapsto 3 \\ 4 &\mapsto 6 \\ 5 &\mapsto 1 \\ 6 &\mapsto 2 \\ 7 &\mapsto 5. \end{aligned}$$

By the definition of relabeling in (6.1), we have

$$\mathbf{y}_a^\sigma = \begin{pmatrix} 0 & 1 & 0 & 0 & 1 & 0 & 0 \\ 1 & 0 & 1 & 1 & 0 & 1 & 0 \\ 0 & 1 & 0 & 1 & 0 & 1 & 1 \\ 0 & 1 & 1 & 0 & 0 & 0 & 0 \\ 1 & 0 & 0 & 0 & 0 & 0 & 0 \\ 0 & 1 & 1 & 0 & 0 & 0 & 0 \\ 0 & 0 & 1 & 0 & 0 & 0 & 0 \end{pmatrix} = \mathbf{y}_b$$

and $C \circ \sigma = C$. Let $\mathbf{y}_a = (y_{ij})_{1 \le i,j \le n}$. By definition of the SBM in (8.2), we have

$$\begin{aligned} &\Pr(\mathbf{Y}_n = \mathbf{y}_b; w, C) = \\ &= \Pr(\mathbf{Y}_n = \mathbf{y}_a^\sigma; w, C) \\ &= \prod_{1 \le i \ne j \le n} w((C\circ\sigma)(i), (C\circ\sigma)(j))^{y_{\sigma(i)\sigma(j)}} \times \\ &\qquad \times (1 - w((C\circ\sigma)(i), (C\circ\sigma)(j)))^{1 - y_{\sigma(i)\sigma(j)}} \\ &= \prod_{1 \le i \ne j \le n} w((C\circ\sigma)(\sigma^{-1}(i)), (C\circ\sigma)(\sigma^{-1}(j)))^{y_{ij}} \times \\ &\qquad \times (1 - w((C\circ\sigma)(\sigma^{-1}(i)), (C\circ\sigma)(\sigma^{-1}(j))))^{1-y_{ij}} \\ &= \prod_{1 \le i \ne j \le n} w(C(i), C(j))^{y_{ij}} (1 - w(C(i), C(j)))^{1-y_{ij}} \\ &= \Pr(\mathbf{Y}_n = \mathbf{y}_a; w, C). \end{aligned}$$

8.9.2 Exercise 8.2

Let $\mathbf{Y}_N$ be distributed according to the SBM with block structure C and weight function w as in (8.2),

$$\Pr(\mathbf{Y}_N = \mathbf{y}; w, C) = \prod_{1 \le i \ne j \le N} w(C(i), C(j))^{y_{ij}} (1 - w(C(i), C(j)))^{1-y_{ij}}.$$

To see that $\mathbf{Y}_N$ is relatively exchangeable with respect to C, let $S \subseteq [N]$ be any subset of vertices and let $\sigma : S \to S$ be such that $C|_S \circ \sigma = C|_S$. By independence of the entries Y_{ij} in $\mathbf{Y}_N$, the marginal distribution of $\mathbf{Y}_N|_S$ is

$$\Pr(\mathbf{Y}_N|_S = \mathbf{y}; w, C|_S) = \prod_{i,j\in S: i\neq j} w(C(i),C(j))^{y_{ij}}(1-w(C(i),C(j)))^{1-y_{ij}}.$$

Thus, for any $\sigma : S \to S$ such that $C|_S \circ \sigma = C|_S$, we have (writing $C' = C|_S$)

$$\begin{aligned}
&\Pr(\mathbf{Y}_N|_S = \mathbf{y}^\sigma; w, C') = \\
&= \prod_{i,j\in S: i\neq j} w(C'(i),C'(j))^{y_{\sigma(i)\sigma(j)}}(1-w(C'(i),C'(j)))^{1-y_{\sigma(i)\sigma(j)}} \\
&= \prod_{i,j\in S: i\neq j} w((C'\circ\sigma)(\sigma^{-1}(i)),(C'\circ\sigma)(\sigma^{-1}(j)))^{y_{ij}} \times \\
&\qquad \times (1-w((C'\circ\sigma)(\sigma^{-1}(i)),(C'\circ\sigma)(\sigma^{-1}(j)))^{1-y_{ij}} \\
&= \prod_{i,j\in S: i\neq j} w(C'(i),C'(j))^{y_{ij}}(1-w(C'(i),C'(j)))^{1-y_{ij}} \\
&= \Pr(\mathbf{Y}_N|_S = \mathbf{y}; w, C').
\end{aligned}$$

Since S and σ were chosen arbitrarily subject to the constraint $C|_S \circ \sigma = C|_S$, the proof is complete by the definition of relative exchangeability in (8.3).

8.9.3 Exercise 8.3

For any permutation $\sigma : [N] \to [N]$ and $\mathbf{y} \in \{0,1\}^{N\times N}$, we compute

$$\begin{aligned}
&\Pr(\mathbf{Y}_N = \mathbf{y}^\sigma; \phi) = \\
&= \sum_{\mathbf{c}:[N]\to[K]} \Pr(C = \mathbf{c})\Pr(\mathbf{Y}_N = \mathbf{y}^\sigma; \phi, \mathbf{c}) \\
&= \sum_{\mathbf{c}:[N]\to[K]} \Pr(C = \mathbf{c}^\sigma)\Pr(\mathbf{Y}_N = \mathbf{y}^\sigma; \phi, \mathbf{c}^\sigma) \\
&= \sum_{\mathbf{c}:[N]\to[K]} \Pr(C = \mathbf{c})\Pr(\mathbf{Y}_N = \mathbf{y}; \phi, \mathbf{c}) \\
&= \Pr(\mathbf{Y}_N = \mathbf{y}; \phi),
\end{aligned}$$

since the orbit of $\mathbf{c}^\sigma$ over all $\mathbf{c} : [N] \to [K]$ coincides with the set $\mathbf{c} : [N] \to [K]$. It follows that $\mathbf{Y}_N$ is exchangeable.

8.9.4 Exercise 8.4

To see that $\mathbf{Y}^*$ is relatively exchangeable, let $S \subset \mathbb{N}$ with $|S| = n$ and take any permutation $\sigma : S \to S$ that fixes $X|_S$. Writing $X|_S = (X_1|_S, \ldots, X_r|_S)$, we have

$$
\begin{aligned}
&\Pr(\mathbf{Y}|_S = \mathbf{y}^\sigma; X) = \\
&= \int_{[0,1]^S} \prod_{i,j\in S} \phi(X|_{\{i,j\}}, u_i, u_j)^{y_{\sigma(i),\sigma(j)}} (1 - \phi(X|_{\{i,j\}}, u_i, u_j))^{1-y_{\sigma(i),\sigma(j)}} \prod_{k\in S} du_k \\
&= \int_{[0,1]^S} \prod_{i,j\in S} \phi(X|_{\{\sigma^{-1}(i),\sigma^{-1}(j)\}}, u_{\sigma^{-1}(i)}, u_{\sigma^{-1}(j)})^{y_{ij}} \times \\
&\qquad \times (1 - \phi(X|_{\{\sigma^{-1}(i),\sigma^{-1}(j)\}}, u_{\sigma^{-1}(i)}, u_{\sigma^{-1}(j)}))^{1-y_{ij}} \prod_{k\in S} du_k \\
&= \int_{[0,1]^S} \prod_{i,j\in S} \phi(X|_{\{i,j\}}, u_i, u_j)^{y_{ij}} (1 - \phi(X|_{\{i,j\}}, u_i, u_j))^{1-y_{ij}} \prod_{k\in S} du_{\sigma(k)} \\
&= \int_{[0,1]^S} \prod_{i,j\in S} \phi(X|_{\{i,j\}}, u_i, u_j)^{y_{ij}} (1 - \phi(X|_{\{i,j\}}, u_i, u_j))^{1-y_{ij}} \prod_{k\in S} du_k \\
&= \Pr(\mathbf{Y}|_S = \mathbf{y}; X),
\end{aligned}
$$

as required.

Chapter 9

Edge exchangeable models

With the exception of the Caron–Fox model (Section 7.3), all of the models discussed so far are tailored to contexts in which the vertices are the units. Even the Caron–Fox model, though not defined explicitly in terms of vertex selection, is most naturally interpreted in terms of t-selection for the point process $\mathbf{Y}$, as in (7.14), or p-sampling of the induced array defined in (7.6); see Section 7.3.7 and [148]. An important distinction between vertex selection, simple random vertex sampling, t-selection, and p-sampling and the alternative sampling schemes presented in Section 3.6 is that the latter depend on the network structure while the former do not. Vertex selection, simple random vertex sampling, t-selection, and p-sampling can all be performed without referring to the network, but the edges in a network cannot be sampled independently of the network, nor can paths. The network must exist, and have edges or paths between its vertices, in order for edge or path sampling to be possible. A vertex that does not participate in any edges or paths cannot appear in a network obtained by edge or path sampling. With this observation, it is no surprise that the correct approach to modeling network data obtained by edge, hyperedge, or more generic 'relational' sampling differs from what is appropriate in more conventionally assumed vertex sampling contexts. We explore some of these differences over the next two chapters. For further details see [53, 54].

9.1 Scenario: Monitoring phone calls

Consider a network constructed by sampling entries from a phone call database, as in Section 3.6.1.1. Associated to each call is the ordered pair (s,r) indicating the *sender* s and *receiver* r. As shown in Table 9.1, database entries also contain information about the time of the call, topic discussed, etc., but we disregard this extra information here. We focus only on the structure induced by regarding each phone call (s,r) as a directed edge $s \to r$ and representing a sequence of calls, e.g.,

$$X_1 = (a,b), \quad X_2 = (c,a), \quad X_3 = (d,e), \quad X_4 = (a,c),$$

as the network-like structure shown in Figure 9.1. In this representation, the vertex labels a,b,c,d,e identify the phone numbers involved in each call, while the edge labels $1,2,3,4$ identify each phone call $(a,b),(c,a),(d,e),(a,c)$, respectively, according to when in the sequence it was observed.

Table 9.1 *Database of phone calls. Each row contains information about a single phone call: caller and receiver (identified by phone number), time of call, topic discussed, etc.*

Caller	Receiver	Time of Call	Topic Discussed	...
555-7892 (a)	555-1243 (b)	15:34	Business	...
550-9999 (c)	555-7892 (a)	15:38	Birthday	...
555-1200 (d)	445-1234 (e)	16:01	School	...
555-7892 (a)	550-9999 (c)	15:38	Sports	...
555-1243 (b)	555-1200 (d)	16:17	Business	...
$\vdots$	$\vdots$	$\vdots$	$\vdots$	$\ddots$

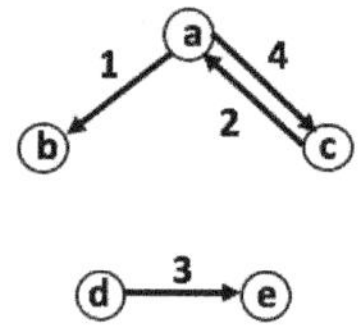

Figure 9.1 *Network depiction of phone call sequence* $X_1 = (a,b)$, $X_2 = (c,a)$, $X_3 = (d,e)$, $X_4 = (a,c)$ *obtained from the first 4 rows of Table 9.1, with label i assigned to edge representing* X_i *for each* $i = 1,2,3,4$.

9.2 Edge-centric view

Imagine sampling 4 phone calls uniformly at random from the database in Table 9.1. Then the observed calls X_1,X_2,X_3,X_4 form an exchangeable sequence of ordered pairs. In the network representation of Figure 9.1, exchangeability of the sequence X_1,X_2,X_3,X_4 induces exchangeability on the network, in the sense that any two realizations that are isomorphic up to edge relabeling have equal probability; see Figure 9.2 for illustration. Because there is a one-to-one correspondence between edge sequences $X_1,\ldots,X_n$ and their vertex-edge labeled network representation, as in Figure 9.1, the class of models for such data is determined by de Finetti's theorem [60] for exchangeable sequences. (See Section 6.6.1 for a connection between de Finetti's theorem and the Aldous–Hoover theorem for vertex exchangeable models.) The key observation at this stage is that exchangeability is defined with respect to relabeling of the edges, not the vertices as was assumed in Chapter 6, and that this shift toward exchangeability of edges arises naturally by considering how the network is observed via edge sampling. This immediately suggests a different way to analyze networks than the conventional networks-as-graphs perspective (Section 1.2). It is also clear at this point why the common practice of delabeling the edges and viewing the network as the graph in Figure 9.3 imposes an expressly vertex-centric, and therefore misleading, perspective on network data observed under the conditions of Section 9.1.

The above scenario, in which a network is observed by sampling phone calls (i.e.,

Figure 9.2 *(Left) Network representation of phone call sequence* $X_1 = (a,b)$*,* $X_2 = (c,a)$*,* $X_3 = (d,e)$*,* $X_4 = (a,c)$*. (Right) Representation of the network on left after reordering of its edges* X_2, X_3, X_4, X_1*. Any such reordering has equal probability under an exchangeable model.*

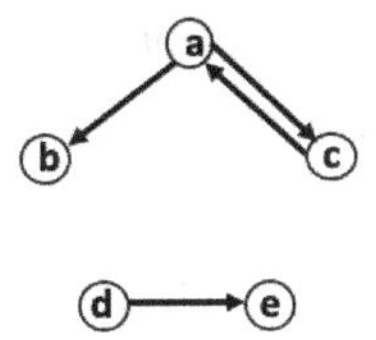

Figure 9.3 *Graphical representation of phone call sequence* $X_1 = (a,b)$*,* $X_2 = (c,a)$*,* $X_3 = (d,e)$*,* $X_4 = (a,c)$ *in Figure 9.1 after delabeling edges.*

interactions/edges) instead of callers (i.e., vertices), fits into Crane and Dempsey's [54] edge-centric perspective for modeling interaction networks. As we see throughout this chapter and the next, the edge-centric perspective is best suited to network data for which the relations or interactions are the units of observation, as in networks built from phone calls, email correspondence, scientific coauthorship, movie actor collaborations, path sampling of Internet topology, and many other kinds of *interaction data.*

Harkening back to Chapter 6, there is a tendency to regard the vertex labels a,b,c,d,e in Figure 9.3 as arbitrary 'names' which serve no additional purpose except to uniquely identify different vertices in the network. Regarded this way, it may seem natural to choose a model for the vertex-labeled representation in Figure 9.3 that is invariant under renaming the vertices, as shown in Figure 9.4. This line of reasoning is sometimes given in favor of vertex exchangeable network models: *since the vertex names are arbitrary, the distribution of the data should be invariant to arbitrary renaming of the vertices.* But such reasoning is flawed. Remember, vertex exchangeability is more than distributional invariance with respect to arbitrary relabeling of the sampled vertices. It is invariance with respect to arbitrary relabeling of all vertices, sampled and unsampled. In particular, it implies that those vertices which have been sampled can be 'exchanged' with vertices which have not been sampled. So while it is true that the vertex labels (i.e., the 'names') in Figure 9.1 serve only to identify individual vertices, vertex exchangeability implies that the vertices themselves (regardless of their 'names') can be interchanged, i.e., are representative

Figure 9.4 *Network representation of phone call sequence* $X_1 = (a,b)$, $X_2 = (c,a)$, $X_3 = (d,e)$, $X_4 = (a,c)$ *from Figure 9.3 and its transformation under renaming vertices. A vertex exchangeable model assigns equal probability to both realizations.*

copies of one another. This latter implication of representative vertex sampling does not hold in Section 9.1 and many other scenarios involving interaction data.

In Section 9.1, the identity of each vertex is determined not by its 'name' but rather by how it relates to other vertices through its phone call activity. It follows that the identity of each sampled vertex is embedded in the data, so that once the identities are given—by articulating how different vertices interact with one another—they cannot be changed arbitrarily. So even though the vertex 'names' a,b,c,d,e are inconsequential, the fact remains that each vertex has a unique identity; and while their names may be arbitrary, their identities are not. To see this explicitly, note that the two networks shown in Figure 9.4 represent two different sequences of calls. On the left are the calls $(a,b),(c,a),(d,e),(a,c)$ from Figure 9.3 and on the right are $(c,b),(e,c),(a,d),(c,e)$. With respect to observing phone calls from the database, the left-hand side and right-hand side of Figure 9.4 are different observations, and we have no reason to assume that both have the same probability of being chosen by sampling from the database. Given that the calls are sampled uniformly without replacement from the database, however, it is reasonable to assume that the sequence of calls determining, e.g., the right-hand network of Figure 9.4, is exchangeable. Thus, for example, the sequence $X_1' = (c,b), X_2' = (e,c), X_3' = (a,d), X_4' = (c,e)$ leading to the network on the right-hand side of Figure 9.4 has the same probability as the reordering of calls by, say, X_2', X_3', X_4', X_1'.

In summary, once the *structure* of the data is accounted for, as in Figure 9.1, the vertex *names* are inconsequential, but the vertex *identities* are not. In other words, the names can be *disregarded*, not *reassigned*, but only after the essential network structure (i.e., pattern of edges) has been recorded. Even though individually the two observations X_1, X_2, X_3, X_4 and X_1', X_2', X_3', X_4' may occur with different probabilities, for the purposes of inference they both convey the same information about the database of calls, because they both produce the same pattern of edges, as shown in Figure 9.5. This viewpoint leads naturally to the concept of edge exchangeability.

Remark 9.1 *Along with Section 1.2, the previous section is the most important of the whole book. Some of the points are subtle and warrant reflection, especially by readers predisposed to the networks-as-graphs mindset. The reader is urged to reread Sections 1.2 and 9.2. Can you think of another context (outside of network analysis)*

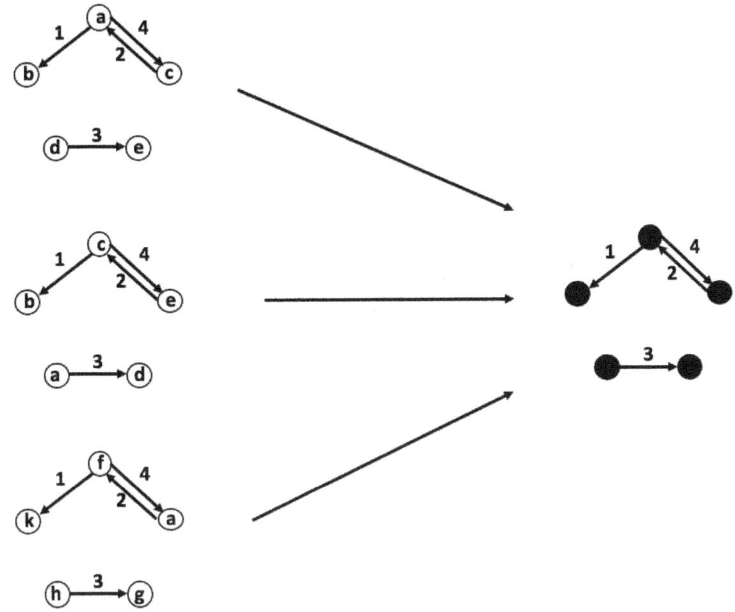

Figure 9.5 *Phone call networks (on left) containing the same sufficient information upon removing vertex labels, as shown in the edge-labeled graph on the right.*

in which a change in perspective provides substantial insight and greatly simplifies an otherwise difficult or intractable problem?

9.3 Edge exchangeability

In Section 9.1 each phone call is represented as a caller-receiver pair $(s,r) \in \mathscr{P} \times \mathscr{P}$ in some (at most countable) set $\mathscr{P}$ representing the population of all phone numbers in a database. Sampling uniformly (with or without replacement) from the database gives an exchangeable sequence $X_1, X_2, \ldots$ in $\mathscr{P} \times \mathscr{P}$, so that each finite segment $(X_1, \ldots, X_n)$ satisfies

$$(X_1, \ldots, X_n) =_{\mathscr{D}} (X_{\sigma(1)}, \ldots, X_{\sigma(n)}) \quad \text{for all permutations } \sigma : [n] \to [n]. \tag{9.1}$$

In particular, for any specific realization $(s_1,t_1), \ldots, (s_n,t_n)$ in $\mathscr{P} \times \mathscr{P}$, the distribution of $(X_i)_{1 \leq i \leq n}$ satisfies

$$\Pr(X_i = (s_i,t_i), i = 1, \ldots, n) = \Pr(X_i = (s_{\sigma(i)}, t_{\sigma(i)}),\ i = 1, \ldots, n) \tag{9.2}$$

for all permutations $\sigma : [n] \to [n]$, for all $n \geq 1$. This condition is reflected in Figure 9.2, which shows two vertex-edge labeled networks whose probabilities are equal under an exchangeable model.

The distribution of the edge-labeled graph obtained by disregarding the names of the vertices, as on the right-hand side of Figure 9.5, is determined by the sequence of calls $X_1, X_2, \ldots$ as follows. First, note that the edge-labeled graph on the right-hand side of Figure 9.5 corresponds to the *set* (i.e., equivalence class) of all call sequences $X_1, X_2, \ldots$ whose induced network structures are the same after disregarding vertex labels. (The three vertex-edge labeled graphs on the left-hand side of Figure 9.5 are members of this equivalence class, as are many other graphs that are not listed.) In particular, the sufficient information in any realization $\mathbf{x} = (x_1, x_2, \ldots)$ of $X_1, X_2, \ldots$ is given by the equivalence class

$$\mathscr{E}_{\mathbf{x}} = \{(\rho(x_1), \rho(x_2), \ldots) \mid \rho : \mathscr{P} \to \mathscr{P} \text{ is a bijection}\}, \tag{9.3}$$

where $\rho(x) = (\rho(s), \rho(t))$ is the image of $x = (s,t)$ under ρ. The operation in (9.3) has the effect of disregarding the observed vertex names by taking the data to be the set of all caller-receiver sequences which give rise to the same edge-labeled graph, as in Figure 9.5. The notation '$\mathscr{E}_{\mathbf{x}}$' in (9.3) is to be understood as the 'edge-labeled network' ($\mathscr{E}$) 'induced by $\mathbf{x}$' (subscript $\mathbf{x}$).[1]

Let $\mathbf{Y}_n$ denote the random edge-labeled network induced by an exchangeable sequence $\mathbf{X}_n = (X_i)_{1 \le i \le n}$, i.e., $\mathbf{Y}_n = \mathscr{E}_{\mathbf{X}_n}$. Then the distribution of $\mathbf{Y}_n$ is given by aggregating the probabilities of all sequences that produce the same edge-labeled graph through (9.3), i.e.,

$$\Pr(\mathbf{Y}_n = \mathscr{E}_{\mathbf{x}}) = \sum_{\mathbf{x}' \in \mathscr{E}_{\mathbf{x}}} \Pr(\mathbf{X}_n = \mathbf{x}').$$

Now, for any sequence $\mathbf{x} = (x_i)_{1 \le i \le n}$ (with each x_i consisting of a pair (s_i, t_i)) and any permutation $\sigma : [n] \to [n]$, let $\mathbf{x}^\sigma = (x_{\sigma(i)})_{1 \le i \le n}$ be the reordering of $\mathbf{x}$ according to σ. Exchangeability of $(X_i)_{1 \le i \le n}$ implies

$$\Pr((X_i)_{1 \le i \le n} = \mathbf{x}) = \Pr((X_i)_{1 \le i \le n} = \mathbf{x}^\sigma), \quad \mathbf{x} = (x_1, \ldots, x_n) \in \mathscr{P} \times \mathscr{P},$$

for all permutations $\sigma : [n] \to [n]$. Since reordering $\mathbf{X}_n$ by σ has the effect of relabeling the edges of the induced edge-labeled network $\mathbf{Y}_n$, it follows that

$$\begin{aligned}
\Pr(\mathbf{Y}_n = \mathscr{E}_{\mathbf{x}^\sigma}) &= \sum_{\mathbf{x}' \in \mathscr{E}_{\mathbf{x}^\sigma}} \Pr(\mathbf{X}_n = \mathbf{x}') \\
&= \sum_{\mathbf{x}' \in \mathscr{E}_{\mathbf{x}}} \Pr(\mathbf{X}_n = \mathbf{x}'^{\sigma^{-1}}) \\
&= \sum_{\mathbf{x}' \in \mathscr{E}_{\mathbf{x}}} \Pr(\mathbf{X}_n = \mathbf{x}') \\
&= \Pr(\mathbf{Y}_n = \mathscr{E}_{\mathbf{x}}).
\end{aligned}$$

This relation gives rise to *edge exchangeability*, whereby the distribution of a random edge-labeled graph is invariant under arbitrary relabeling of its edges, as illustrated

[1] I use the terms 'edge-labeled graph' and 'edge-labeled network' interchangeably throughout this chapter.

Figure 9.6 *Relabeling of two edge-labeled graphs. An edge exchangeable model assigns equal probability to both outcomes.*

in Figure 9.6. Interaction propensity processes make up an important family of edge exchangeable models that are analogous to graphons in the vertex exchangeable setting.

9.4 Interaction propensity processes

For a set $\mathscr{P}$, assume data is collected as a $(\mathscr{P} \times \mathscr{P})$-valued sequence $x_1, x_2, \ldots$, or equivalently as a function $\mathbf{x} : \mathbb{N} \to \mathscr{P} \times \mathscr{P}$, $i \mapsto x_i$. For any bijection $\rho : \mathscr{P} \to \mathscr{P}$, we write $\rho\mathbf{x}$ to denote the composition of $\mathbf{x}$ and the map $(a,b) \mapsto (\rho(a), \rho(b))$. Thus, $\rho\mathbf{x} : \mathbb{N} \to \mathscr{P} \times \mathscr{P}$ corresponds to the sequence $i \mapsto \rho(x_i)$ which 'renames' the vertices of $\mathbf{x}$ according to ρ. With this convention, the *edge-labeled network* induced by $\mathbf{x} : \mathbb{N} \to \mathscr{P} \times \mathscr{P}$ (as defined in (9.3) and illustrated in Figure 9.5) can be re-expressed as

$$\mathscr{E}_{\mathbf{x}} = \{\mathbf{x}' : \mathbb{N} \to \mathscr{P} \times \mathscr{P} \mid \rho\mathbf{x}' = \mathbf{x} \text{ for some bijection } \rho : \mathscr{P} \to \mathscr{P}\}, \tag{9.4}$$

i.e., the equivalence class of all edge sequences that are isomorphic up to their induced edge structure. For example, each of the vertex-edge labeled graphs on the left-hand side of Figure 9.5 corresponds to a sequence $\mathbf{x}'$ as follows. Referring to their relative position in Figure 9.5:

$$\begin{aligned}
\text{(Top Left)} \quad & \mathbf{x}'_1 = (a,b),(c,a),(d,e),(a,c) \\
\text{(Middle Left)} \quad & \mathbf{x}'_2 = (c,b),(e,c),(a,d),(c,e) \\
\text{(Bottom Left)} \quad & \mathbf{x}'_3 = (f,k),(a,f),(h,g),(f,a).
\end{aligned}$$

For population $\mathscr{P} = \{a,b,c,\ldots\}$, we define $\rho : \mathscr{P} \to \mathscr{P}$ by

$$\begin{aligned}
\rho : & \mathscr{P} \to \mathscr{P} \\
& a \mapsto c \\
& b \mapsto b \\
& c \mapsto e \\
& d \mapsto a \\
& e \mapsto d,
\end{aligned}$$

from which we obtain $\rho\mathbf{x}'_1 = \mathbf{x}'_2$, showing that both $\mathbf{x}'_1$ and $\mathbf{x}'_2$ are associated to the same edge-labeled graph according to (9.4). By a similar argument, we can show that $\mathbf{x}'_2$ and $\mathbf{x}'_3$ are equivalent in the sense of (9.4), as are $\mathbf{x}'_1$ and $\mathbf{x}'_3$. Together, the set of all such $\mathbf{x}'$ determines the 'shape' given by the edge-labeled graph on the right-hand side of Figure 9.5.

Exercise 9.1 *Show that the definitions of $\mathscr{E}_{\mathbf{x}}$ in* (9.3) *and* (9.4) *are equivalent.*

For $S \subseteq \mathbb{N}$, we write $\mathfrak{E}_S$ to denote the set of all edge-labeled networks with edges labeled in S. (Every $\mathbf{y} \in \mathfrak{E}_S$ is an edge-labeled network with each edge labeled uniquely by an element in S. For example, the structure $\mathscr{E}_{\mathbf{x}}$ corresponding to (9.4) is labeled by $\mathbb{N}$.) Notice that the set labeling the population $\mathscr{P}$ plays a diminished role in $\mathscr{E}_{\mathbf{x}}$ because it is quotiented out by the equivalences ρ. Without loss of generality, we assume $\mathscr{P} = \mathbb{N}$ wherever it appears.

For any permutation $\sigma : \mathbb{N} \to \mathbb{N}$, we write $\mathbf{x}^\sigma : \mathbb{N} \to \mathscr{P} \times \mathscr{P}$ as the reordering of $\mathbf{x}$ according to σ, so that $\mathbf{x}^\sigma(i) = \mathbf{x}(\sigma(i))$. To avoid confusion, it is important to distinguish the notations:

$$\begin{aligned} \rho\mathbf{x} &: \text{vertices renamed according to bijection } \rho \\ \mathbf{x}^\sigma &: \text{edges relabeled according to permutation } \sigma. \end{aligned}$$

The composition by ρ on the left in $\rho\mathbf{x}$ *renames* the *vertices* according to ρ. The superscript σ in $\mathbf{x}^\sigma$ *relabels* the *edges* by σ. An edge-labeled graph $\mathscr{E}_{\mathbf{x}}$ is obtained in (9.4) by aggregating all $\mathbf{x}'$ that are equivalent to $\mathbf{x}$ up to renaming by some ρ. An edge exchangeable model (Definition 9.1 below) is defined as a distributional invariance with respect to relabeling edges by arbitrary σ.

As defined here, every edge-labeled graph $\mathbf{y} \in \mathfrak{E}_{\mathbb{N}}$ corresponds to an equivalence class as in (9.4) for some $\mathbf{x} : \mathbb{N} \to \mathscr{P} \times \mathscr{P}$. For any edge-labeled graph $\mathbf{y} \in \mathfrak{E}_{\mathbb{N}}$, let $\mathbf{y}^\sigma$ be the structure obtained by relabeling the edges of $\mathbf{y}$ according to σ. Since $\mathbf{y}$ corresponds to an equivalence class, we formally define $\mathbf{y}^\sigma$ by taking any $\mathbf{x}$ such that $\mathbf{y} = \mathscr{E}_{\mathbf{x}}$ and putting $\mathbf{y}^\sigma = \mathscr{E}_{\mathbf{x}^\sigma}$, where $\mathbf{x}^\sigma = (x_{\sigma(i)})_{i\geq 1}$ is as defined above. A random edge-labeled graph $\mathbf{Y}$ is *edge exchangeable* if its distribution is invariant under this relabeling operation.

Exercise 9.2 *Verify that the above definition of* $\mathbf{y}^\sigma$ *does not depend on the choice of representative* $\mathbf{x}$ *for* $\mathbf{y}$.

Definition 9.1 (Edge exchangeability [54]) *A random edge-labeled graph* $\mathbf{Y}$ *on* $\mathfrak{E}_S$ *is* edge exchangeable *if* $\mathbf{Y}^\sigma =_{\mathscr{D}} \mathbf{Y}$ *for all permutations* $\sigma : S \to S$.

For countable set $\mathscr{P}$, we define the $(\mathscr{P} \times \mathscr{P})$*-simplex* as the set of all probability distributions on $\mathscr{P} \times \mathscr{P}$, i.e.,

$$\mathscr{F}_{\mathscr{P}\times\mathscr{P}} = \left\{ (f_{(s,t)})_{s,t\in\mathscr{P}\times\mathscr{P}} : f_{(s,t)} \geq 0 \quad \text{and} \quad \sum_{s,t\in\mathscr{P}} f_{(s,t)} = 1 \right\}. \tag{9.5}$$

Given any $f \in \mathscr{F}_{\mathscr{P}\times\mathscr{P}}$, we define ε_f as the probability distribution of a random edge-labeled graph $\mathbf{Y}$ induced by an i.i.d. sequence $\mathbf{X} = (X_1, X_2, \ldots)$, with each X_i distributed as

$$\Pr(X_i = (s,t); f) = f_{(s,t)}, \quad (s,t) \in \mathscr{P} \times \mathscr{P}. \tag{9.6}$$

In particular, $\mathbf{Y} \sim \varepsilon_f$ is a random edge-labeled graph obtained by putting $\mathbf{Y} = \mathscr{E}_{\mathbf{X}}$, for $\mathscr{E}_{\mathbf{X}}$ defined in (9.4) and $\mathbf{X}$ i.i.d. according to (9.6). We call $\mathbf{Y} \sim \varepsilon_f$ the *interaction propensity process directed by* f. The following is immediate by the construction.

Theorem 9.1 *For any* $f \in \mathscr{F}_{\mathscr{P} \times \mathscr{P}}$, $\mathbf{Y} \sim \varepsilon_f$ *is edge exchangeable.*

Exercise 9.3 *Prove Theorem 9.1.*

In the next section we see that interaction propensity processes play an analogous role for edge exchangeable models as the ϕ-process does for vertex exchangeable models. In particular, the interaction propensity processes comprise the ergodic distributions for edge exchangeable networks labeled by $\mathbb{N}$. This analogy should not, however, suggest that interaction propensity processes exhibit the same behavior as ϕ-processes. There are a couple of important differences.

1. Interaction propensity processes allow for the occurrence of multiple edges. In fact, multiple edges between two vertices will occur with probability 1 in any large enough sample of edges from (9.6). This is a consequence of the strong law of large numbers for i.i.d. sequences: assuming $f_{(s,t)} > 0$ for some $(s,t) \in \mathscr{P} \times \mathscr{P}$, the limiting relative frequency of (s,t) in $X_1, X_2, \ldots$ satisfies

   $$n^{-1} \sum_{i=1}^{n} \mathbf{1}(X_i = (s,t)) \approx f_{(s,t)} > 0 \quad \text{for large } n \text{ with probability 1.}$$

   In particular, the number of occurrences of (s,t) satisfies $\sum_{i=1}^{n} \mathbf{1}(X_i = (s,t)) \approx n f_{(s,t)} \to \infty$ as $n \to \infty$. So whereas each edge appears at most once in a vertex exchangeable network, each edge appears either 0 or infinitely many times, with probability 1, in an outcome of the interaction propensity process.

2. New vertices appear in $\mathscr{E}_{\mathbf{X}} \sim \varepsilon_f$ in size-biased order according to their overall frequency of occurrence in the network. Specifically, for any $s \in \mathscr{P}$, the probability that s is the sender of a given call equals $\sum_{t \in \mathscr{P}} f_{(s,t)}$, and the probability that s is the receiver of a given call equals $\sum_{t \in \mathscr{P}} f_{(t,s)}$. Thus, although the identities of the vertices are disregarded in the edge-labeled representation $\mathscr{E}_{\mathbf{X}}$, whichever vertices have been observed are known to be atypical (in the sense of being a size-biased pick from $\mathscr{P}$).

In the context of Section 9.1, uniform sampling of calls makes it more likely to observe callers who appear more often in the database. For example, a caller who appears 2,000 times in the database is twice as likely to be chosen as another caller who only appears 1,000 times, explaining why the observed vertices in edge exchangeable networks are a size-biased sample of the population of all vertices. This last point demonstrates why vertex exchangeability is incompatible with the way in which the data in Section 9.1 and other interaction networks are typically observed. Whereas vertex exchangeability treats observed vertices as representative of the population of all vertices, edge exchangeability treats the observed edges as representative of the population of all edges. And when the observed edges are representative

Figure 9.7 *Edge-labeled network obtained from sequence* $X_1 = (0,0)$, $X_2 = (0,0)$, $X_3 = (0,0)$, $X_4 = (0,0)$, ... *in the modified interaction propensity process construction of Section 9.5.*

of the population of edges, the observed vertices are not representative of the population of vertices: they are instead a size-biased sample. This observation is made formal with the following characterization of edge exchangeable random graphs.

9.5 Characterizing edge exchangeable random graphs

The Aldous–Hoover theorem (Theorem 6.3) states that the distribution of any vertex exchangeable random graph with infinitely many vertices can be expressed as a mixture of graphon processes. Except for a minor technical modification, interaction propensity processes stand in relation to edge exchangeable random graphs in the same way that graphon models stand in relation to vertex exchangeable random graphs. The need for this technical modification (described below) arises because the vertices in an edge-labeled graph cannot be identified independently of the edges in which they participate. The identity of each vertex is determined by how it relates to other vertices through the observed edge relations. (Contrast this with a vertex-labeled graph, in which each vertex can be identified by reference to its label $1, 2, \ldots$ without any mention of its edge relations to other vertices.)

This observation features into the forthcoming characterization of edge exchangeable network models because ordinarily when sampling an infinite i.i.d. sequence from a countable set S, each element $s \in S$ will appear either 0 times or infinitely many times with probability 1 by the strong law of large numbers. But in edge exchangeable networks, the countable set of edges $S \times S$ (as pairs of vertices) become de-identified once the network is represented as the edge-labeled structure associated to the equivalence class in (9.4). This additional step allows for the infinite occurrence of certain edge *types*, which manifest themselves in the edge-labeled representation through the singular occurrence of vertices called *blips*. Thus, in an edge exchangeable graph, vertices appear 1 or infinitely many times with probability 1, and the following modification is intended to address this possibility.[2]

To set the stage, consider the edge-labeled graph in Figure 9.7, which has infinitely many loops at otherwise isolated vertices. Since relabeling the edges by any permutation changes nothing, a distribution which assigns probability 1 to the outcome in Figure 9.7 is edge exchangeable, but such a distribution cannot be expressed in terms of the interaction propensity process of Section 9.4 because each of the

[2]Notice that I do not write that a vertex can appear 0 times because in the edge-labeled representation such vertices cannot be assumed to 'exist' independently of any edge in which they appear. This philosophical point should not distract from the main results of this section.

edges would have to be encoded by a loop with positive probability assignment (i.e., a pair (s,s) for some s with $f_{(s,s)} > 0$); and by the strong law of large numbers any such loop appears infinitely often with probability 1.

Exercise 9.4 *Prove or explain why the edge exchangeable distribution that assigns probability 1 to the network consisting of infinitely many isolated loops (as in Figure 9.7) cannot be described by the interaction propensity process in Section 9.4.*

Evidently, there are edge exchangeable distributions on $\mathfrak{E}_{\mathbb{N}}$ which cannot be described by the interaction propensity processes of the preceding section. We account for these additional cases by putting $\mathscr{P} = \mathbb{N}$ and defining the *ranked* $\mathbb{N} \times \mathbb{N}$*-simplex* as

$$\mathscr{F}^{\cong}_{\mathbb{N}\times\mathbb{N}} = \mathscr{F}_{\mathbb{N}_{\geq -1}\times\mathbb{N}_{\geq -1}} / \cong, \tag{9.7}$$

where

$$\mathscr{F}_{\mathbb{N}_{\geq -1}\times\mathbb{N}_{\geq -1}} = \{(f_{(i,j)})_{i,j\geq -1} : f_{(i,j)} \geq 0 \text{ and } \sum_{i,j\geq -1} f_{(i,j)} = 1\}$$

and $\cong$ is an equivalence relation on $\mathscr{F}_{\mathbb{N}_{\geq -1}\times\mathbb{N}_{\geq -1}}$ defined below. We can think of $\mathscr{F}_{\mathbb{N}_{\geq -1}\times\mathbb{N}_{\geq -1}}$ in the same way that we thought of $\mathscr{F}_{\mathscr{P}\times\mathscr{P}}$ in the interaction propensity process—for any $f \in \mathscr{F}_{\mathbb{N}_{\geq -1}\times\mathbb{N}_{\geq -1}}$ and $i,j \geq -1$, $f_{(i,j)}$ is the probability of observing an edge from i to j—but with the understanding that $\mathscr{F}_{\mathbb{N}_{\geq -1}\times\mathbb{N}_{\geq -1}}$ is augmented by the additional possibilities of $f_{(0,i)}$, $f_{(i,0)}$, $f_{(-1,i)}$, and $f_{(i,-1)}$ for $i \geq 1$, and $f_{(0,0)}$, $f_{(-1,-1)}$, $f_{(0,-1)}$, and $f_{(-1,0)}$. In this modified presentation, the labels 0 and -1 serve as placeholders for special edge 'types', such as the isolated loops in Figure 9.7. In general, when a 0 or -1 occurs in the interaction propensity process associated to $f \in \mathscr{F}_{\mathbb{N}_{\geq -1}\times\mathbb{N}_{\geq -1}}$, the edge-labeled graph representation maps those elements to a new vertex which has never appeared before and will never appear again in the network. This is explained further in (I)–(III) and Figure 9.8 below.

Given $f = (f_{(i,j)})_{i,j\geq -1} \in \mathscr{F}_{\mathbb{N}_{\geq -1}\times\mathbb{N}_{\geq -1}}$ we define ε_f as the distribution of $\mathbf{Y}$ obtained by first taking $\mathbf{X} = (X_1, X_2, \ldots)$ i.i.d. from f as in (9.6), next converting $\mathbf{X}$ into a new sequence $\mathbf{X}^* = (X_1^*, X_2^*, \ldots)$ by relabeling each instance of 0 and -1 with a unique non-positive integer, and finally defining $\mathbf{Y} = \mathscr{E}_{\mathbf{X}^*}$, for $\mathscr{E}_{\mathbf{X}^*}$ just as in (9.4). In the following special case the mapping $\mathbf{X} \mapsto \mathbf{X}^*$ gives

$$\begin{array}{llllll}
\mathbf{X}: & X_1 = (3,1), & X_2 = (1,0), & X_3 = (2,3), & X_4 = (0,-1), & X_5 = (0,2) \\
\downarrow & \downarrow & \downarrow & \downarrow & \downarrow & \downarrow \\
\mathbf{X}^*: & X_1^* = (3,1), & X_2^* = (1,0), & X_3^* = (2,3), & X_4^* = (-1,-2), & X_5^* = (-3,2).
\end{array} \tag{9.8}$$

Notice that each occurrence of a positive integer $1,2,3$ is unchanged in passing from $\mathbf{X}$ to $\mathbf{X}^*$, but each occurrence of a non-positive integer 0 or -1 is replaced by the largest non-positive integer which has not previously appeared in the updated sequence. It is through this process that the vertices corresponding to 0 and -1 become isolated in the edge-labeled graph determined by $X_1^*, \ldots, X_5^*$. Since it is possible that different $f, f' \in \mathscr{F}_{\mathbb{N}_{\geq -1}\times\mathbb{N}_{\geq -1}}$ produce the same distribution on $\mathfrak{E}_{\mathbb{N}}$, we define the equivalence relation $f \cong f'$ to mean that $\varepsilon_f = \varepsilon_{f'}$ under the above operation. (Specifically, $f \cong f'$ indicates that for $\mathbf{X}$ i.i.d. according to $\Pr(X_i = \cdot; f)$ in (9.6) and $\mathbf{X}'$

i.i.d. according to $\Pr(X_i' = \cdot; f')$ in (9.6) satisfy $\mathscr{E}_{\mathbf{X}^*} =_{\mathscr{D}} \mathscr{E}_{\mathbf{X}'^*}$.) We write $\mathscr{F}_{\mathbb{N}\times\mathbb{N}}^{\cong}$ for the set of all equivalence classes of $\mathscr{F}_{\mathbb{N}_{\geq -1}\times\mathbb{N}_{\geq -1}}$ under $\cong$, expressed symbolically as $\mathscr{F}_{\mathbb{N}\times\mathbb{N}}^{\cong} = \mathscr{F}_{\mathbb{N}_{\geq -1}\times\mathbb{N}_{\geq -1}}/\cong$ in (9.7).

The constituents of any $f \in \mathscr{F}_{\mathbb{N}\times\mathbb{N}}^{\cong}$ play the following roles:

(I) $i, j \geq 1$: since positive labels are unaffected by the operation $\mathbf{X} \mapsto \mathbf{X}^*$, each $f_{(i,j)}$, $i, j \geq 1$, is the probability that a particular interaction (i, j) occurs in sequence $\mathbf{X}$. Each occurrence of such an interaction is associated to a distinct edge between the corresponding vertices in $\mathscr{E}_{\mathbf{X}^*}$.

(II) $i \geq 1$, $j \in \{0, -1\}$ or $i \in \{0, -1\}$, $j \geq 1$: each occurrence of $(i, 0)$ in $\mathbf{X}$ will be replaced in $\mathbf{X}^*$ by (i, z) for a unique non-positive label $z = 0, -1, \ldots$. For example, $f \in \mathscr{F}_{\mathbb{N}\times\mathbb{N}}^{\cong}$ corresponding to $f_{(i,0)} = 1$ gives the sequence

$$\begin{array}{lllll} \mathbf{X}: & (i,0), & (i,0), & (i,0), & \ldots \\ \downarrow & \downarrow & \downarrow & \downarrow & \ldots \\ \mathbf{X}^*: & (i,0), & (i,-1), & (i,-2), & \ldots \end{array}$$

with probability 1. In this way, each instance of $(i, 0)$ or $(i, -1)$ corresponds to an edge from i to a unique vertex, i.e., a 'blip' which appears only once in $\mathbf{X}^*$ and thus only once in $\mathscr{E}_{\mathbf{X}^*}$. If $f_{(i,0)} > 0$, then $(i, 0)$ will appear infinitely often in $\mathbf{X}$ with probability 1, meaning that in $\mathscr{E}_{\mathbf{X}^*}$ there will be an infinite, isolated 'star' emerging from the vertex corresponding to i. (The same description holds for $(0, j)$ and $(-1, j)$, with the only change being that each occurrence of $(0, j)$ and $(-1, j)$ produces an edge from an isolated vertex toward the vertex corresponding to j.)

(III)(a) $i = j = 0$ or $i = j = -1$: since each non-positive entry is replaced by a unique non-positive entry in $\mathbf{X}^*$, each occurrence of $(0, 0)$ or $(-1, -1)$ produces an isolated loop at an otherwise isolated vertex in $\mathscr{E}_{\mathbf{X}^*}$. For example, $f \in \mathscr{F}_{\mathbb{N}\times\mathbb{N}}^{\cong}$ corresponding to $f_{(0,0)} = 1$ produces the sequence

$$\begin{array}{lllll} \mathbf{X}: & (0,0), & (0,0), & (0,0), & \ldots \\ \downarrow & \downarrow & \downarrow & \downarrow & \ldots \\ \mathbf{X}^*: & (0,0), & (-1,-1), & (-2,-2), & \ldots \end{array}$$

with probability 1. Its network representation $\mathscr{E}_{\mathbf{X}^*}$ is shown in Figure 9.7.

(III)(b) $(i, j) = (0, -1)$ or $(i, j) = (-1, 0)$: since both labels are non-positive, each occurrence of $(0, -1)$ corresponds to an edge between two distinct, otherwise isolated vertices in $\mathscr{E}_{\mathbf{X}^*}$. In particular, each occurrence of $(0, -1)$ is replaced by (z, z') for $z \neq z'$ and z, z' not appearing anywhere else in the sequence. For example, $f \in \mathscr{F}_{\mathbb{N}\times\mathbb{N}}^{\cong}$ corresponding to $f_{(0,-1)} = 1$ produces

$$\begin{array}{lllll} \mathbf{X}: & (0,-1), & (0,-1), & (0,-1), & \ldots \\ \downarrow & \downarrow & \downarrow & \downarrow & \ldots \\ \mathbf{X}^*: & (0,-1), & (-2,-3), & (-4,-5), & \ldots \end{array}$$

with probability 1.

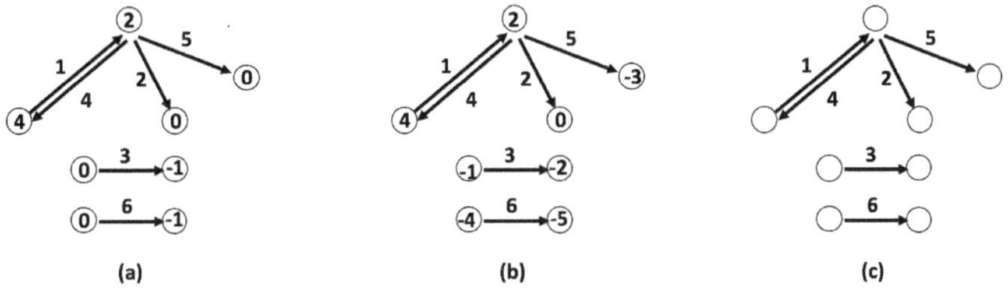

Figure 9.8 *Illustration of different edge types (I)–(III) from the construction in Section 9.5 based on sequence* $X_1 = (4,2)$, $X_2 = (2,0)$, $X_3 = (0,-1)$, $X_4 = (2,4)$, $X_5 = (2,0)$, $X_6 = (0,-1)$. *(a) The vertex-edge labeled graph structure with original labels* $X_1,\ldots,X_6$. *(b) The vertex-edge labeled network structure in (a) with vertex labels transformed according to the operation* $\mathbf{X} \mapsto \mathbf{X}^*$ *in* (9.8). *(c) The edge-labeled graph obtained by removing vertex labels from the vertex-edge labeled graph in (b). In the sequence,* X_1 *and* X_4 *are of type (I),* X_2 *and* X_5 *are of type (II),* X_3 *and* X_6 *are of type (III)(b). There are no edges of type (III)(a) in this example. See Figure 9.7 for a network with edges of type (III)(a).*

To better understand why cases (II) and (III) must be treated differently from case (I) by the passage from $\mathbf{X}$ to $\mathbf{X}^*$, consider the extreme case of $f \in \mathscr{F}^{\cong}_{\mathbb{N}\times\mathbb{N}}$ corresponding to $f_{(0,0)} = 1$, so that $\mathbf{X}$ i.i.d. from (9.6) has $X_i = (0,0)$ for all $i \geq 1$ with probability 1. The edge-labeled network resulting from this process is shown in Figure 9.7. The distribution of $\mathscr{E}_{\mathbf{X}^*}$ obtained in this way is clearly edge exchangeable, but this distribution cannot be represented as one of the interaction propensity processes in Section 9.4; see Exercise 9.4 above.

The same argument that applies in Exercise 9.4 also applies to networks comprised entirely of edge types (II) and (III)(b). In particular, for an edge to appear in the interaction propensity process, there must be some (s,t), $s,t \geq 1$, for which $f_{(s,t)} > 0$. But if $f_{(s,t)} > 0$, then (s,t) will appear infinitely often in $\mathbf{X}$ with probability 1 by the strong law of large numbers. Thus, if an edge appears once in $\mathscr{E}_{\mathbf{X}}$ it must appear infinitely many times with probability 1, meaning that the isolated loops of Figure 9.7 or edges of type (II) and (III)(b) cannot occur in a realization of the interaction propensity process from Section 9.4. Thus, exchangeability implies that edges of *type* (II) and (III) will occur infinitely often (if they occur at all), but each *instance* of that type occurs only once.

Theorem 9.2 (Crane–Dempsey [54]) *Let* $\mathbf{Y}$ *be an edge-exchangeable random graph in* $\mathfrak{E}_{\mathbb{N}}$. *Then there exists a unique probability measure* φ *on* $\mathscr{F}^{\cong}_{\mathbb{N}\times\mathbb{N}}$ *such that* $\mathbf{Y} =_{\mathscr{D}} \mathscr{E}_{\mathbf{X}^*}$, *for* $\mathbf{X}^*$ *obtained by first taking* $f \sim \varphi$ *and, given* f, *generating* $\mathbf{X}$ *i.i.d. as in* (9.6) *and applying the operation* $\mathbf{X} \mapsto \mathbf{X}^*$ *in* (9.8). *We write* $\mathbf{Y} \sim \varepsilon_{\varphi}$ *to denote the distribution of* $\mathbf{Y}$.

Although the representation in Theorem 9.2 could, on its own, be useful for modeling edge exchangeable networks using ideas from Bayesian nonparametrics, it offers an immediate conceptual insight into the sampling behavior of edge exchangeable networks. For each $i \geq 1$, let $f_i^{\bullet} = \sum_{j \geq 0} f_{(i,j)} + f_{(j,i)}$ be the sum of the weights over all edges incident to i. Then $f_i^{\bullet}$ is the probability that vertex i is contained in any given edge $X_1, X_2, \ldots$, leading to the following observation.

We call a vertex in $\mathbf{y} \in \mathfrak{E}_{\mathbb{N}}$ *recurrent* if it appears in more than one edge of $\mathbf{y}$.

Theorem 9.3 ([54]) *Recurrent vertices in an infinite edge exchangeable graph appear in size-biased order according to their expected relative degrees* $f_i^{\bullet}$.

Theorem 9.3 shows that the behavior of vertices in edge exchangeable networks, in which vertices arrive in size-biased order, is incompatible with the behavior of vertices in vertex exchangeable networks, in which the vertices can be interpreted as arriving in exchangeable random order. This theorem thus clarifies why vertex exchangeability is often expressly violated, and therefore untenable, in applications for which the sampling scheme depends on the network, as in the edge sampling scheme of Section 9.1 and a number of others from Section 3.6. Finally, I note that the breakdown of edge types in (I)–(III) above, and the analogous characterization for graphex models in Section 7.3, seems to reflect a universal property of exchangeable processes. The interested reader can consult [9, 16, 98] for a general introduction to exchangeable processes and further discussion of this phenomenon, another instance of which arises in the characterization of jump types in exchangeable graph-valued Markov processes [48, 49]. I discuss these latter processes in Chapter 11.

9.6 Vertex components models

The vertex components model is a subclass of interaction propensity processes whose propensities $f_{(s,t)}$ are expressed as the product of vertex-specific propensities as follows. Let

$$\Delta_1 = \left\{ (f_1, f_2, \ldots) : f_i \geq 0 \text{ and } \sum_{i \geq 1} f_i = 1 \right\}$$

denote the infinite simplex and let φ be any joint probability distribution on $\Delta_1 \times \Delta_1$. In this way, $(f^{\text{out}}, f^{\text{in}}) = ((f_i^{\text{out}}, f_i^{\text{in}}))_{i \geq 1} \sim \varphi$ describes the relative frequency with which each vertex makes outgoing calls f_i^{out} and receives incoming calls f_i^{in}. From $(f^{\text{out}}, f^{\text{in}})$, we define $f \in \mathscr{F}^{\cong}_{\mathbb{N} \times \mathbb{N}}$ by first reordering f^{out} in decreasing order so that $f_1^{\text{out}} \geq f_2^{\text{out}} \geq \cdots$, then relabeling f^{in} consistently with f^{out}, and finally putting

$$f_{(i,j)} = f_i^{\text{out}} f_j^{\text{in}}, \quad i, j \geq 1.^3 \tag{9.9}$$

(We relabel $(f^{\text{out}}, f^{\text{in}})$ for the purpose of identifiability in the induced distributions. Our choice to reorder f^{out} instead of f^{in} is a matter of convention and does not affect the distribution ε_f determined by f.)

Interaction propensity processes associated to $f = (f_{(i,j)})_{i,j \geq 1}$ defined in (9.9)

[3] If for $(f_i^{\text{out}}, f_i^{\text{in}})$ and $(f_{i'}^{\text{out}}, f_{i'}^{\text{in}})$ there is a tie ($f_i^{\text{out}} = f_{i'}^{\text{out}}$), then when reordering we assign the smaller label to whichever has the larger second component. For example, if $(f_i^{\text{out}}, f_i^{\text{in}})$ and $(f_{i'}^{\text{out}}, f_{i'}^{\text{in}})$ are to appear as the n and $n+1$ elements of the reordered sequence, then we assign label n to $(f_i^{\text{out}}, f_i^{\text{in}})$ if $f_i^{\text{in}} \geq f_{i'}^{\text{in}}$. If both $f_i^{\text{out}} = f_{i'}^{\text{out}}$ and $f_i^{\text{in}} = f_{i'}^{\text{in}}$ then the labels n and $n+1$ can be assigned arbitrarily to either.

offer a simple, if not simplistic, description of network formation. Conditional on the random out- and in-degree propensities $(f^{\text{out}}, f^{\text{in}})$, each edge in $\mathbf{Y} \sim \varepsilon_f$ is an independent choice of vertices, one from f^{out} and one from f^{in}. For modeling draws from a phone call database, for example, the vertex components model assumes that each observed call behaves as if the sender and receiver were chosen conditionally independently from f^{out} and f^{in}, respectively. This assumption is very likely violated in the scenario of Section 9.1 and in most conceivable networks applications involving interaction processes. Incorporating dependence between caller and receiver in a meaningful and tractable way is a worthwhile open problem with potential implications beyond the specific scenario of Section 9.1.

Research Problem 9.1 *Identify a natural parametric class of edge exchangeable models, akin to the Hollywood model in Section 9.7 below, which incorporates dependence between vertices on either end of an edge. For example, construct the interaction propensities* $(f_{(i,j)})_{i,j\geq 1}$ *from a collection of vertex components* $f = (f_i)_{i\geq 1}$ *and conditional vertex components* $f_{\cdot|i} = (f_{j|i})_{j\geq 1}$ *for every* $i \geq 1$ *so that*

$$f_{(i,j)} = f_i f_{j|i}, \quad i,j \geq 1.$$

What properties does this model exhibit? If possible, compare its empirical properties to those of the Hollywood model.

9.6.1 *Stick-breaking constructions for vertex components*

Since in general there need not be an upper bound on the total number of vertices that can appear in a network, the vertex components f^{out} and f^{in} may be infinite in length. (Recall from Section 6.4 that allowing for the possibility that the population network has countably many vertices does not imply that the population is assumed to be infinite, but rather that the population could be any finite size.) Since an infinite length vector $(f^{\text{out}}, f^{\text{in}})$ cannot be stored in a computer, the potentially infinite population size poses practical issues for fitting such models.

Stick-breaking representations for random elements of Δ_1, see, e.g., [94, 137], offer a computationally tractable way to simulate from and estimate certain vertex components models by constructing the sequences of interactions and the vertex components simultaneously. In a nutshell, although the number of vertices in the population is unbounded, only a finite number can appear in any sample of finitely many edges. The behavior of any such finite sample can be described using only the propensities of the (finitely many) sampled vertices.

Since the vertex labels in any realization of the interaction propensity process are forgotten by the induced edge-labeled graph $\mathscr{E}_{\mathbf{X}^*}$, we can reconstruct a representative of $\mathscr{E}_{\mathbf{X}^*}$, i.e., a member of its equivalence class (9.4), by labeling vertices in the order that they arrive. And since the propensity of a vertex is irrelevant until it appears in the network, the propensity of the vertex labeled i need not be generated until the ith vertex appears. We describe this protocol by letting $\{g_i\}_{i\geq 1}$ be a collection of probability densities on $[0,1]$, putting $S_1 = 1$, and sampling $W_1 \sim g_1$. We then continue inductively for each $n = 1,2,\ldots$ as follows. As above let $\mathbf{X}_n$ be the collection of all pairs $X_1 = (S_1, T_1), \ldots, X_n = (S_n, T_n)$ chosen up to and including stage n. Given $\mathbf{X}_n$ and $W_1, \ldots, W_{V_n}$, where $V_n = \max(\mathbf{X}_n)$ is the largest vertex label appearing in $\mathbf{X}_n$, choose S_{n+1} according to

$$\Pr(S_{n+1} = r \mid \mathbf{X}_n, W_1, \ldots, W_{V_n}) = \begin{cases} W_r, & r = 1, \ldots, V_n, \\ 1 - \sum_{j=1}^{V_n} W_j, & r = V_n + 1. \end{cases} \tag{9.10}$$

If $S_{n+1} = V_n + 1$, then choose $W_{V_n+1} \sim g_{V_n+1}(\cdot/(1 - \sum_{j=1}^{V_n} W_j))$. (In the 'stick-breaking' interpretation, $W_1, W_2, \ldots$ are the sizes broken off of a stick of unit length. The division by $1 - \sum_{j=1}^{V_n} W_j$ is the normalization by the length of the 'stick' from which the $(n+1)$st piece is to be broken off.) The process continues by drawing T_{n+1} given $S_1, T_1, \ldots, S_n, T_n, S_{n+1}$ as in (9.10), with V_n updated according to whether or not S_{n+1} was chosen as a previously unseen vertex or not.

In words, the construction in (9.10) chooses the next vertex to be one that already exists with probability W_r, for $r = 1, \ldots, V_n$. Otherwise, with probability $1 - \sum_{r=1}^{V_n} W_r$, a new (previously unseen) vertex is assigned the next available label $V_n + 1$. When a new vertex is chosen, it is assigned a random weight W_{V_n+1} as a fraction of the remaining mass $1 - \sum_{j=1}^{V_n} W_j$ drawn according to density g_{V_n+1}.

From the sequence $(S_1, T_1), (S_2, T_2), \ldots$ constructed above, $\mathbf{X}_n : [n] \to \mathbb{N} \times \mathbb{N}$ is the interaction process defined by $\mathbf{X}_n(i) = (S_i, T_i)$, $i = 1, \ldots, n$, and $\mathscr{E}_{\mathbf{X}_n}$ is the edge-labeled network induced by $\mathbf{X}_n$ for each $n \in \mathbb{N}$, as in (9.4). As computed in [54], the joint density of $\mathscr{E}_{\mathbf{X}_n}$ and W is given by

$$\Pr(\mathscr{E}_{\mathbf{X}_n} = \mathbf{y}, (W_1, \ldots, W_{v(\mathbf{y})}) \in (dw_i)_{1 \le i \le v(\mathbf{y})}; \{g_i\}_{i \ge 1}) = \tag{9.11}$$

$$= \prod_{j=1}^{v(\mathbf{y})} \left(1 - \sum_{i=1}^{j-1} w_i\right) g_j \left(\frac{w_j}{1 - \sum_{i=1}^{j-1} w_i}\right) w_j^{D_n(j)-1} dw_1 \cdots dw_{v(\mathbf{y})},$$

where $D_n(j)$ is the number of times the vertex with weight w_j appears in $\mathbf{y}$ and $\sum_{i=1}^{0} w_i = 0$ by convention. The distribution of $\mathscr{E}_{\mathbf{X}_n}$ can be recovered by integrating over the vertex components W_i in (9.11).

Research Problem 9.2 *The above stick-breaking formulation of edge exchangeable models invites further study. Just about any question, statistical, mathematical, or computational, about these models is currently open. A special case of this construction is described in the next section. Before tackling this problem, the reader should be aware of the extensive work on stick-breaking constructions in the Bayesian nonparametrics literature; for an overview, see Ishwaran and James [94] and followup work by other authors.*

9.7 Hollywood model

The Hollywood model is the canonical two-parameter family of edge exchangeable network models. It corresponds to the vertex components model in (9.11) with W generated from the Poisson–Dirichlet distribution [46, 72, 73, 130]. The definition given below first invokes the Poisson–Dirichlet distribution, but the model's alternative description by the Hollywood process (Section 9.7.1) is more easily digested.

The reader interested in learning more about the Poisson–Dirichlet distribution and its many appearances in probability, statistics, and science is referred to [46].

Remark 9.2 *The upcoming description only allows edges with exactly two vertices, as in our running phone call example. In the intended semantics of the Hollywood model, as conceived in [54], each edge in $\mathbf{Y}_n$ corresponds to a movie whose actors are represented by the vertices incident to that edge. To bring this interpretation to fruition, we must extend the model below to allow for edges of arbitrary finite size. I discuss the extension to arbitrary edge sizes in Chapter 10.*

Let φ be the distribution defined on the diagonal of $\Delta_1 \times \Delta_1$ by taking $f \sim \mathrm{PD}(\alpha, \theta)$ and putting $(f^{\mathrm{out}}, f^{\mathrm{in}}) = (f, f)$, where $\mathrm{PD}(\alpha, \theta)$ denotes the *Poisson–Dirichlet distribution* with parameter (α, θ) satisfying either

- (finite population) $\alpha < 0$ and $\theta = -\kappa\alpha$ for some positive integer $\kappa = 1, 2, \ldots$ or
- (infinite population) $0 \leq \alpha \leq 1$ and $\theta > -\alpha$.

The significance of these parameters will become more apparent in Sections 9.7.1–9.7.2. Until then, we observe that the split parameter space of the Hollywood model accounts for both bounded and unbounded population sizes, with the region $0 \leq \alpha \leq 1$ and $\theta > -\alpha$ giving rise to sequences of edge-labeled networks $(\mathbf{Y}_n)_{n\geq 1}$ for which $v(\mathbf{Y}_n) \to \infty$ almost surely (a.s.) as $n \to \infty$, and the region $\alpha < 0$ and $\theta = -\kappa\alpha$ for some positive integer $\kappa = 1, 2, \ldots$ giving rise to sequences of edge-labeled networks $(\mathbf{Y}_n)_{n\geq 1}$ for which $v(\mathbf{Y}_n) \to \kappa$ a.s. as $n \to \infty$.

In the infinite population regime, the Hollywood model is a special case of the vertex components model with $W = (W_i)_{i\geq 1}$ chosen from the Griffiths–Engen–McCloskey (GEM) distribution with parameter (α, θ) on Δ_1. The GEM distribution with parameter (α, θ) is the distribution of $(W_1, W_2, \ldots)$ obtained by a size-biased reordering of $(W_1', W_2', \ldots)$ from the Poisson–Dirichlet distribution with parameter (α, θ). See [72] for further details on the GEM distribution and its relationship to the Poisson–Dirichlet distribution.

In the finite population regime, the Hollywood model corresponds to the vertex components model with $W = (W_1, \ldots, W_\kappa)$ chosen from the symmetric Dirichlet distribution with parameter $(\alpha, \ldots, \alpha)$ on the $(\kappa - 1)$-dimensional simplex

$$\Delta_{\kappa-1} = \{(s_1, \ldots, s_\kappa) : 0 \leq s_i \leq 1, s_1 + \cdots + s_\kappa = 1\}.$$

The symmetric Dirichlet distribution has density

$$f(w_1, \ldots, w_\kappa; \alpha) = \frac{\Gamma(\kappa\alpha)}{\Gamma(\alpha)^\kappa} \prod_{i=1}^{\kappa} w_i^{\alpha-1} dw_i, \quad (w_1, \ldots, w_\kappa) \in \Delta_{\kappa-1},$$

where $\Gamma(\cdot)$ is the gamma function.

An important practical feature of edge exchangeability is its ability to account for the empirical properties of sparsity and power law degree distribution, as can be seen through the connection between the Hollywood process and the two-parameter Chinese restaurant process, discussed in the next section. I now present a more explicit sequential construction of the Hollywood model according to the (binary) Hollywood process.

9.7.1 *The Hollywood process*

Let (α,θ) be in the parameter space of the Hollywood model and initiate $\mathbf{Y}_0$ as the empty edge-labeled graph with 0 edges and 0 vertices. We construct a sequence of edge-labeled networks $(\mathbf{Y}_n)_{n\geq 1}$ by sampling edges $\mathbf{X} = (S_1,T_1),(S_2,T_2),\ldots$, putting $\mathbf{X}_n = ((S_1,T_1),\ldots,(S_n,T_n))$, and defining $\mathbf{Y}_n = \mathscr{E}_{\mathbf{X}_n}$ as in (9.4) for each $n \geq 1$. As above, we write $v(\mathbf{Y}_n)$ to denote the number of vertices in $\mathbf{Y}_n$, which equals the number of distinct labels observed among $S_1,T_1,\ldots,S_n,T_n$, and $e(\mathbf{Y}_n) = n$ to denote the number of edges in $\mathbf{Y}_n$. We initialize both $v(\cdot)$ and $e(\cdot)$ at $v(\mathbf{Y}_0) = e(\mathbf{Y}_0) = 0$.

Given $\mathbf{X}_{n-1} = \mathbf{x}_{n-1} = (s_1,t_1,\ldots,s_{n-1},t_{n-1}), n \geq 1$, we write $D(i;\mathbf{x}_{n-1})$ to denote the number of times i appears in the sequence $\mathbf{x}_{n-1}$, i.e., $D(i;\mathbf{x}_{n-1})$ is the degree of the vertex corresponding to i in $\mathbf{y}_{n-1} = \mathscr{E}_{\mathbf{x}_{n-1}}$, and we choose S_n,T_n successively as follows.

1. Select S_n randomly according to

$$\Pr(S_n = i \mid \mathbf{x}_{n-1}) \propto \begin{cases} D(i;\mathbf{x}_{n-1}) - \alpha, & i = 1,\ldots,v(\mathbf{y}_{n-1}), \\ \theta + \alpha v(\mathbf{y}_{n-1}), & i = v(\mathbf{y}_{n-1}) + 1. \end{cases} \tag{9.12}$$

2. Update $\mathbf{x}_{n-1} \mapsto \mathbf{x}_n^* = (s_1,t_1,\ldots,s_{n-1},t_{n-1},s_n)$ to include the newly chosen vertex $S_n = s_n$.
3. Choose T_n according to the distribution $\Pr(T_n = i \mid \mathbf{x}_n^*)$ in (9.12), with

$$v(\mathbf{y}_n^*) = \begin{cases} v(\mathbf{y}_{n-1}), & s_n = 1,\ldots,v(\mathbf{y}_{n-1}), \\ v(\mathbf{y}_{n-1}) + 1, & s_n = v(\mathbf{y}_{n-1}) + 1. \end{cases}$$

For any edge-labeled graph $\mathbf{y} \in \mathfrak{E}_{[n]}$, let $v(\mathbf{y})$ be the number of (non-isolated) vertices in $\mathbf{y}$ and let $N_k(\mathbf{y})$ be the number of vertices in $\mathbf{y}$ with total degree k.[4] This construction gives the following closed-form formula for the distribution of $\mathbf{Y}_n = \mathscr{E}_{\mathbf{X}_n}$, for each $n \geq 1$:

$$\Pr(\mathbf{Y}_n = \mathbf{y};\alpha,\theta) = \alpha^{v(\mathbf{y})}\frac{(\theta/\alpha)^{\uparrow v(\mathbf{y})}}{\theta^{\uparrow(2n)}}\prod_{k=2}^{\infty}\exp\{N_k(\mathbf{y})\log(1-\alpha)^{\uparrow(k-1)}\}, \tag{9.13}$$

for $\mathbf{y} \in \mathfrak{E}_{[n]}$, where $x^{\uparrow j} = x(x+1)\cdots(x+j-1)$ is the rising factorial. The reader should verify that (9.13) is the correct distribution of $\mathbf{Y}_n$ based on the above construction. Note, in particular, that although the update probabilities in (9.12) depend on the sequence of pairs $\mathbf{X}_{n-1}$, the conditional distribution of $\mathbf{Y}_n$ given $\mathbf{X}_{n-1}$ depends on $\mathbf{X}_{n-1}$ only through $\mathbf{Y}_{n-1} = \mathscr{E}_{\mathbf{X}_{n-1}}$.

Exercise 9.5 *Show that the Hollywood distribution in* (9.13) *is edge exchangeable.*

Alternatively, the Hollywood model with $0 \leq \alpha < 1$ and $\theta > -\alpha$ can be constructed from the stick-breaking construction of the GEM distribution by taking g_j to be the density of the Beta distribution with parameter $(1-\alpha,\theta+j\alpha)$ for each $j \geq 1$. We recover (9.13) by marginalizing over W in (9.11). See [94] for more on stick-breaking.

---

[4]By the construction in (9.12), all vertices appearing in the sequence $(s_1,t_1),\ldots,(s_n,t_n)$ are 'non-isolated'.

Example 9.1 *Let* (α, θ) *be in the parameter space of the Hollywood model and suppose* $n = 4$*. Then the sequence of edges* $(1,2)$*,* $(3,1)$*,* $(2,1)$*,* $(2,4)$ *occurs as a realization of the Hollywood process description* (9.12) *as follows.*

- *First, choose* $(1,2)$ *with probability*

$$\frac{\theta}{\theta} \times \frac{\theta+\alpha}{\theta+1}.$$

Since both 1 *and* 2 *are newly labeled vertices at the time of first arrival, the probabilities of their selection are both given by the second line in* (9.12)*. (Notice that the first vertex to arrive is always new and is labeled '1' with probability* $\theta/\theta = 1$*, where we adopt the convention that* $0/0 = 1$ *in case* $\theta = 0$*.)*

- *Given* $\mathbf{x}_1 = (S_1, T_1) = (1,2)$*, we have* $D(1;\mathbf{x}_1) = D(2;\mathbf{x}_1) = 1$ *and* $(S_2, T_2) = (3,1)$ *occurs with probability*

$$\frac{\theta+2\alpha}{\theta+2} \times \frac{1-\alpha}{\theta+3},$$

with the left-hand piece giving the probability of choosing a new vertex labeled 3 *and the right-hand piece giving the conditional probability of choosing* $T_2 = 1$ *given* S_1, T_1, S_2*, as in the top line of* (9.12)*.*

- *Continuing in this way, we have* $\mathbf{x}_2 = ((S_1,T_1),(S_2,T_2)) = ((1,2),(3,1))$ *so that* $D(1;\mathbf{x}_2) = 2$ *and* $D(2;\mathbf{x}_2) = D(3;\mathbf{x}_2) = 1$*. Thus,* $(S_3, T_3) = (2,1)$ *occurs with probability*

$$\frac{1-\alpha}{\theta+4} \times \frac{2-\alpha}{\theta+5},$$

since both labels '1' and '2' appear previously in $\mathbf{x}_2$*, with label '2' appearing* 1 *time (and receiving weight* $1-\alpha$*) and label '1' appearing* 2 *times (and receiving weight* $2-\alpha$*).*

- *Finally, with* $\mathbf{x}_3 = ((1,2),(3,1),(2,1))$*, we choose* $(S_4, T_4) = (2,4)$ *with conditional probability*

$$\frac{2-\alpha}{\theta+6} \times \frac{\theta+3\alpha}{\theta+7}.$$

Multiplying these probabilities gives a total probability of

$$\alpha^4 \frac{(\theta/\alpha)(\theta/\alpha+1)(\theta/\alpha+2)(\theta/\alpha+3)}{\theta(\theta+1)\cdots(\theta+7)}(1-\alpha)^2(1-\alpha+1)^2,$$

which simplifies to (9.13) *with* $v(\mathbf{y}_4) = 4$ *and degree distribution* $(2,0,2)$*, in agreement with* (9.13)*.*

9.7.2 Role of parameters in the Hollywood model

By (9.12), $\alpha > 0$ increases the probability of observing previously unseen vertices but decreases the probability of observing a vertex again after its initial appearance. Thus, α values near 1 make it more likely that new edges involve previously unseen vertices, but less likely that previously seen vertices appear in subsequent edges. On the other hand, $\alpha < 0$ corresponds to a finite population size, so that each newly observed vertex decreases the number of unseen vertices and increases the probability that subsequent edges involve previously seen vertices.

In the $0 < \alpha < 1$ regime, larger values of θ increase the probability of seeing previously unobserved vertices in new edges, but the effect of θ diminishes as $n \to \infty$. In Section 9.7.3, we see that $0 < \alpha < 1$ is directly related to the sparsity and power law behavior of $(\mathbf{Y}_n)_{n\geq 1}$ constructed from the Hollywood process.

9.7.3 *Statistical properties of the Hollywood model*

The update rule of the Hollywood process in (9.12) is related to that of the two-parameter Chinese restaurant process (CRP); see [46, 54, 132] for a detailed description. The main difference between Hollywood process and CRP is that the former process groups subsequent arrivals into pairs which form the edges of a network. With this observation, the number of vertices $v(\mathbf{Y}_n)$ in the Hollywood process with n edges corresponds to the number of blocks in a random partition of $[2n]$ generated by the CRP. The distributional properties of the CRP and Ewens–Pitman distribution [46, 69, 132] are well-known, allowing us to immediately deduce the following sparsity and power law properties of the Hollywood model. (Refer to Section 4.2 for the initial discussion of sparsity and power law degree distribution. See [54] for a more precise derivation of the forthcoming properties of the Hollywood model.)

When $0 < \alpha < 1$, $v(\mathbf{Y}_n)$ satisfies

$$E(v(\mathbf{Y}_n)) \sim \frac{\Gamma(\theta+1)}{\alpha\Gamma(\theta+\alpha)}(2n)^{\alpha} \quad \text{as } n \to \infty, \tag{9.14}$$

where '$a_n \sim b_n$ as $n \to \infty$' indicates that $\lim_{n\to\infty} a_n/b_n = 1$ and $\Gamma(t) = \int_0^\infty u^{t-1}e^{-u}du$ is the gamma function; see [132, p. 69]. From this it follows that the sequence $(\mathbf{Y}_n)_{n\geq 1}$ obtained from the Hollywood model with parameter (α, θ) is sparse with probability 1 provided that $1/2 < \alpha < 1$. Moreover, by [132, Lemma 3.11] the degree distribution of $\mathbf{Y}_n$ defined by $\deg(\mathbf{Y}_n) = (\deg_k(\mathbf{Y}_n))_{k\geq 1}$, where $\deg_k(\mathbf{Y}_n)$ is the number of vertices appearing exactly k times in $\mathbf{Y}_n$, satisfies

$$\deg_k(\mathbf{Y}_n) \sim \frac{\alpha(1-\alpha)^{\uparrow(k-1)}}{k!}(2n)^{\alpha}S_{\alpha} \quad \text{a.s. for every } k \geq 1 \text{ as } n \to \infty$$

for some strictly positive random variable S_α. By this relationship, the asymptotic proportion of vertices with degree $k \geq 1$ is seen to exhibit a power law degree distribution with exponent $\alpha + 1$,

$$\frac{\deg_k(\mathbf{Y}_n)}{v(\mathbf{Y}_n)} \to \frac{\alpha(1-\alpha)^{\uparrow(k-1)}}{k!} \sim k^{-(\alpha+1)} \quad \text{a.s. as } n \to \infty. \tag{9.15}$$

I summarize these properties here for completeness. See [54] for further details.

Theorem 9.4 (Sparsity and Power Law in the Hollywood model [54]) *For* $\mathbf{Y} = (\mathbf{Y}_n)_{n\geq 1}$ *be a realization of the Hollywood process with parameter* (α, θ)*. Then*

- $(\mathbf{Y}_n)_{n\geq 1}$ *is sparse with probability 1, provided* $1/2 < \alpha < 1$, *and*
- $(\mathbf{Y}_n)_{n\geq 1}$ *exhibits power law degree distribution with exponent* $\alpha + 1$ *with probability 1, provided* $0 < \alpha < 1$.

Research Problem 9.3 *Simulation results suggest that the power law degree distribution might be preserved under thresholding multiple edges in the Hollywood model to single edges, but so far this remains unproven. See [54, Section 4.5] and [95] for further discussion of this and other unexplored technical aspects of the Hollywood model.*

9.7.4 *Prediction from the Hollywood model*

The sequential description of the Hollywood model in (9.12) allows for predictive inferences about future interactions. For a concrete example, suppose we are interested in whether the next pull from a phone call database involves at least one previously unobserved caller. The probability of this outcome can be computed explicitly using the update rule of the Hollywood process in (9.12):

$$\Pr(\text{new caller in } (n+1)\text{st call} \mid \mathbf{y}_n) = 1 - \left(\frac{2n - v(\mathbf{y}_n)\alpha}{\theta + 2n}\right)\left(\frac{2n+1-v(\mathbf{y}_n)\alpha}{\theta+2n+1}\right),$$

where the product on the right-hand side is the probability of choosing a previously observed vertex for both S_{n+1} and T_{n+1}. In practice, we can estimate this probability by plugging in estimates for α and θ based on $\mathbf{y}_n$.

Depending on the objective of the analysis, the Hollywood model offers a number of other possibilities for statistical inference, many of which have not yet been explored in detail. See [54, Sections 5 and 6] for a preliminary illustration of some potential avenues of inquiry.

9.8 Contexts for edge sampling

The edge exchangeable models presented in this chapter mark our first major departure from the standard 'networks-as-graphs' perspective of earlier chapters. When edges are the units of observation, a statistical network model $(\{\mathscr{M}_n\}_{n\geq 1}, \{\Sigma_{m,n}\}_{n\geq m\geq 1})$ (in the sense of Chapter 5) is defined as a set $\mathscr{M}_n$ of candidate distributions on $\mathfrak{E}_{[n]}$ for every finite sample of $n \geq 1$ edges and a system $\{\Sigma_{m,n}\}_{n\geq m\geq 1}$ of (random) edge sampling schemes. Much like the vertex exchangeable models of Chapter 6, the edge exchangeable models of this chapter are most naturally set in the context of edge selection.

To make this notion precise, we define the *edge selection map*

$$\begin{aligned} \mathbf{S}_{m,n} : \mathfrak{E}_{[n]} &\to \mathfrak{E}_{[m]} \\ \mathscr{E}_{\mathbf{x}} &\mapsto \mathscr{E}_{\mathbf{x}|_{[m]}} \end{aligned} \tag{9.16}$$

as follows. For $\mathbf{y} \in \mathfrak{E}_{[n]}$, let $\mathbf{x} = (x_1, \ldots, x_n)$ be a representative of the equivalence class $\mathbf{y}$, so that $\mathbf{y} = \mathscr{E}_{\mathbf{x}}$ as defined in (9.4). We define $\mathbf{S}_{m,n}\mathbf{y} = \mathscr{E}_{\mathbf{x}|_{[m]}}$, where $\mathbf{x}|_{[m]} =$

$(x_1,\ldots,x_m)$ is the projection of $\mathbf{x}$ onto its first m coordinates. (The reader should verify that this definition of $\mathbf{S}_{m,n}$ is well-defined, i.e., does not depend on the choice of representative $\mathbf{x} \in \mathbf{y}$.)

Exercise 9.6 *Show that the projective Hollywood model* $(\{\mathscr{M}_n\}_{n\geq 1}, \{\mathbf{S}_{m,n}\}_{n\geq m\geq 1})$ *is coherent in the sense of Definition 5.2, for*

$$\mathscr{M}_n = \{Pr(\mathbf{Y}_n = \cdot;\alpha,\theta) \text{ in } (9.13)\} \text{ for each } n \geq 1 \quad \text{and} \tag{9.17}$$
$$\mathbf{S}_{m,n} \text{ as defined in } (9.16) \text{ for } n \geq m \geq 1.$$

As in Section 3.9, for any injection $\psi : [m] \to [n]$, we define the *ψ-selection map* $\mathbf{S}^{\psi}_{m,n} : \mathfrak{E}_{[n]} \to \mathfrak{E}_{[m]}$ as follows. For $\mathbf{y} \in \mathfrak{E}_{[n]}$, let $\mathbf{x} = (x_1,\ldots,x_n)$ be a representative of the equivalence class $\mathbf{y}$, so that $\mathbf{y} = \mathscr{E}_{\mathbf{x}}$ as defined in (9.4). For any injection ψ : $[m] \to [n]$, we define $\mathbf{S}^{\psi}_{m,n}\mathbf{y} = \mathscr{E}_{\mathbf{x}^{\psi}}$, where $\mathbf{x}^{\psi} = (x_{\psi(1)},\ldots,x_{\psi(m)})$ is the subsequence of $\mathbf{x}$ obtained by ψ-selection. With this definition of ψ-selection, we define a *random edge sampling scheme* $\Sigma_{m,n}$ as a ψ-selection map $\mathbf{S}^{\psi}_{m,n} : \mathfrak{E}_{[n]} \to \mathfrak{E}_{[m]}$ chosen randomly in a way that possibly depends on $\mathbf{Y}_n$. To recapitulate the definition of consistency under subsampling (Definition 3.2) in terms of network models defined on $\mathfrak{E}_{[n]}$, we say that random edge-labeled graphs $\mathbf{Y}_n$ and $\mathbf{Y}_m$ on $\mathfrak{E}_{[n]}$ and $\mathfrak{E}_{[m]}$, respectively, are *consistent under subsampling from* $\Sigma_{m,n}$, or $\Sigma_{m,n}$-consistent, if $\Sigma_{m,n}\mathbf{Y}_n =_{\mathscr{D}} \mathbf{Y}_m$.

Exercise 9.7 *For any probability distribution φ on $\mathscr{F}^{\cong}_{\mathbb{N}\times\mathbb{N}}$, let $\mathbf{Y} \sim \varepsilon_\varphi$, for ε_φ defined as the φ-mixture of interaction propensity processes in Theorem 9.2, and let $\mathbf{Y}_n = \mathbf{Y}|_{[n]}$ and $\mathbf{Y}_m = \mathbf{Y}|_{[m]}$ for $n \geq m \geq 1$. Prove that $\mathbf{Y}_m$ and $\mathbf{Y}_n$ are $\Sigma_{m,n}$-consistent for any random edge sampling scheme $\Sigma_{m,n}$ that is independent of $\mathbf{Y}_n$.*

I leave it as a relevant open problem to study how models for edge-labeled networks behave under more general edge sampling schemes $\{\Sigma_{m,n}\}_{n\geq m\geq 1}$.

Research Problem 9.4 *Analyze (mathematically, empirically, or computationally) the behavior of the Hollywood model and the interaction propensity process under different edge sampling contexts.*

9.9 Relative edge exchangeability

The concept of relative exchangeability from Chapter 8 extends to edge exchangeable models by a straightforward modification of Definition 8.2. The resulting theory of relative edge exchangeability then follows by the close association between edge exchangeable networks $\mathbf{Y}$ and exchangeable sequences $\mathbf{X}$ through the interaction propensity process (Theorem 9.2). For brevity, I sketch the main ideas here and leave the details to the interested reader.

Definition 9.2 (Relative edge exchangeability) *Let $\mathbf{Y}$ be a random edge-labeled network in $\mathfrak{E}_{\mathbb{N}}$ and let $\mathbf{z} \in \mathfrak{E}_{\mathbb{N}}$ be fixed. We say that $\mathbf{Y}$ is* relatively edge exchangeable with respect to **z**, *or simply* **z**-edge exchangeable, *if $\mathbf{Y}|^{\sigma}_{S} =_{\mathscr{D}} \mathbf{Y}|_S$ for all permutations $\sigma : S \to S$ such that $\mathbf{z}|^{\sigma}_{S} = \mathbf{z}|_S$, for all $S \subseteq \mathbb{N}$.*

Fix $\mathbf{x} : \mathbb{N} \to \mathbb{N}\times\mathbb{N}$ and, for each $i,j \geq 1$ let $f^{(i,j)} = (f^{(i,j)}_{rs})_{r,s\geq -1} \in \mathscr{F}^{\cong}_{\mathbb{N}\times\mathbb{N}}$. For

$f = (f^{(i,j)})_{i,j\geq 1}$, let $\mathbf{X} = (X_1, X_2, \ldots)$ be a sequence of independent random variables with distribution given by

$$\Pr(X_k = (r,s); f, \mathbf{x}) = f_{rs}^{\mathbf{x}(k)}, \quad r,s \geq -1, \tag{9.18}$$

for every $k \geq 1$. (Notice that the superscript $\mathbf{x}(k)$ in $f_{rs}^{\mathbf{x}(k)}$ records the dependence of X_k on $\mathbf{x}(k)$. In particular, the distribution in (9.18) reads $\Pr(X_k = (r,s); f, \mathbf{x}) = f_{rs}^{(i,j)}$ whenever $\mathbf{x}(k) = (i,j)$.) Given $\mathbf{X}$, we put $\mathbf{Y} = \mathscr{E}_{\mathbf{X}^*}$ for $\mathbf{X}^*$ defined by the operation $\mathbf{X} \mapsto \mathbf{X}^*$ in (9.8).

Exercise 9.8 *Show that the distribution of* $\mathbf{Y} = \mathscr{E}_{\mathbf{X}^*}$ *constructed from* $\mathbf{X}$ *with distribution* (9.18) *is relatively edge exchangeable with respect to* $\mathbf{z} = \mathscr{E}_{\mathbf{X}}$.

By the close connection between edge exchangeable networks and exchangeable sequences of random variables through the interaction propensity process (Sections 9.4–9.5 and Theorem 9.2), the theory of relative edge exchangeability does not seem to stray too far from the theory of relatively exchangeable sequences, as in [58, 59]. But even if the theory of relative edge exchangeability does not reveal anything especially surprising, it may still be worthwhile to work out the details for potential use in applications. Questions remain, however, about the practical implications of relative edge exchangeability; see Problem 9.5 below. I conclude this section with a conjecture about the structure of relatively edge exchangeable networks. For the definition of ultrahomogeneity in the following conjecture, refer to Section 8.3.3 and/or [59].

Conjecture 9.1 *Let* $\mathbf{z} \in \mathfrak{E}_{\mathbb{N}}$ *be an ultrahomogeneous edge-labeled network and let* $\mathbf{Y}$ *be* $\mathbf{z}$*-edge exchangeable. Then there exists a probability measure* φ *on the space of collections* $(f^{(i,j)})_{i,j\geq 1}$, *with* $f^{(i,j)} \in \mathscr{F}_{\mathbb{N}\times\mathbb{N}}^{\cong}$ *for each* $i,j \geq 1$, *such that* $\mathbf{Y} =_{\mathscr{D}} \mathbf{Y}^*$, *for* $\mathbf{Y}^* = \mathscr{E}_{\mathbf{X}^*}$ *constructed from* $\mathbf{X}$ *distributed according to the distribution in* (9.18) *conditional on* $f \sim \varphi$ *and* $\mathbf{x} : \mathbb{N} \to \mathbb{N} \times \mathbb{N}$ *chosen to be any representative of the equivalence class associated to* $\mathbf{z}$ *through* (9.4).

Research Problem 9.5 *At the present time, it is unclear how relative edge exchangeability fits into the practice of network analysis. In particular, under what circumstances might the invariance condition in Definition 9.2 be natural for a statistical network model?*

9.10 Thresholding and its unintended consequences

Moving away from the specific context of this chapter for a moment, I call attention here to the unrealized impact of rampant thresholding in network analysis. For example, Figure 9.9(a) shows the complete karate club dataset [161], for which each edge records a social interaction between 2 of the 34 members in a university karate club. The karate club dataset is canonical in network community detection because of the known separation of club members according to their allegiance to one of the club's two leaders.[5] Many community detection techniques have been tested on the

[5]The karate club network has become so widely cited in the networks literature that it is now the subject of parody; see Section 1.6.3 for further discussion. For our purposes here, however, we invoke the

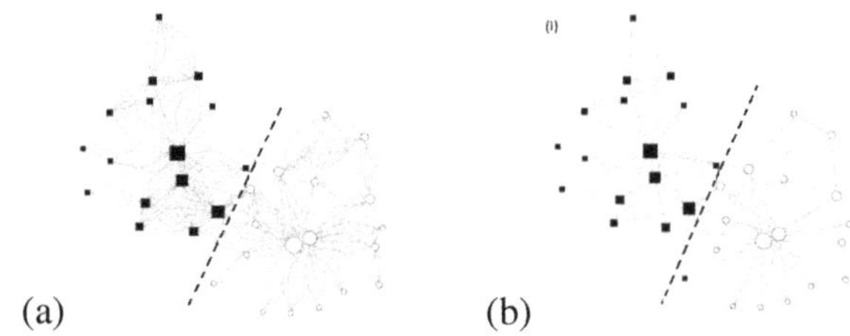

Figure 9.9 *(Left) Interaction network of Zachary karate club. (Right) Projection of the network in (a) by removing edge multiplicities. In both pictures, the dashed line separates vertices into the true communities, as identified by Zachary [161], with vertices left of the line corresponding to one community and vertices right of the line corresponding to the other. The color of the vertices, as black or white, shows the classification given by either a simple method that accounts for multiple edges (in (a)) or more complicated approaches, such as degree-corrected stochastic blockmodels, which ignore multiplicities (in (b)). The classification in (a) coincides with Zachary's analysis while the classification in (b) does not. (Notice the one incorrectly specified black vertex on the right-hand side of the dashed line in (b).)*

karate club network after projecting multiple occurrences of each edge to a single occurrence, as shown in Figure 9.9(b). As a result of thresholding, it is common, e.g., in [19, 99], for the leading community detection methods to correctly classify all but one of the karate club members. Because it is so commonplace to misclassify this one particular vertex, the incorrect classification shown in Figure 9.9(b) is often taken as good enough for demonstrating the usefulness of a given method.

Easily lost amid the decision to threshold multiple edges is that the 'real' community structure among these karate club members is best understood in terms of the multigraph in Figure 9.9(a), which reflects the frequency of social interactions outside the club, and not the thresholded graph in Figure 9.9(b), which only records whether or not two club members have interacted outside the club regardless of the frequency of those interactions. Because there is no logical reason to expect the same community structure to persist after arbitrarily projecting multiple edges to a single edge, misclassification should be expected since the act of removing multiple edges does not guarantee that the fundamental structure of the data is preserved. With this observation, the fact that only one vertex is misclassified in Figure 9.9(b) should not be interpreted as evidence that the community detection method producing this classification 'works'. It should instead be seen as a matter of blind luck! It turns out that all vertices can be correctly classified by applying a very simple method to the complete network with multiple edges; see [51] for more discussion.

This example illustrates in a manageable and well-understood context how data processing, and in particular thresholding, can have unintended consequences for inferences from network data. Despite these drawbacks, the practice of thresholding remains widespread in network analysis.

Research Problem 9.6 *Needless data processing, such as thresholding multiple*

karate club network as a base case for network community detection, which allows us to demonstrate the main pitfalls of thresholding in network analysis.

edges as in the above karate club example and the analysis in [96], can have adverse consequences on statistical inference from network data. Figure 9.9 demonstrates one specific pitfall of thresholding multiple edges in community detection. There are likely many unknown consequences of this and other data processing techniques, such as decomposing multi-way interactions into pairwise edges (to be discussed further in Chapter 10). It is an important open problem to better understand the practical, computational, and theoretical consequences of data processing on network analysis.

Returning to the topic of edge exchangeability and the Hollywood model, I note that the characterization of edge exchangeable models by the interaction propensity process (Theorem 9.2) implies that any edge exchangeable network is guaranteed to exhibit multiple edges between some of its vertices, unless its distribution is concentrated on edges involving 'blips'. (Refer to Section 9.5 for the definition of 'blip'.) Theorem 5.4 in [54] demonstrates one immediate consequence of removing multiple edges from an edge-labeled multigraph constructed by the Hollywood model: whereas Theorem 9.4 shows that the Hollywood model (without thresholding) is sparse if and only if $1/2 < \alpha < 1$, the projection of $(\mathbf{Y}_n)_{n\geq 1}$ to a sequence of simple graphs by removing multiple edges is sparse for all $0 < \alpha < 1$. Therefore, thresholding not only throws away data but also alters the observed asymptotic behavior of the network.

Under the paradigmatic 'networks-as-graphs' mindset, it is the convention, even when dealing with interaction networks, to threshold multiple occurrences of edges to obtain a $\{0,1\}$-valued adjacency matrix as in (2.1). A practical reason for projecting multiple edges seems to be that many network models are unable to accommodate the natural occurrence of multiple edges. In other cases, it seems to arise out of the faulty association between 'networks' and 'graphs' that has overtaken much of the network science literature; see Chapter 1, in particular Section 1.2, for more discussion on this point. In addition to handling multiple occurrences of edges in a natural way, the relationally exchangeable models presented in Chapter 10 accommodate networks formed from interactions involving more than two vertices, as in networks of scientific coauthorships, movie collaborations, and Internet paths.

9.11 Comparison: Edge exchangeability v. graphex

The perspective presented in this chapter flows logically from the study of interaction data, to its representation as an edge-labeled network, to the concept of edge exchangeability. Edge exchangeability offers an alternative to more common network models, such as graphons, exponential random graph models, and stochastic blockmodels, which cannot account for the most basic properties observed in modern network data, namely sparsity and power law degree distribution. The fact that edge exchangeable models easily replicate these basic features of network data is merely empirical justification that the framework may be viable for a range of applications in network science and statistics. But the main argument in favor of edge exchangeability, and the more general concept of relational exchangeability in the next chapter, is that it respects the role of the edges (i.e., interactions) as the units of observation in many network datasets [54].

To conclude this chapter, I briefly discuss some similarities and differences between the Caron–Fox/graphex models from Section 7.3 and the Crane–Dempsey/edge exchangeable models presented here. Since other authors have already highlighted some connections between the two frameworks, see, e.g., the discussion accompanying [32], I focus here on key differences between the two approaches, as I have previously emphasized in [47].

Sparsity and power law degree distribution

Both the Caron–Fox and Crane–Dempsey models are able to reproduce sparsity and power law degree distributions. In fact, the range of power law distributions, with exponents in the range $(1,2)$, seems to be identical in both model classes. From this point of view, both approaches successfully account for these two minimal empirical properties that are believed to be found in many real-world networks, as highlighted in Sections 1.7.1 and 7.1.

Probabilistic symmetry

The concept of edge exchangeability arises naturally in Crane and Dempsey's framework by considering network data that is constructed from a representative sample of edges (i.e., interactions). The concept of exchangeability invoked in graphex models, however, is less clearly connected to the way in which real-world networks are typically observed. As I discussed in Section 7.3.4, the 'exchangeability' in graphex models refers to invariance under measure-preserving transformations of the associated point process. If the entire point pattern (including the 'arrival times') is observed, then exchangeability of the point process can be understood in terms of sampling edge patterns over a fixed duration of time, as noted in Section 7.3.7. But more often this temporal information is not available, forcing the data to be modeled as the unlabeled structure induced by the edge patterns, as in (7.6). The distribution induced on these unlabeled structures does not seem to be exchangeable in any recognizable sense of the term (Exercise 7.3). And even in the motivating scenario of Section 7.3.1, the point process representation merely labels the vertices according to the time at which they first arrived in the system. The exchangeability condition of the point process does not allow these temporal labels to serve any additional purpose. Edge exchangeability can also model such edge patterns but without the need to associate labels to the vertices.

In [32], the authors present the Caron–Fox model as an answer to the question posed in [128], "Is there a notion of probabilistic symmetry whose ergodic measures [...] describe useful statistical models for sparse graphs with network properties?" Given the roundabout way in which probabilistic symmetry arises in the Caron–Fox model, through the invariance of a latent point process, and the lack of a clearcut motivating context for this approach, the extent to which Caron and Fox's proposal addresses the misspecification issues highlighted in [128] remains unclear, as noted previously in [126]. Throughout Section 7.3, I have been as charitable as possible to the Caron–Fox model, first by motivating the approach with a concrete scenario (Section 7.3.1) and then by giving a natural sampling interpretation to the exchangeable point process construction in terms of t-sampling (Section 7.3.7). The latter gives a more direct interpretation of the proposed exchangeable point processes as models for networks, and therefore improves upon the more stylized and artificial p-sampling interpretation of the induced $\{0, 1\}$-valued arrays.

Edge exchangeable models, on the other hand, do exhibit a novel probabilistic symmetry (i.e., edge exchangeability) which seems to be natural for modeling interaction networks under a presumed exchangeable edge sampling scheme, as illustrated in Section 9.1. Because of its straightforward description in terms of sampling interaction networks, I believe that the edge exchangeable framework in [54] does satisfactorily answer the question posed in [128], at least in its intended context of modeling interaction data. The extent to which answering this question will have any far-reaching implications in the practice of network analysis, however, remains to be seen.

Sampling interpretation/projectivity

The sampling interpretation of edge exchangeable models is clear: the observed edges are a representative sample of the population of all edges (i.e., interactions). The sampling interpretation of graphex models is less clear, except when considered in the full context of the associated point process, in which case the exchangeability criterion for the point process can be interpreted in terms of representative sampling of edge patterns over a given length of time (Section 7.3.7). This offers an uncontroversial interpretation of sampling from graphex models which is apparently distinct from the sampling interpretation of edge exchangeable models. When considering instead the discrete structure induced by the point process, as in (7.6), the natural sampling interpretation for the point process invokes the interpretation of 'p-sampling' [148], which for most applications is as unrealistic as selection and simple random vertex sampling. See Section 7.3.7 for further discussion.

9.12 Further reading

The class of edge exchangeable network models introduced by Crane and Dempsey [52, 54] has since been studied in follow-up work by other authors [95, 124]. Janson [95] examined some technical properties of edge exchangeable models, with specific focus on the Crane–Dempsey Hollywood model from Section 9.7. Ng and Silva [124] initiate a study of dynamic edge exchangeable models for time-varying networks. For other related work on link prediction in interaction networks, see Williamson [153]. More recently, the Crane–Dempsey model has been recast as a model in physics [41]. A translation of the main results of [52, 54] into the language of contemporary Bayesian nonparametrics can be found in [30].

9.13 Solutions to exercises

9.13.1 Exercise 9.1

Fix $\mathbf{x} : \mathbb{N} \to \mathscr{P} \times \mathscr{P}$ and define

$$\mathscr{E}_{\mathbf{x}} = \{\rho\,\mathbf{x} \mid \rho : \mathscr{P} \to \mathscr{P} \text{ is a bijection}\} \quad \text{and}$$
$$\mathscr{E}'_{\mathbf{x}} = \{\mathbf{x}' : \mathbb{N} \to \mathscr{P} \times \mathscr{P} \mid \rho\,\mathbf{x}' = \mathbf{x} \text{ for some bijection } \rho : \mathscr{P} \to \mathscr{P}\}.$$

We need to show $\mathscr{E}_{\mathbf{x}} = \mathscr{E}'_{\mathbf{x}}$. First note that for every bijection ρ, there is an inverse bijection ρ^{-1} so that $\rho^{-1}(\rho\,\mathbf{x}) = \mathbf{x}$, and thus $\rho\,\mathbf{x} \in \mathscr{E}'_{\mathbf{x}}$ and $\mathscr{E}_{\mathbf{x}} \subseteq \mathscr{E}'_{\mathbf{x}}$. Conversely, let $\mathbf{x}' \in \mathscr{E}'_{\mathbf{x}}$ so that $\mathbf{x}' = \rho^{-1}\mathbf{x}$, where ρ^{-1} is the inverse of ρ and is also a bijection. Thus $\rho^{-1}\mathbf{x} \in \mathscr{E}_{\mathbf{x}}$ and $\mathscr{E}'_{\mathbf{x}} \subseteq \mathscr{E}_{\mathbf{x}}$. This completes the proof.

9.13.2 Exercise 9.2

Let $\mathbf{y} = \mathscr{E}_{\mathbf{x}}$ for some $\mathbf{x} : \mathbb{N} \to \mathscr{P} \times \mathscr{P}$. Let $\mathbf{x}'$ be any representative of the equivalence class $\mathscr{E}_{\mathbf{x}}$, so that by Exercise 9.1 there is a bijection ρ such that $\rho\,\mathbf{x}' = \mathbf{x}$. Then, given any permutation $\sigma : \mathbb{N} \to \mathbb{N}$, we must prove that $\mathscr{E}_{\mathbf{x}^\sigma} = \mathscr{E}_{\mathbf{x}'^\sigma}$. By the definition of $\mathscr{E}_{\mathbf{x}}$ in (9.3),

$$\begin{aligned}\mathscr{E}_{\mathbf{x}'^\sigma} &= \{\rho'(\mathbf{x}'^\sigma) \mid \rho' : \mathscr{P} \to \mathscr{P} \text{ is a bijection}\} \\ &= \{(\rho'(x'_{\sigma(i)}))_{i\geq 1} \mid \rho' : \mathscr{P} \to \mathscr{P} \text{ is a bijection}\}.\end{aligned}$$

Since $\rho\,\mathbf{x}' = \mathbf{x}$ and ρ^{-1} is a bijection, we have $\mathbf{x}' = \rho^{-1}\mathbf{x}$ and $\rho(\rho^{-1}\mathbf{x}^\sigma) = \mathbf{x}^\sigma \in \mathscr{E}_{\mathbf{x}'^\sigma}$. And since $\mathscr{E}_{\mathbf{x}'^\sigma}$ is an equivalence class containing $\mathbf{x}^\sigma$, we must have $\mathscr{E}_{\mathbf{x}'^\sigma} = \mathscr{E}_{\mathbf{x}^\sigma}$, completing the proof.

9.13.3 Exercise 9.3

Fix $f \in \mathscr{F}_{\mathscr{P}\times\mathscr{P}}$, let $\mathbf{X} = (X_1, X_2, \ldots)$ be i.i.d. according to (9.6), and put $\mathbf{Y} = \mathscr{E}_{\mathbf{X}}$ as defined in (9.4). Then by Exercise 9.2 we have $\mathbf{Y}^\sigma = \mathscr{E}_{\mathbf{X}^\sigma}$ for every permutation $\sigma : \mathbb{N} \to \mathbb{N}$, and $\mathbf{X}$ is exchangeable because it is i.i.d.; whence, $\mathbf{X}^\sigma =_{\mathscr{D}} \mathbf{X}$ and

$$\mathbf{Y}^\sigma = \mathscr{E}_{\mathbf{X}^\sigma} =_{\mathscr{D}} \mathscr{E}_{\mathbf{X}} = \mathbf{Y} \quad \text{for all permutations } \sigma : \mathbb{N} \to \mathbb{N},$$

thus proving that $\mathbf{Y} \sim \varepsilon_f$ is edge exchangeable.

9.13.4 Exercise 9.4

Suppose that there is an edge exchangeable network $\mathbf{Y}$ whose distribution on $\mathfrak{E}_{\mathbb{N}}$ assigns probability 1 to the network $\mathbf{y}^\circ$ in Figure 9.7 and such that the distribution of $\mathbf{Y}$ can be expressed as one of the interaction propensity processes ε_f in Section 9.4. Then by definition of ε_f there must be some $f = (f_{(i,j)})_{i,j\geq 1} \in \mathscr{F}_{\mathbb{N}\times\mathbb{N}}$ such that $\mathbf{Y} =_{\mathscr{D}} \mathscr{E}_{\mathbf{X}}$ for $\mathbf{X} = (X_1, X_2, \ldots)$ drawn i.i.d. from (9.6). First notice that since all edges in $\mathbf{Y}$ are loops, f must satisfy $f_{ij} \equiv 0$ for all $i \neq j$. But now, if $f_{ii} > 0$ for any $i \geq 1$, then the limiting frequency of occurrences of the pair (i,i) in $\mathbf{X}$ will be $f_{ii} > 0$ with probability 1, and thus (i,i) will appear infinitely often in $\mathbf{X}$ with probability 1. But since each loop appears exactly once in $\mathbf{y}^\circ$, it follows that $f_{(i,i)} \equiv 0$ for all $i \geq 1$. Thus, any f for which $\Pr(\mathbf{Y} = \mathbf{y}^\circ; f) = 1$ must have $f_{ij} \equiv 0$ for all $i, j \geq 1$. There can be no such f in $\mathscr{F}_{\mathbb{N}\times\mathbb{N}}$ because all $f' \in \mathscr{F}_{\mathbb{N}\times\mathbb{N}}$ satisfy $\sum_{i,j\geq 1} f'_{(i,j)} = 1$, but $\sum_{i,j\geq 1} f_{(i,j)} = 0$. A contradiction.

9.13.5 Exercise 9.5

We can immediately deduce that the Hollywood model is edge exchangeable by noticing that (9.13) depends on $\mathbf{y}$ only through its degree distribution, which is invariant to edge relabeling.

9.13.6 Exercise 9.6

To prove coherence, we must show that $\mathbf{S}_{m,n}\mathscr{M}_n = \mathscr{M}_m$ for all $n \geq m \geq 1$, with $\{\mathscr{M}_n\}_{n\geq 1}$ defined in (9.17) and $\mathbf{S}_{m,n}\mathscr{M}_n$ defined in (5.6). For this, it is enough to show that the Hollywood model is consistent under selection, i.e., $\mathbf{Y}_n$ and $\mathbf{Y}_m$ distributed according to (9.13) satisfy $\mathbf{Y}_m =_{\mathscr{D}} \mathbf{S}_{m,n}\mathbf{Y}_n$ for any choice of (α, θ). This can be proven the hard way, by calculating the marginal distribution of $\mathbf{S}_{m,n}\mathbf{Y}_n$ and showing that it coincides with the distribution of $\mathbf{Y}_m$, as we did for the p_1 model in (3.10). An easier way is to leverage the generative description of the Hollywood model through the Hollywood process in Section 9.7.1. From this, consistency under selection is automatic by (4.4). Thus, for every (α, θ) in the parameter space of the Hollywood model, $\mathbf{S}_{m,n}\mathbf{Y}_n$ is distributed according to the Hollywood model on $\mathfrak{E}_{[m]}$ with parameter (α, θ). It follows that $\mathscr{M}_m = \mathbf{S}_{m,n}\mathscr{M}_n$, as desired.

9.13.7 Exercise 9.7

The solution follows the same program as that of Exercise 6.2, without any change in syntax and with the appropriate change of interpretation from vertex sampling to edge sampling.

9.13.8 Exercise 9.8

By Definition 9.2, we need to show that $\mathbf{Y}\,|_S^\sigma =_{\mathscr{D}} \mathbf{Y}\,|_S$ for all permutations $\sigma : S \to S$ for which $\mathbf{z}|_S^\sigma = \mathbf{z}|_S$, for all $S \subseteq \mathbb{N}$. Without loss of generality, we take $S = [n]$ for arbitrary $n \geq 1$. By the definition of restriction and relabeling for edge-labeled graphs, it is enough to establish

$$\mathbf{Y}\,|_{[n]}^\sigma =_{\mathscr{D}} \mathscr{E}_{\mathbf{X}^*\,|_{[n]}^\sigma} =_{\mathscr{D}} \mathscr{E}_{\mathbf{X}^*\,|_{[n]}} =_{\mathscr{D}} \mathbf{Y}\,|_{[n]}$$

for all $\sigma : [n] \to [n]$ such that

$$\mathbf{z}|_{[n]}^\sigma = \mathscr{E}_{\mathbf{X}\,|_{[n]}^\sigma} = \mathscr{E}_{\mathbf{X}\,|_{[n]}} = \mathbf{z}|_{[n]}.$$

The first and third equalities in both displays follow by definition. And by the construction of $\mathbf{X}^*$ from $\mathbf{X}$ in (9.8), it is enough to show that $\mathbf{X}$ is relatively exchangeable with respect to $\mathbf{x}$, in which case the middle inequality would follow. But $\mathbf{X}$ is $\mathbf{x}$-exchangeable by its construction in (9.18). The proof is completed upon confirming that this holds for arbitrary $\mathbf{x} \in \mathbf{z}$.

Alternatively, we could prove this by calculating the distribution of $\mathbf{Y}$ induced by (9.18) and showing that it is relatively exchangeable with respect to $\mathscr{E}_{\mathbf{x}}$ for every $\mathbf{x} \in \mathbf{z}$.

Chapter 10

Relationally exchangeable models

In broadening the scope of network analysis beyond the conventional paradigm, the framework of edge exchangeability put forward in Chapter 9 takes a small step toward realizing the broader vision of complex data analysis laid out in Chapter 1. This chapter expands upon the previous toward a theory for network data constructed from a sample of arbitrary relations. In Section 9.1, for example, an edge-labeled network is constructed out of pairwise interactions (i.e., phone calls) between callers and receivers sampled from a phone call database. More generally, networks can be built from repeated observations of any relational structure, including pairwise interactions (i.e., edges) as in Figure 9.5, multiway interactions (i.e., hyperedges), paths (i.e., ordered multisets), or even networks (i.e., graphs) or arbitrary relational structures $(X_1, \ldots, X_r)$ as in Section 8.5. Networks constructed from an exchangeable sequence of such relations are called *relationally exchangeable* [53]. I focus here on a few special cases of relationally exchangeable network models.

10.1 Sampling multiway interactions (hyperedges)

Many real-world networks are built from interactions that involve more than two entities. In conventional approaches to network analysis, these multiway interactions are often decomposed into their constituent pairwise edges. But as we have already noted the potential pitfalls of thresholding in network analysis (Section 9.10), we should seek to avoid needless data processing whenever possible. Both examples discussed below and the subsequent theory for hyperedge-labeled networks (Section 10.2) fit within the emerging edge-centric paradigm of Section 9.2.

10.1.1 Collaboration networks

Consider a network that represents a sample of movie actor collaborations taken from the Internet Movie Database (IMDb). Each IMDb entry corresponds to a different movie and includes information such as title, year, cast of actors, description of plot, etc. For each movie, we observe the set of actors $\{a_1, \ldots, a_k\}$ starring in that movie and ignore all other meta-data, such as genre, year, director, etc. Table 10.1 illustrates such an observation.

In this scenario, we assume $M_1, \ldots, M_N$ are sampled uniformly without replacement from among all movies in the database. Associated to each M_i is a set

Table 10.1 *Database of movies and actors. Each row contains the set of actors in the corresponding movie.*

Movie title	Starring cast
Rocky (1976)	Sylvester Stallone, Bert Young, Carl Weathers, ...
Rounders (1998)	Matt Damon, Ed Norton, John Turturro, ...
Groundhog Day (1993)	Bill Murray, Andie McDowell, Chris Elliott, ...
A Bronx Tale (1993)	Robert DeNiro, Chazz Palminteri, Joe Pesci, ...
Over the Top (1987)	Sylvester Stallone, Robert Loggia, ...
⋮	⋮

Figure 10.1 *(a) Hypergraph structure formed by the sequence of interactions in* (10.1). *A binary interaction is represented by a line, as in the line labeled* 2 *connecting a and c, a three-way interaction by a triangle, as in the triangles labeled* 1 *and* 3 *connecting a, c, d and b, d, e, respectively, and four-way interactions are represented by a square/rectangle, as in the rectangle labeled* 4 *connecting a, d, f, g. (b) The graph formed by decomposing each hyperedge into its pairwise edges and disregarding the higher-order structure (e.g., triangle, square, etc.) induced by each hyperedge. For example, in this representation the hyperedge* $\{a,b,c\}$ *labeled '1' in (a) decomposes into three binary edges ab, ac, and bc, without any indication that these three edges occurred as part of a single hyperedge in the generating process.*

$\{M_i(1),\ldots,M_i(K_i)\}$, where each $M_i(j)$ identifies a different actor and $K_i \geq 1$ is the number of actors in the ith sampled movie. The structure induced by the sampled movies $M_1,\ldots,M_N$ can be represented as a network. For example, Figure 10.1(a) shows the network structure associated to the sequence

$$M_1 = \{a,b,c\}, \quad M_2 = \{a,c\}, \quad M_3 = \{b,d,e\}, \quad M_4 = \{a,d,f,g\}. \tag{10.1}$$

In Figure 10.1(a), each movie is represented by a *hyperedge* whose vertices correspond to the actors in that movie, with line representing a movie with 2 actors, a triangle a movie with 3 actors, a square a movie with 4 actors, and so on. To maintain the integrity of the data structure, the analysis should not decompose hyperedges into their pairwise interactions, e.g., by breaking down $\{a,c,d\}$ into its constituent

Table 10.2 *Database of statistics articles. Each row contains the list of authors of the corresponding article. Note: this table is for illustration only. It was not obtained by sampling from SSRN or arXiv.*

Article title	Authors
A nonparametric view of network models ...	Bickel, Chen
Edge exchangeable models for interaction networks	Crane, Dempsey
Snowball sampling	Goodman
Latent space approaches to social network analysis	Hoff, Raftery, Handcock
⋮	⋮

pairs ac, ad, cd as in Figure 10.1(b). I discuss the implications of such decomposition further in Section 10.2.

10.1.2 Coauthorship networks

Similar to movie collaboration networks are networks constructed from scientific coauthorships. For example, the Social Science Research Network (SSRN) is a repository of more than $700,000$ academic articles written by more than $300,000$ authors in 24 disciplines within social science. Suppose N articles are sampled uniformly without replacement from among all articles posted to SSRN in the year 2016. For each article the set of authors is recorded, so that each observation $A_i = \{A_i(1), \ldots, A_i(K_i)\}$ is a finite set consisting of the authors of the ith sampled article. An example of articles about network analysis is shown in Table 10.2. The structure induced by the sampled articles can be represented by a hypergraph as in Figure 10.1(a).

10.2 Representing multiway interaction networks

Aside from models for random hypergraphs and multilayer networks, e.g., [103], the vast majority of statistical network models (and all of the models discussed in the preceding chapters) are tailored specifically to networks with binary edges.[1] The lack of models for hypergraph data goes hand-in-hand with the common practice of decomposing multiway interactions into their pairwise components, so that a three-way interaction abc decomposes as ab, ac, and bc, a four-way interaction $abcd$ decomposes as ab, ac, ad, bc, bd, cd, and so on. For example, the graphical representation of (10.1) shown in Figure 10.1(b) preserves the pairwise structure in the data, but not its higher-order structure. And if the data is processed further by removing multiple occurrences of the same interaction, as in Figure 9.9(b), then even more structure is lost. As Chapter 9.10 already highlights the adverse consequences of thresholding, the potential ramifications of decomposing hyperedges into their pairwise compo-

[1] As noted in Problem 7.3, it seems that the completely random measure approach of Section 7.3 could be extended to account for multiway interactions, but I leave this as a topic of future study.

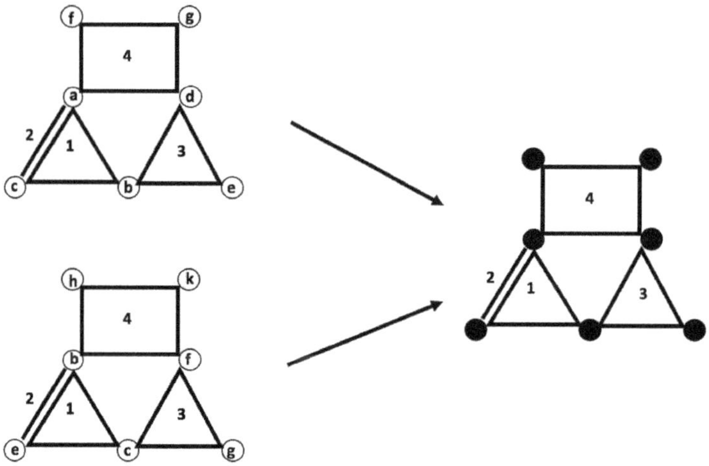

Figure 10.2 *Analogous illustration to Figure 9.5 from Chapter 9. (Left) On top is the structure whose vertices and hyperedges are labeled according to the observation in* (10.1). *On the bottom is the structure whose vertices and hyperedges are labeled according to the observation in* (10.2). *Both of these observations induce the same hyperedge-labeled structure, as shown on the right. (Right) Hyperedge-labeled structure (from* (10.3)*) induced by the equivalence class of the sequence of hyperedges in* (10.1) *(and also* (10.2)*).*

nents seem inevitable. Practical efforts in network analysis would benefit greatly from a better understanding of these implications; see Problem 9.6.

The class of relationally exchangeable models presented in this chapter refine the edge exchangeable models from Chapter 9, providing a straightforward way to analyze the multiway interaction networks that arise in the scenarios of Section 10.1. Building on the edge-centric perspective put forward in Section 9.2, the models in this chapter facilitate network analysis from the point of view of the interactions (or relations) rather than the vertices making up those interactions. As discussed in Section 9.2, the names of actors convey no additional information beyond what is already contained in the network structure induced by their interactions. Thus, for example, the sequence

$$M'_1 = \{b,c,e\}, \quad M'_2 = \{b,e\}, \quad M'_3 = \{c,f,g\}, \quad M'_4 = \{b,f,h,k\}, \tag{10.2}$$

induces the same essential structure as (10.1), as can be seen by comparing their induced hyperedge-labeled networks, e.g., on the right-hand side of Figure 10.2.

10.3 Hyperedge exchangeability

Sampling interactions uniformly from the IMDb as in Section 10.1.1 or from SSRN as in Section 10.1.2 produces networks that are exchangeable with respect to relabeling of their hyperedges. Whereas the pairwise interaction data in Section 9.1 takes the form of a sequence $X_1, X_2, \ldots$ of ordered pairs in $\mathscr{P} \times \mathscr{P}$, the movie collaboration and scientific coauthorship data in Section 10.1 take the form of a sequence $X_1, X_2, \ldots$ in the set $\text{fin}(\mathscr{P})$ of all finite multisets of an at most countable population $\mathscr{P}$. (Each $a \in \text{fin}(\mathscr{P})$ has the form $\{a_1, \ldots, a_k\}$ for some finite $k \geq 1$.) Sampling uniformly from the database results in an exchangeable $\text{fin}(\mathscr{P})$-valued sequence $X_1, X_2, \ldots$, for which each finite initial segment satisfies

$$(X_1, \ldots, X_n) =_{\mathscr{D}} (X_{\sigma(1)}, \ldots, X_{\sigma(n)}) \quad \text{for all permutations } \sigma : [n] \to [n].$$

Any bijection $\rho : \mathscr{P} \to \mathscr{P}$ acts on $\text{fin}(\mathscr{P})$ elementwise. Specifically, for any $x = \{x_1, \ldots, x_k\} \in \text{fin}(\mathscr{P})$ and any bijection $\rho : \mathscr{P} \to \mathscr{P}$, we write $\rho x = \{\rho(x_1), \ldots, \rho(x_k)\} \in \text{fin}(\mathscr{P})$ to denote the action of 'renaming' the population $\mathscr{P}$ according to ρ. For any sequence $\mathbf{x} = (x_1, \ldots, x_n)$ in $\text{fin}(\mathscr{P})$, with $x_i = \{x_{i,1}, \ldots, x_{i,k_i}\} \subseteq \mathscr{P}$ for each i, the induced edge-labeled hypergraph is defined analogously to (9.3) by

$$\mathscr{E}_{\mathbf{x}} = \{\rho\,\mathbf{x} = (\rho(x_1), \ldots, \rho(x_n)) \mid \rho : \mathscr{P} \to \mathscr{P} \text{ a bijection}\}, \tag{10.3}$$

or equivalently by

$$\mathscr{E}_{\mathbf{x}} = \{\mathbf{x}' : \mathbb{N} \to \text{fin}(\mathscr{P}) \mid \rho\,\mathbf{x}' = \mathbf{x} \text{ for some bijection } \rho : \mathscr{P} \to \mathscr{P}\}$$

as in (9.4); see Exercise 9.1. The equivalence class $\mathscr{E}_{\mathbf{x}}$ in (10.3) corresponds to a hyperedge-labeled graph. For example, the hyperedge-labeled graph on the right-hand side of Figure 10.2 identifies the equivalence class $\mathscr{E}_{\mathbf{x}}$ built from the sequence in (10.1), of which the vertex-hyperedge labeled networks on the left-hand side of Figure 10.2 are both members.

Following the same program as in Section 9.3, we derive the analogous notion of *hyperedge exchangeability* for edge-labeled hypergraphs constructed from exchangeable sequences $\mathbf{X} = (X_1, X_2, \ldots)$ in $\text{fin}(\mathscr{P})$. In this case, the random edge-labeled hypergraph $\mathscr{E}_{\mathbf{X}}$ is *hyperedge exchangeable* if its distribution is invariant under relabeling of its hyperedges, as illustrated in Figure 10.3. The interaction propensity processes from Section 9.4 extend in a straightforward way to the present setting.

10.3.1 Interaction propensity process

For a countable set $\mathscr{P}$, we define the $\text{fin}(\mathscr{P})$-simplex by

$$\mathscr{F}_{\text{fin}(\mathscr{P})} = \left\{ (f_x)_{x \in \text{fin}(\mathscr{P})} : f_x \geq 0 \text{ and } \sum_{x \in \text{fin}(\mathscr{P})} f_x = 1 \right\}.$$

From any $f \in \mathscr{F}_{\text{fin}(\mathscr{P})}$, the *interaction propensity process directed by* f is the distribution of a random edge-labeled hypergraph $\mathscr{E}_{\mathbf{X}}$ constructed from a sequence $\mathbf{X} = (X_1, X_2, \ldots)$ sampled i.i.d. according to

$$\Pr(X_i = x; f) = f_x, \quad x \in \text{fin}(\mathscr{P}). \tag{10.4}$$

Writing ε_f to denote the distribution of $\mathscr{E}_{\mathbf{X}}$, we define ε_φ as the mixture of ε_f with respect to choosing f from a probability measure φ on $\mathscr{F}_{\text{fin}(\mathscr{P})}$ and sampling $\mathbf{X}$

Figure 10.3 *Two edge-labeled hypergraphs equivalent up to relabeling of their hyperedges. Both realizations are assigned equal probability by any hyperedge exchangeable network model.*

conditionally i.i.d. given f as in (10.4). In particular, ε_φ is the distribution of $\mathbf{Y}$ obtained by first taking $f \sim \varphi$, then, given f, generating $\mathbf{X}$ to be conditionally i.i.d. from (10.4), and finally putting $\mathbf{Y} = \mathscr{E}_\mathbf{X}$ as in (10.3). The mixture distribution ε_φ is expressed formally as

$$\varepsilon_\varphi(\cdot) = \int_{\mathscr{F}_{\mathrm{fin}(\mathscr{P})}} \varepsilon_f(\cdot)\varphi(df).$$

The same basic observations for edge exchangeable models carry over to hyperedge exchangeable networks generated from the interaction propensity process. In particular, vertices arrive in size-biased order according to their limiting average frequency of occurrence in interactions: that is, for each $s \in \mathscr{P}$, the element s appears with frequency $f_\bullet^s = \sum_{x \in \mathrm{fin}(\mathscr{P}): s \in x} f_x$. So again the sampled vertices are atypical in that they tend to be more 'active' or 'popular' relative to the other vertices in the population. In the movie collaboration example of Section 10.1.1, for example, popular actors appear in more movies than D-listers, and thus are more likely to appear in any given uniformly sampled entry of the database. An actor who has been cast in 100 movies, for instance, is 10 times more likely to be observed in any given sampled movie than an actor who has been cast in only 10 movies. Similarly for the SSRN coauthorship network in Section 10.1.2: authors who write more papers are more likely to be observed when sampling uniformly from among all articles.

Although the IMDb and SSRN datasets arise similarly as networks constructed from sequences of multiway interactions, we might anticipate some differences in their empirical properties based on the circumstances under which they are obtained. The IMDb contains a wide range of movies, including many produced in Hollywood. As many Hollywood movies have a few starring roles which tend to be filled by a select group of famous actors, there are likely 'hubs' in these networks due to the recurrence of a few lead actors in a substantial fraction of movies. On the other hand, although academic articles do not customarily recruit famous researchers in the same way that Hollywood movies recruit famous actors, there is plenty of evidence suggesting that journals give preferential treatment to well-known authors, their own editors and associate editors, and researchers at prestigious institutions [125]. Unfortunately, the leading statistical journals, such as the *Annals of Statistics*, are not immune to these biases.[2] But since the SSRN is a repository, its articles are not subject to peer review and, therefore, the structure of the network of SSRN coauthorships likely exhibits a wider variety of vertex configurations than articles published in the academic literature. An empirical study of the differences between the network of coauthorships associated to SSRN/arXiv articles and those in the published literature would make for an interesting applied research project. These specific observations lie beyond the scope of our discussion here, but all are important to keep in mind when modeling data for a given application.

10.3.2 Characterization of hyperedge exchangeable network models

In much the same way that the interaction propensity processes are ergodic for edge exchangeable networks, the interaction propensity processes of the preceding section are ergodic for hyperedge exchangeable networks.[3] Recall that the generic representation of edge exchangeable models (Section 9.5) required a technical modification to account for the occurrence of 'blips'. For edge exchangeable networks, each edge consists of at most two distinct vertices, and the blips can be handled by the labels 0 and -1, which serve as placeholders in any interaction involving a vertex that appears in exactly one edge of the network. Hyperedge exchangeable networks also admit blips, but since hyperedges can be of any finite size the corresponding representation includes all non-positive integers $0,-1,-2,\ldots$, subject to the identifiability restriction that if $-k$ appears in an interaction for some $k \geq 1$ then $-k+1,-k+2,\ldots,0$ must also appear.

Toward the representation, we let $\mathscr{F}^{\cong}_{\text{fin}(\mathbb{N})} = \mathscr{F}_{\text{fin}(\mathbb{Z})}/\cong$ be the quotient space of the simplex

$$\begin{aligned}\mathscr{F}_{\text{fin}(\mathbb{Z})} = \{(f_x)_{x\in\text{fin}(\mathbb{Z})} : f_x \geq 0,\ \sum_{x\in\text{fin}(\mathbb{Z})} f_x = 1,\ f_x > 0, \text{ and}\\ -k \in x \text{ implies } -k+1,\ldots,0 \in x\},\end{aligned}$$

where $f \cong f'$ if and only if $\varepsilon_f = \varepsilon_{f'}$.[4] Any $f = (f_x) \in \mathscr{F}_{\text{fin}(\mathbb{Z})}$ determines a distribution ε_f on the space of hypergraphs with edges labeled in $\mathbb{N}$ by first taking $\mathbf{X} = (X_1, X_2, \ldots)$ i.i.d. as in (10.4). From $\mathbf{X}$, we construct a new sequence $\mathbf{X}^* = (X_1^*, X_2^*, \ldots)$ by replacing every occurrence of a non-positive integer with a unique non-positive integer. (Two occurrences of the same non-positive integer within a given interaction are assigned the same unique label for that hyperedge only.) Finally, $\mathbf{Y} = \mathscr{E}_{\mathbf{X}^*}$ is defined as in (10.3), with $\mathbf{Y} \sim \varepsilon_f$ denoting the distribution of $\mathbf{Y}$ obtained in this way.

[2]See http://www.harrycrane.com/AOS-publishing-stats.xlsx for recent publication data showing that a disproportionate fraction of *Annals of Statistics* articles are either authored or co-authored by members of its own editorial board.

[3]For our purposes here the term 'ergodic' essentially means that every edge exchangeable network model can be expressed as a mixture of interaction propensity processes, as in Theorem 9.2. We observe a similar outcome for hyperedge exchangeable networks.

[4]This definition of $\mathscr{F}^{\cong}_{\text{fin}(\mathbb{N})}$ is exactly analogous to that of $\mathscr{F}^{\cong}_{\mathbb{N}\times\mathbb{N}}$ in Section 9.5, to which the reader is referred for more details.

For example, suppose

$$X_1 = \{0,2,4\}, \quad X_2 = \{1,2\}, \quad X_3 = \{-1,0,4,5\}, \quad X_4 = \{-3,-2,-1,0\}.$$

Then $\mathbf{X}^*$ is constructed by relabeling each non-positive element with the largest non-positive label which has not yet been assigned:

$$\begin{array}{cccccc} \mathbf{X}: & \{0,2,4\}, & \{1,2\}, & \{-1,0,4,5\}, & \{-3,-2,-1,0\}, & \cdots \\ \downarrow & \downarrow & \downarrow & \downarrow & \downarrow & \cdots \\ \mathbf{X}^*: & \{0,2,4\}, & \{1,2\}, & \{-2,-1,4,5\}, & \{-6,-5,-4,-3\} & \cdots. \end{array}$$

In this example, X_4^* represents a movie with four actors who have never acted before and will never act again (in the IMDb setting), or an article with four authors who have never published before and will never publish again (in the SSRN setting). Dempsey and I call these vertices *blips*, as their appearance in the data is just a 'blip on the radar'. Blips certainly do occur in real-world interaction processes—many academic papers, for example, are the sole publication of one or more authors—but in practice the behavior of blips is hard to account for since one cannot determine whether a given vertex is a blip based on only a finite sample. (Due to random chance, an author who appears in the database numerous times might only be observed once, even in a relatively large sample.)

Theorem 10.1 (Crane–Dempsey [53]) *Let* $\mathbf{Y}$ *be an edge exchangeable hypergraph with hyperedges labeled in* $\mathbb{N}$. *Then there exists a unique probability distribution* φ *on* $\mathscr{F}^{\cong}_{fin(\mathbb{N})}$ *such that* $\mathbf{Y} \sim \varepsilon_\varphi$, *for* $\varepsilon_\varphi(\cdot)$ *defined by*

$$Pr(\mathbf{Y} \in \cdot; \varphi) = \varepsilon_\varphi(\cdot) = \int_{\mathscr{F}^{\cong}_{fin(\mathbb{N})}} \varepsilon_f(\cdot)\varphi(df).$$

Remark 10.1 *Our choice to allow for multisets in the above presentation allows us to handle a wider range of interaction networks. In the setting of Section 10.1, multiple occurrences of the same actor in a given movie does happen from time to time. (For example, Eddie Murphy plays multiple roles in* The Nutty Professor, The Nutty Professor II, *and* Coming to America.*) In coauthorship networks, however, each author's name appears at most once on each article, and so there is no multiplicity. In this case, the probability distribution* φ *in Theorem 10.1 would assign all of its mass to the subset of* $\mathscr{F}^{\cong}_{fin(\mathbb{N})}$ *for which each vertex appears at most once in a hyperedge with probability 1. The reader is referred to [53, 54] for more discussion of these technical aspects of hyperedge exchangeable models.*

10.4 Scenario: Traceroute sampling of Internet topology

Some of the earliest interest in network science arose from questions about the structure of paths sampled from the Internet. To avoid technicalities here, we think of the Internet as the physical structure of routers (vertices) and wires connecting routers (edges). Information is transmitted over the Internet by passing a message along this physical structure of routers and wires. Given the size and complexity of the Internet, it is natural to ask: What does the Internet look like?

```
traceroute to                    (128.135.10.17), 64 hops max, 72
byte packets
 1  fios_quantum_gateway (192.653.22.69) 2.557 ms  3.073 ms  3.881
ms
 2  lo0-100.nyp-sec-4513 -319.comcast-xxg .net ( 158.19.2130 )  5.677
ms  15.916 ms  15.397 ms
 3  b3319.  nyp-sèc -21.'comcast-xxg .net (100.41.209.120)  16.386
ms  10.418 ms  16.390 ms
 4  * * *
 5  0.ae3.br2.nyc4.alter.net (140.222.231.133)  13.368 ms  9.816
ms  13.792 ms
 6  204.255.168.110 (204.255.168.110)  15.426 ms  28.583 ms
10.595 ms
 7  be2061.ccr42.jfk02.atlas.cogentco.com (154.54.3.69)  13.331 ms
15.715 ms  15.677 ms
 8  be2890.ccr22.cle04.atlas.cogentco.com (154.54.82.245)  34.393
ms  30.023 ms  26.264 ms
 9  be2718.ccr42.ord01.atlas.cogentco.com (154.54.7.129)  32.745
ms  29.979 ms  28.970 ms
10  be2522.agr21.ord01.atlas.cogentco.com (154.54.81.62)  38.000
ms  32.219 ms  36.152 ms
11  te0-0-2-0.nr11.b010917-1.ord01.atlas.cogentco.com
(154.24.4.38)  48.702 ms  33.046 ms  29.155 ms
12  38.104.103.10 (38.104.103.10)  28.439 ms  35.973 ms  31.425 ms
13  192.170.192.19 (192.170.192.19)  34.475 ms  31.105 ms  28.312
ms
14  192.170.192.27 (192.170.192.27)  29.062 ms  71.833 ms  28.353
ms
15  * * *
16                   (128.135.10.17)  31.506 ms  30.992 ms
35.041 ms
```

Figure 10.4 *Output of traceroute path between IP addresses 192.653.22.69 and 128.135.10.17.*

Research Problem 10.1 *In light of the competing vertex-centric and edge-centric points of view put forward in Chapters 6 and 9, respectively, ponder the meaning of the question* 'What does the Internet look like?' *while keeping in mind that there is no canonical or unique way to think about what it means to 'look at' the Internet. How does this question relate to issues of sampling and network modeling? How does the perspective from which the Internet (or any other network) is viewed affect judgments about what that network 'looks like'? See [112] for some early considerations of this problem in the context of network sampling. Earlier discussion in Section 1.3 is also relevant to these questions.*

In keeping with our general theme, we note that what the Internet 'looks like' depends on the point of view from which it is 'looked at', i.e., the angle from which we 'shine our flashlight' in the analogy of Section 1.3. Here we consider the implications of analyzing the Internet network when paths are sampled via *traceroute*. Given a *source* s (usually the IP address from which the search is initiated) and a *target* t, traceroute returns the path (i.e., 'traces the route') of IP addresses visited in accessing t from s. An example of the output from traceroute sampling is shown in Figure 10.4.

Given the population $\mathscr{P}$ of all IP addresses and $s,t \in \mathscr{P}$, let $\mathbf{trace}(s,t)$ denote the path from s to t obtained by traceroute sampling. Any such path is a finite ordered set $(s,x_1,\ldots,x_k,t)$ in $\mathscr{P}$, interpreted here as a message transmitted from s to x_1, then x_1 to x_2, and so on until finally transmitting x_k to t. Writing path$(\mathscr{P})$ to denote the space of finite paths in $\mathscr{P}$, which is in correspondence with ordered multisets of $\mathscr{P}$, we can obtain a snapshot of the Internet topology by sampling sources and targets $(S_1,T_1),\ldots,(S_n,T_n)$ according to some protocol and then observing $\mathbf{X}=(X_1,\ldots,X_n)$ with each $X_i = \mathbf{trace}(S_i,T_i)$ given by traceroute sampling of the path between S_i and T_i.[5] From the collection of paths $X_1,\ldots,X_n$, we assemble a 'path-labeled network' representation, just as we have constructed edge- and hyperedge-labeled networks previously.

10.4.1 Representing the data

It is conventional to analyze the structure induced by the paths $X_1 = \mathbf{trace}(S_1,T_1),\ldots,$ $X_n = \mathbf{trace}(S_n,T_n)$ by decomposing each path $(s,x_1,\ldots,x_k,t)$ into its constituent edges $(s,x_1),(x_1,x_2),\ldots,(x_k,t)$, assembling a graph from these edges, and then studying the features of that graph. But just as we observed in earlier discussions about networks assembled from interaction sequences (in Chapter 9 and again in Section 10.3), decomposing paths into binary edges neither respects the structure of the data nor reflects the context in which the network was observed. Much like decomposing a multiway interaction $\{a,b,c\}$ into its pairwise components ab, ac, and bc is unfaithful to the data structures of Section 10.1, so is decomposing a path (s,a,b,c,t) into its four components sa, ab, bc, ct unfaithful in the context of path sampling via traceroute. Decomposing the network by disassociating paths from their constituent edges may give a misleading perspective on the network's structure vis-à-vis the prevailing sampling scheme; see Problem 9.6.

For a concrete illustration, consider the networks of paths shown in Figure 10.5. The network in Figure 10.5(a) represents a single path (s,a,b,c,t) from s to t; the network in Figure 10.5(b) represents two paths, (s,a) and (a,b,c,t); and the network in Figure 10.5(c) represents three paths, (s,a), (a,b,c,t), and (s,d,e,t). Even though the networks in Figures 10.5(a) and 10.5(b) involve the same edge traversals, as shown in Figures 10.6(a) and 10.6(b), the two observations differ in that the path from s to t in Figure 10.5(a) reflects a single observation (s,a,b,c,t), and thus permits a direct interpretation of the relationship between s and t via traceroute, whereas the two paths in Figure 10.5(b) compose to give a path from s to t which is forced to pass through a. In particular, the fact that the link sa occurs in Figure 10.5(a) cannot be thought of independently of the endpoint t in the path of which it is part, but the occurrence of sa in Figure 10.5(b) does not depend on any relationship between s and t. This distinction is not captured in the vertex-labeled networks of Figure 10.6, as disassociating the edges from the paths in which they occur removes this relevant information about the observation mechanism that produces the network. Figure 10.5(c) shows a third possibility in which the direct path from s to t passes

[5] Although an IP address will not appear more than once in a sample from traceroute, other kinds of path sampling might include repeated occurrences of the same vertex and so we allow for repeated vertices in the paths of path$(\mathscr{P})$.

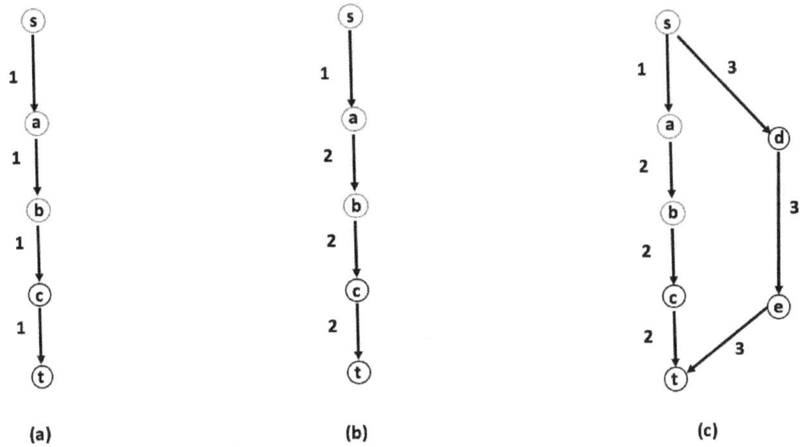

Figure 10.5 *Networks assembled from collections of paths obtained by traceroute sampling. The label on each edge indicates the path to which each edge corresponds. For example, two edges labeled '2' should be interpreted as being part of the same path. (a) A single path from s to t given by (s,a,b,c,t). (b) A network built from two paths, with the first path (s,a) from s to a and the second path (a,b,c,t) from a to t. (c) A network built from three paths, with the first two paths as in (b) and the third (s,d,e,t) from s to t going through d and e.*

through d and e, and thus differs from the composite path (through a) suggested by Figure 10.5(b).

To avoid the ambiguity caused by decomposing paths into their constituent edges, we associate a sample of paths $X_1,\dots,X_n$ to a *path-labeled network*, just as we have previously for samples of pairwise and multiway interactions. As before, the passage from vertex-path labeled structure (as in Figure 10.5(a)-(c)) to path-labeled network (as in Figure 10.7(a)-(c)) reflects the representation of the data as the equivalence class of all collections of paths that induce the same essential structure. The justification for the path-labeled network representation follows the same rationale as in the preceding section and in Chapter 9. Refer to Figures 9.5 and 10.2 for the analogous construction of edge- and hyperedge-labeled networks as equivalence classes of vertex-edge labeled structures. The concept of *path exchangeability* arises by considering how such networks behave when the sources and targets are sampled in an exchangeable way.

10.4.2 *Path exchangeability*

Given a bijection $\rho : \mathscr{P} \to \mathscr{P}$ and a path $a = (a_0,a_1,\dots,a_n) \in \text{path}(\mathscr{P})$, we write $\rho a = (\rho(a_0),\dots,\rho(a_n))$ to denote the action that ρ induces on $\text{path}(\mathscr{P})$ by renaming elements of $\mathscr{P}$. Similarly to the definition of edge- and hyperedge-labeled networks in Chapter 9 and Section 10.3, respectively, the *path-labeled network* associated to a

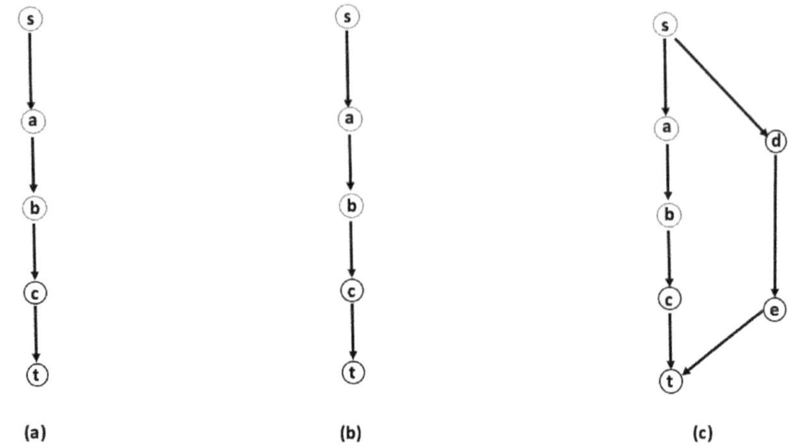

Figure 10.6 *Networks obtained by removing the edge labels (i.e., disassociating edges from their paths) in Figure 10.5. Observe that the representations in (a) and (b), which correspond to two different observations in Figures 10.5(a) and 10.5(b), coincide upon removing pathwise information.*

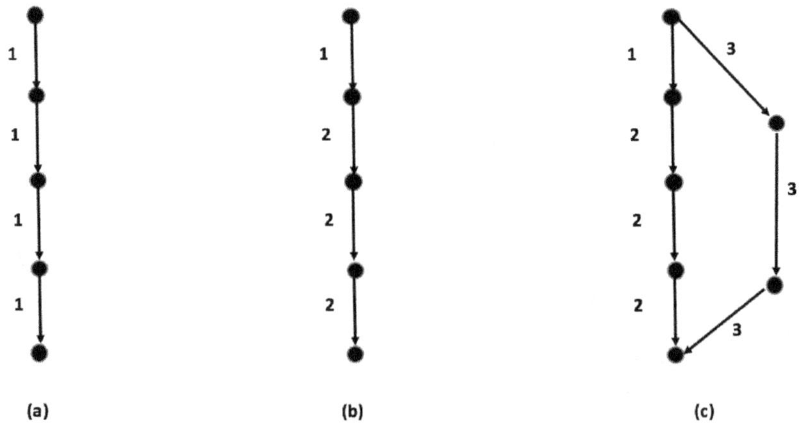

Figure 10.7 *Path-labeled networks associated to each of the observations in Figure 10.5 through the correspondence in display* (10.5). *Note that the path-labeled networks in (a) and (b) are different, capturing the fact that Figures 10.5(a)–(b) represent different observations. The vertex-labeled networks in Figures 10.6(a)–(b) do not make this distinction.*

sequence of paths $\mathbf{x} = (x_1, x_2, \ldots)$ in $\mathrm{path}(\mathscr{P})$ is given by

$$\mathscr{E}_{\mathbf{x}} = \{\rho\mathbf{x} \mid \rho : \mathscr{P} \to \mathscr{P} \text{ is a bijection}\}, \tag{10.5}$$

where as usual we overload notation and write $\rho\mathbf{x} = (\rho(x_1), \rho(x_2), \ldots)$ for the componentwise application of ρ to $\mathbf{x}$. An exchangeable sequence of paths $X_1, X_2, \ldots$ thus gives rise to a *path exchangeable network* in the same way that exchangeable sequences of pairs and exchangeable sequences of multisets give rise to edge exchangeable and hyperedge exchangeable networks in those preceding discussions. If $\mathbf{X} = (X_1, X_2, \ldots)$ is exchangeable in $\mathrm{path}(\mathscr{P})$, then $\mathbf{Y} = \mathscr{E}_{\mathbf{X}}$ is path exchangeable in the sense that its distribution is invariant under arbitrary relabeling of its paths. At this point, the analog to the interaction propensity processes in Sections 9.4 and 10.3 immediately suggests itself: Take any $(f_x)_{x \in \mathrm{path}(\mathscr{P})}$ with $f_x \geq 0$ and $\sum_{x \in \mathrm{path}(\mathscr{P})} f_x = 1$ and construct $\mathbf{Y} \sim \varepsilon_f$ by first taking $\mathbf{X} = (X_1, X_2, \ldots)$ i.i.d. from

$$\Pr(X_i = x; f) = f_x, \quad x \in \mathrm{path}(\mathscr{P}), \tag{10.6}$$

and then putting $\mathbf{Y} = \mathscr{E}_{\mathbf{X}}$ as in (10.5).

Once again, the generic form of path exchangeable networks arises by modifying the interaction propensity process to allow for blips. Let $\mathscr{F}^{\cong}_{\mathrm{path}(\mathbb{N})} = \mathscr{F}_{\mathrm{path}(\mathbb{Z})}/\cong$ be the equivalence class obtained by putting $f \cong f'$ whenever they both determine the same distribution as follows. Let $f = (f_x)_{x \in \mathrm{path}(\mathbb{Z})}$ be indexed by paths labeled by both negative and positive integers and take $\mathbf{X} = (X_1, X_2, \ldots)$ to be i.i.d. from the distribution in (10.6). From $\mathbf{X}$, define $\mathbf{X}^* = (X_1^*, X_2^*, \ldots)$ by replacing each occurrence of a non-positive label with a unique non-positive label and then put $\mathbf{Y} = \mathscr{E}_{\mathbf{X}^*}$. For example,

$$\begin{array}{cccccc}
\mathbf{X}: & (3,0,2), & (3,1,4), & (2,0), & (2,0,1,-1), & \ldots \\
\downarrow & \downarrow & \downarrow & \downarrow & \downarrow & \ldots \\
\mathbf{X}^*: & (3,0,2), & (3,1,4), & (2,-1), & (2,-2,1,-3) & \ldots.
\end{array}$$

The representation theorem of path exchangeable networks has the same form as Theorems 9.2 and 10.1.

Theorem 10.2 (Crane–Dempsey [53]) *Let $\mathbf{Y}$ be an infinite path exchangeable random network. Then there exists a unique probability distribution φ on $\mathscr{F}^{\cong}_{path(\mathbb{N})}$ such that $\mathbf{Y} \sim \varepsilon_\varphi$, for ε_φ defined as the φ-mixture of interaction propensity processes ε_f,*

$$\varepsilon_\varphi(\cdot) = \int_{\mathscr{F}^{\cong}_{path(\mathbb{N})}} \varepsilon_f(\cdot)\varphi(df).$$

10.4.3 Relational exchangeability

By now the basic structure of relationally exchangeable networks ought to be clear. In general, one can take $\mathscr{R}$ to be any class of relations and let $\mathscr{R}(\mathscr{P})$ be the set of all such relations indexed by the elements in a population $\mathscr{P}$. In Chapter 9, $\mathscr{R}$ consists of pairs (i.e., $\mathscr{R}(\mathscr{P}) = \mathscr{P} \times \mathscr{P}$); in Section 10.3, $\mathscr{R}$ consists of finite multisets (i.e.,

$\mathscr{R}(\mathscr{P}) = \text{fin}(\mathscr{P})$); and throughout Section 10.4, $\mathscr{R}$ consists of paths (i.e., $\mathscr{R}(\mathscr{P}) = \text{path}(\mathscr{P})$). With the relevant set of relations specified, the definition of the interaction propensity process follows without any change: let $(f_x)_{x\in\mathscr{R}(\mathscr{P})}$ satisfy $f_x \geq 0$ and $\sum_{x\in\mathscr{R}(\mathscr{P})} f_x = 1$ and draw relations $X_1, X_2, \ldots$ i.i.d. according to

$$\Pr(X_i = x; f) = f_x, \quad x \in \mathscr{R}(\mathscr{P}).$$

Given $\mathbf{X} = (X_1, X_2, \ldots)$, construct a *relationally exchangeable network* (in $\mathscr{R}$) as the equivalence relation $\mathscr{E}_{\mathbf{X}}$ determined by $\mathbf{X}$, just as in (9.3), (10.3), and (10.5) above. The representation of relationally exchangeable networks then follows similarly by taking $\mathscr{P} = \mathbb{Z}$ and replacing any occurrence of a non-positive vertex by a unique 'blip', as in the passage from $\mathbf{X} \mapsto \mathbf{X}^*$ in previous special cases. The realization that many networks are more naturally represented and analyzed as relationally-labeled structures seems to have been first discovered and studied in [52, 53, 54], to which the reader is referred for further details about the general case. I conclude this chapter by discussing the general class of relationally exchangeable Hollywood models.

10.5 General Hollywood model

The Hollywood model introduced in Section 9.7 (cf. [54]) extends to a more general class of models for random hyperedge- and path-labeled networks.[6] For this extension, we let $\text{fin}_k(\mathbb{N})$ be the set of all finite multisets of $\mathbb{N}$ with cardinality $k \geq 1$ and define $\text{fin}(\mathbb{N}) = \cup_{k\geq 1} \text{fin}_k(\mathbb{N})$ as the set of all finite multisets of $\mathbb{N}$. For each $k \geq 1$, we define a distribution $f^{(k)} = (f_s^{(k)})_{s\in\text{fin}_k(\mathbb{N})}$ on $\text{fin}_k(\mathbb{N})$ and, given a distribution $\nu = (\nu_k)_{k\geq 1}$ on the positive integers, we define $f = (f_s)_{s\in\text{fin}(\mathbb{N})}$ by

$$f_s = \nu_k f_s^{(k)}, \quad s \in \text{fin}_k(\mathbb{N}), \quad k \geq 1, \tag{10.7}$$

where $k = |s|$ is the cardinality of s. By the law of total probability, any probability distribution $(f_s)_{s\in\text{fin}(\mathbb{N})}$ on $\text{fin}(\mathbb{N})$ can be uniquely expressed as in (10.7).

To generate a draw from the hyperedge-labeled interaction propensity process directed by f, we first take an i.i.d. sequence $X_1, X_2, \ldots$ of finite subsets of $\mathbb{N}$ with distribution

$$\Pr(X_i = s; f) = f_s, \quad s \in \text{fin}(\mathbb{N}),$$

and then construct a hyperedge-labeled network $\mathscr{E}_{\mathbf{X}}$, as in (10.3), by associating each X_j to a hyperedge with label j; see Figure 10.2 for illustration. The resulting network $\mathscr{E}_{\mathbf{X}}$ is hyperedge exchangeable.

The *Hollywood model with parameter* (α, θ) *on hyperedge-labeled networks* is defined by first drawing $f^{(1)} \in \Delta_1$ according to the GEM distribution with parameter (α, θ) (see Section 9.7) and then, given $f^{(1)} = (f_i)_{i\geq 1}$, putting

$$f_s = \nu_k \prod_{i=1}^{k} f_{s_i}, \quad s = (s_1, \ldots, s_k) \in \text{fin}(\mathbb{N}). \tag{10.8}$$

[6]The description of the Hollywood model presented in Chapters 9 and 10 first appeared, in whole or in part, in an article published by the American Statistical Association [54].

In the Hollywood model, edges correspond to ordered multisets, so that vertices can appear multiple times in a single edge and the order in which vertices appear in an edge matters. Neither property can be relaxed without some theoretical repercussions, but I defer that discussion to [54].

Using the same notation as in Section 9.7, we let $v(\mathbf{Y}_n)$ denote the number of vertices in $\mathbf{Y}_n$ and $e(\mathbf{Y}_n) = n$ denote the number of hyperedges/paths in $\mathbf{Y}_n$. In the intended semantic interpretation of the Hollywood process as a description of movie formation, every edge in $\mathbf{Y}_n$ corresponds to a movie, with the vertices incident to edge i corresponding to the actors in movie $i = 1,\ldots,n$. The special case of network data with binary edges in Chapter 9 is easily handled by setting $\nu_2 = 1$ in (10.8).

As an alternative to the above definition in terms of the GEM distribution, we generalize the binary Hollywood process construction from Section 9.7.1 as follows. Given $\mathbf{Y}_{n-1} = \mathbf{y}$, for $n \geq 1$, choose the number of roles K_n in the next movie independently according to ν. Now, given $\mathbf{Y}_{n-1} = \mathbf{y}$ and $K_n = k$, choose the k actors in order of their prominence, first filling the lead role, then the second role, and so on until all k roles have been filled. Let $N_n(j)$ be the number of unique actors observed up to and including the $(j-1)$st role in movie n. (Thus, e.g., $N_n(1)$ is the number of unique actors observed up to and including the 0th role in movie n, i.e., those actors appearing in movies $1,\ldots,n-1$.) For $j = 1,\ldots,k$, label the actors arbitrarily $1,\ldots,N_n(j)$, with $D_n(i,j)$ denoting the number of roles for which the actor labeled i has been cast up to and including the $(j-1)$st role of movie n. (Remember that an actor is allowed to play more than one role in a given movie.) The actor $v_n(j)$ cast in the jth role of movie n is chosen randomly among the actors labeled $1,\ldots,N_n(j)$ and a previously unseen actor, labeled $N_n(j)+1$, according to

$$\Pr(v_n(j) = i \mid D_n(i,j),\ i = 1,\ldots,N_n(j)) \propto \begin{cases} D_n(i,j) - \alpha, & i = 1,\ldots,N_n(j), \\ \theta + \alpha N_n(j), & i = N_n(j)+1. \end{cases} \tag{10.9}$$

Vertices continue to be chosen according to (10.9) until all k roles of movie n have been filled.

Example 10.1 *Let ν be a probability distribution on the positive integers, $0 < \alpha < 1$, and $\theta > -\alpha$, and suppose $K_1, K_2, \ldots$ are i.i.d. from $\nu = \{\nu_k\}_{k\geq 1}$. Then for $(K_1, K_2, \ldots) = (3,2,4,\ldots)$, suppose we observe X_1, X_2, X_3 as*

- $X_1 = (1,2,1)$ *with probability*

$$\frac{\theta}{\theta} \times \frac{\theta+\alpha}{\theta+1} \times \frac{1-\alpha}{\theta+2},$$

- $X_2 = (3,2)$ *with probability*

$$\frac{\theta+2\alpha}{\theta+3} \times \frac{1-\alpha}{\theta+4}, \quad \textit{and}$$

- $X_3 = (1,4,3,5)$ *with probability*

$$\frac{2-\alpha}{\theta+5} \times \frac{\theta+3\alpha}{\theta+6} \times \frac{1-\alpha}{\theta+7} \times \frac{\theta+4\alpha}{\theta+8}.$$

The Hollywood process (10.9) *assigns the following probability to the hyperedge-labeled graph* $\mathscr{E}_{\mathbf{X}}$ *associated to the observation* $X_1 = (1,2,1)$, $X_2 = (3,2)$, *and* $X_3 = (1,4,3,5)$*:*

$$\nu_3 \times \nu_2 \times \nu_4 \times \alpha^5 \frac{(\theta/\alpha)(\theta/\alpha+1)\cdots(\theta/\alpha+4)}{\theta(\theta+1)\cdots(\theta+8)}(1-\alpha)^3(1-\alpha+1).$$

The distribution of $\mathbf{Y}_n$ constructed from the Hollywood process with parameter (ν,α,θ), as in (10.9), can be written in explicit form by

$$\Pr(\mathbf{Y}_n = \mathbf{y}; \alpha, \theta, \nu) = \tag{10.10}$$
$$\left[\prod_{k\geq 1} \nu_k^{M_k(\mathbf{y})}\right] \alpha^{v(\mathbf{y})} \frac{(\theta/\alpha)^{\uparrow v(\mathbf{y})}}{\theta^{\uparrow m(\mathbf{y})}} \prod_{k=2}^{\infty} \exp\{N_k(\mathbf{y})\log((1-\alpha)^{\uparrow(k-1)})\},$$

where $\mathbf{y}$ is any hyperedge-labeled network with n oriented (i.e., directed) hyperedges, $v(\mathbf{y})$ is the number of non-isolated vertices in $\mathbf{y}$, $N_k(\mathbf{y})$ is the number of vertices in $\mathbf{y}$ with degree $k \geq 1$, $m(\mathbf{y})$ is the total degree of $\mathbf{y}$, $M_k(\mathbf{y})$ is the number of hyperedges of size k in $\mathbf{y}$, and $x^{\uparrow j} = x(x+1)\cdots(x+j-1)$ is the ascending factorial function. By putting $\nu_2 = 1$, $\nu_k = 0$ for all $k \geq 2$, and comparing (10.10) to the binary Hollywood model in Section 9.7, we see that the two coincide. The distribution in (10.10) is called the *Hollywood model with parameter* (ν,α,θ) [54].

The Hollywood model was first introduced under the heading of the *Poisson–Dirichlet model* in [55, Section 6.3]. I also noted the connection between the Hollywood model and the two parameter Poisson–Dirichlet distribution in [45]. From the relationship between the Hollywood model and the Poisson–Dirichlet distribution through the vertex components model (Section 9.6), one can deduce that networks from the general Hollywood model in (10.10) exhibit power law degree distribution with exponent $\gamma = \alpha + 1$ when $0 < \alpha < 1$ and $\theta > -\alpha$; see [54, Section 5.3]. Furthermore, $(\mathbf{Y}_n)_{n\geq 1}$ is sparse provided $1/\mu < \alpha < 1$, where $\mu = \sum_{k\geq 1} k\nu_k$ is the mean interaction size in the general Hollywood model with parameter (α,θ,ν). This last observation refines the analysis in Section 9.7.3, where we saw that the Hollywood model restricted to pairwise interactions is sparse for $1/2 < \alpha < 1$. Since the binary case corresponds to $\nu_2 = 1$, and thus $\mu = 2$, these two analyses agree.

10.6 Markovian vertex components models

The vertex components models described in Sections 9.6 and 10.5 assume that vertices occur in any given interaction independently of one another. As I have already noted in Problem 9.1, it is unreasonable in the scenario of Section 9.1 to assume that the conditional probability of observing a call with receiver r, given that the caller is s, is f_r independently of s. Similarly, when sampling from a movie database as in Section 10.1.1, the occurrence of a certain actor in a sampled movie is likely to affect the conditional probability of another actor's occurrence in that same movie. For example, there are many famous duos, such as Abbott and Costello, Dean Martin and Jerry Lewis, etc., who tend to perform together. Similar synergies exist for scientific coauthorships.

With these observations in mind, it is natural to consider a more robust class of models. In the binary case, Problem 9.1 suggests one possible extension by taking a collection of vertex components $(f_i)_{i\geq 1}$ and conditional probabilities $(f'_{j|i})_{i,j\geq 1}$ and defining the probability of a given pair by

$$f_{(i,j)} = f_i f'_{j|i}, \quad i,j \geq 1.$$

Here, $(f'_{j|i})_{i,j\geq 1}$ describes the conditional probability that the second vertex is j given that the first is i. See Problem 9.1 for more on this approach. To extend this idea to hypergraphs, we let $\nu = \{\nu_k\}_{k\geq 1}$ be a distribution on the positive integers and define

$$f_s = \nu_k f_{s_1} \prod_{j=2}^{k} f_{s_j|s_{j-1}}, \quad s = (s_1, \ldots, s_k) \in \mathrm{fin}(\mathbb{N}).$$

But even this offers just a minor improvement over the ordinary vertex components model, since it only incorporates dependence between vertices that occur in adjacent positions of a directed hyperedge. Although such dependence could make sense for modeling path exchangeable networks (Section 10.4), it seems overly simplistic for modeling the actors and scientific collaboration networks in Section 10.1. Readers are encouraged to explore this and other potential extensions of relationally exchangeable models in future work.

10.7 Contexts for relational sampling

As for edge exchangeable models in Chapter 9, I have defined relationally exchangeable models in terms of the generative interaction propensity process. By this generative description, (4.4) immediately implies consistency under selection as well as coherence of relationally exchangeable models in a selection sampling context. The more general edge sampling contexts discussed throughout Section 9.8 also transfer wholesale to relationally-labeled networks, e.g., hyperedge- and path-labeled networks, without any change in notation. Any of the questions asked of edge sampling in Section 9.8 can just as well be asked of relational sampling and relationally exchangeable networks here. Because of the close parallels to this earlier discussion, I do not rehash the details here. Refer to Section 9.8 for the relevant coverage.

10.8 Concluding remarks and further reading

The mindset of edge and relational exchangeability put forward in Chapters 9 and 10 is an important step toward a complete theory for complex data structures. The significance of the insight afforded by representing interaction data as an equivalence class of structures, as in (9.3), (10.3), and (10.5), has not yet been appreciated by the wider community of statisticians, data scientists, mathematicians, and computer scientists. Given the tendency to resist change, it is likely that the 'networks-as-graphs' point of view, including its extensions to 'networks-as-hypergraphs' and 'multilayer networks', will persist for at least a while longer in network science. But if progress—in the form of clearer and more illuminating insights, not just more articles about the same old models and methods—is to be attained, a change in perspective is not only beneficial but necessary.

If we visualize data analysis as 'shining a flashlight' (as in Section 1.3), then the different viewpoints offered by the vertex-centric 'networks-as-graphs' perspective (Section 1.2) and the alternative 'edge-centric' perspective presented throughout Chapters 9 and 10 correspond to shining the light from different angles. It is expected that the shadow cast by shining the light from one angle will have a different texture than that cast by shining it from a different angle. Evidently, this choice of angle affects which aspects of the data are visible. Contrast, e.g., the inability of vertex exchangeable models to 'see' sparsity and power law with the ease with which edge and relationally exchangeable models are able to model these properties. Since statistical inference takes place in these 'shadows', we ought to be mindful of how we shine our flashlight, both through the models we specify and the data representations we choose. The Crane–Dempsey edge and relationally exchangeable perspective [53, 54] represents a first step out of the dark shadow cast by the 'networks-as-graphs' perspective of network analysis. Future developments in both extending edge and relational exchangeability as well as establishing new network modeling frameworks will be essential to fulfilling the vision of Chapter 1, in which statistical network analysis is viewed as the foundation for structured, complex data analysis.

Chapter 11

Dynamic network models

Social networks, whose edges represent friendships, often change over time due to increased/decreased social activity, formation of new friendships, loss of old friendships, etc. Brain networks, whose edges represent the transmission of an electrical charge between neurons, also change depending on brain function and activity. These are just two examples of networks with dynamic edge patterns. There are many more, with each admitting its own behaviors depending on the domain of application. In some scenarios, both vertex and edge sets change over time; in others, the edges associated to different vertices evolve on different time scales. In this final chapter I focus only on the most basic aspects of modeling dynamic networks whose vertex set stays fixed while its edges vary over time. I leave more nuanced considerations (e.g., dynamic vertex and edge sets, different time scales, etc.) to future developments in their respective fields of study. Because I only emphasize the most basic statistical properties of dynamic network models, the contents of this chapter barely scratch the surface of what is needed to lay a solid foundation for dynamic network analysis. Plenty of questions remain open and invite future exploration.

Dynamic versus evolving networks

The *dynamic* networks studied here should not be confused with the *evolving* (or generative) networks from previous chapters. For example, the preferential attachment model (Section 4.2) describes a network which evolves by the addition of one new vertex at each step. The resulting system of networks $(\mathbf{Y}_n)_{n\geq 1}$ grows in such a way that its existing structure never changes: once the status of an edge Y_{ij} between i and j is determined to be present (or absent), it remains present (or absent) forevermore. By contrast, a *dynamic network* is one whose entire structure varies with respect to time, as shown in Figure 11.1. A dynamic network is thus a collection $\mathbf{Y} = (Y(t))_{t\in T}$ indexed by a set of times T, with each $Y(t)$ representing a network for a (fixed) vertex set. So whereas the components of an evolving network $(\mathbf{Y}_n)_{n\geq 1}$ are related to one another through how they evolve to form a single (limiting) network, the components of a dynamic network are related to one another through their association over time.

Thus, in addition to the usual sampling issues discussed throughout Chapters 3 and 6–10, dynamic network models should also account for the temporal dimension along which the dynamics take place. To streamline this chapter as much as possible, I confine to the 'networks-as-graphs' context, so that each $Y(t)$ in $\mathbf{Y} = (Y(t))_{t\in T}$ is

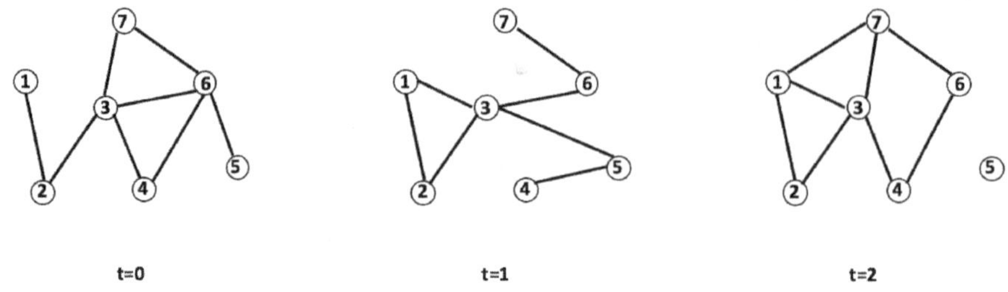

Figure 11.1 *A dynamic network with 7 vertices observed at evenly spaced times* $t = 0, 1, 2$.

regarded as a $\{0,1\}$-valued array indexed by a fixed set of vertices, either $\mathbb{N}$ (if infinite) or $[n]$ (if finite with cardinality n). Network dynamics for edge- and relationally-labeled networks are a topic of active development, and so are not discussed here. See [124] for some recent work in that direction.

Units of observation

In the setting assumed here, dynamic network data is obtained by sampling both a subset of vertices $S \subseteq V$ and a subset of times $T' \subseteq T$ for $\mathbf{Y} = (Y(t))_{t \in T}$ such that each $Y(t)$ is an array in $\{0,1\}^{V \times V}$. Described in this way, the units of observation are determined both by the sample of vertices S and the set of times T' at which the dynamics for that sample are observed, and therefore (by the discussion in Sections 3.7–3.8) the sample size of dynamic network data is jointly determined by both the size of the observed network, i.e., the number of vertices, and the duration of time over which its dynamics are observed. In general, one should bear in mind that the sampled vertices and times of a dynamic network may depend on each other as well as on the underlying network structure; but for the most part we only consider dynamic networks that are observed for a subsample of vertices at a finite collection of times, both of which are selected independently of the network and of each other. In particular, we assume throughout this chapter that a sample of size (t', m), for $t' \leq t$ and $m \leq n$, from a dynamic network $\mathbf{Y} = (Y(s))_{s=0,1,\ldots,t}$ evolving on $\{0,1\}^{n \times n}$ is obtained by restriction of $\mathbf{Y}$ to times $\{0,1,\ldots,t'\}$ and vertices $[m]$, i.e., we observe $(\mathbf{S}_{m,n} Y(s))_{s=0,1,\ldots,t'}$. In Section 11.2.2 I briefly discuss how this setup can be extended to account for random sampling of vertices and times, as was done for singular observations of network data in Sections 3.9, 7.3.7, 8.6, 9.8, and 10.7.

Outline

To convey the above ideas, I first canvass the main considerations of dynamic network analysis (Section 11.2). I then go on to discuss two broad classes of dynamic network models, called *rewiring processes* (Section 11.3) and *Lévy processes* (Section 11.4). While their probabilistic foundation has mostly been established through a series of articles in the mathematical probability literature [44, 48, 49, 50, 57, 58, 59], these models have not yet been developed into viable statistical methodologies for dynamic network analysis. Because of their definition in terms of basic structural assumptions (i.e., exchangeability, projective Markov property, stationary and independent increments), rewiring and Lévy process models are a natural building block for a larger theory of dynamic network analysis. The exposition in this chapter is meant to clarify the main conceptual ideas that are needed in order to make progress in developing such a theory. The reader interested in the technical details is referred to the above series of papers.

11.1 Scenario: Dynamics in social media activity

Consider a network of interactions on a social media platform, such as Twitter, for which each vertex is associated to a different user account and these vertices interact by 'retweeting', 'liking', or 'replying' to content posted by other users.[1] For simplicity, we call any of these actions (i.e., retweet, like, or reply) an *interaction* and construct a dynamic network of Twitter interactions as follows. Assuming Twitter activity has been monitored for each of $T+1$ consecutive days, marked $t = 0, 1, \ldots, T$, we label users $1, \ldots, N$ and record their interactions on day t as an array $Y(t) = (Y_{ij}(t))_{1 \le i,j \le N}$, with

$$Y_{ij}(t) = \begin{cases} 1, & i \text{ and } j \text{ interacted on day } t, \\ 0, & \text{otherwise.} \end{cases}$$

The totality of interactions gives a time-indexed collection of networks $\mathbf{Y} = (Y(t))_{t=0,1,\ldots,T}$.

Suppose we are interested in understanding certain attributes of this dynamic Twitter network. For example, are users who interact on a given day (say, day t) more or less likely to interact again on the next day (say, day $t+1$)? To what extent can future activity be predicted from past activity? And so on. As for many modern social media sites, Twitter currently has a population of hundreds of millions of users, making it practically impossible to analyze the complete dynamics of Twitter interactions over time. We instead infer the dynamics of the population $\mathbf{Y}$ by observing the interaction dynamics for a sample of $n \ll N$ users whose activity we monitor up to some time $T' \le T$. Our objective is to gain insight into the dynamics of the population of all Twitter accounts based on the dynamics of the sample.

11.2 Modeling considerations

The above scenario describes dynamic network data constructed from Twitter interactions among a sample of individuals over some duration of time. When devising a model for $\mathbf{Y}$, the following questions are worth bearing in mind.

[1]The terms 'retweet', 'like', and 'reply' are used here in the same sense as on Twitter. In the modern vernacular, the act of posting content on Twitter is called 'tweeting'. Anytime a user posts a tweet, the content is visible to his/her followers. Other users can broadcast content from any tweet to his/her followers by 'retweeting', can express interest/support for content by 'liking' the tweet, or can reply to a tweet by 'replying'.

1. How does the network structure change with respect to time? To what extent are past interactions $(Y(s))_{s=0,1,\ldots,t-1}$ useful for predicting future interactions $(Y(s))_{s=t,t+1,\ldots}$?
2. If the data reflects a sample from the population, then how are the dynamics of the sampled network related to those of the population network?

Below we consider both of these questions within the context of Markov chain models for $\mathbf{Y}$.

11.2.1 Network dynamics: Markov property

Let $\mathbf{Y} = (Y(t))_{t=0,1,\ldots,T}$ record the dynamics of the population network, whose transition behavior is of primary interest. A preliminary null model for $\mathbf{Y}$ assumes $Y(0), Y(1), \ldots, Y(T)$ are i.i.d. according to one of the previously discussed distributions for $\{0,1\}$-valued arrays (see Chapters 6–8). If such an assumption were appropriate, then the 'dynamic' aspect of $\mathbf{Y}$ adds no complexity beyond what we have already discussed in previous chapters. But for the network of Twitter activity in Section 11.1 it seems reasonable to expect that interactions exhibit non-trivial temporal dependence which ought to be reflected in the model. For example, two users who interact at time t seem more likely than two random users to have a follow-up interaction at some time in the near future. By the same logic, two users who have not interacted for a long period of time would seem unlikely to interact in the near future.

For a venue as complex as Twitter, it is easy to imagine that the entire history of interactions might be informative about future interactions. For example, two users might regularly interact on weekends, but never on weekdays; or one user might only check his account every few days, so that his interactions tend to lag behind the response times of others. But these possibilities are far too application-specific and complicated to discuss here. We instead focus on the basic theoretical implications of *time homogeneous Markov chain* models for dynamic network data.

Let $P(\cdot,\cdot)$ be a *transition probability* on $\{0,1\}^{N\times N}$, i.e., $P(\mathbf{y},\cdot)$ defines a probability distribution on $\{0,1\}^{N\times N}$ for each $\mathbf{y} \in \{0,1\}^{N\times N}$. The distribution of $\mathbf{Y} = (Y(t))_{t=0,1,\ldots}$ is defined by first specifying a distribution for $Y(0)$, called the *initial distribution*, and then generating $(Y(t))_{t=0,1,\ldots}$ sequentially according to the conditional distribution,

$$\Pr(Y(t+1) = \mathbf{y}' \mid Y(t) = \mathbf{y},\ (Y(s))_{s=0,1,\ldots,t}) = P(\mathbf{y},\mathbf{y}'), \quad \mathbf{y},\mathbf{y}' \in \{0,1\}^{N\times N},\ t \geq 0. \tag{11.1}$$

Thus, for any $T \geq 1$, the distribution of $(Y(t))_{t=0,1,\ldots,T}$ is given by

$$\Pr((Y(t))_{t=0,\ldots,T} = (\mathbf{y}_t)_{t=0,\ldots,T}) = \Pr(Y(0) = \mathbf{y}_0) \prod_{t=1}^{T} P(\mathbf{y}_{t-1},\mathbf{y}_t). \tag{11.2}$$

Any model for $(Y_t)_{t=0,\ldots,T}$ specified as in (11.2) is called a *time homogeneous Markov chain model*. Under the *Markov* assumption, the conditional distribution of $Y(t)$ given all past states $(Y(s))_{s<t}$ depends only on the immediately preceding state $Y(t-1)$, as in (11.1). The model is called *time homogeneous* because the conditional

distribution of $Y(t)$ depends only on the previous state $Y(t-1)$ and not on the time t at which the transition from $Y(t-1)$ to $Y(t)$ takes place. These assumptions can be relaxed in several ways, e.g., by assuming the *kth-order Markov property*, by which the conditional distribution of $Y(t)$ depends on the previous k states $Y(t-1),\ldots,Y(t-k)$, or *time inhomogeneity*, by which the conditional distribution of $Y(t)$ depends on both the previous state $Y(t-1)$ and the time t at which the transition takes place. We only consider the first-order, time homogeneous case below.

11.2.1.1 Modeling the initial state

Modeling $(Y(t))_{t=0,1,\ldots,T}$ requires a description of both the initial state $Y(0)$ and the dynamics $Y(t) \mapsto Y(t+1)$ at each time t. Though we are primarily interested in the dynamics, the choice of initial distribution also deserves attention. In principle, the initial state can be described by any of the model classes for $\{0,1\}$-valued arrays discussed throughout Chapters 2, 6, 7, and 8. It is important to realize, however, that in many situations the initial observation $Y(0)$ does not reflect the state of the network at the time it first came into existence; and thus the time at which the initial state $Y(0)$ is observed is itself part of the observation process. In the Twitter interactions of Section 11.1, for example, we do not assume that the observed sequence $(Y(t))_{t=0,1,\ldots,T}$ goes back to the inception of Twitter. And so, when analyzing the data, we should take into account that the process by which users interact on Twitter has been in force since long before our first observation $Y(0)$, and that those past interactions—indeed the very process that we are trying to model—affect the observed network dynamics.

Assuming that no special circumstances distinguish the observed times from all other times—in other words, the observed times have been chosen irrespective of the network dynamics—it is prudent to proceed under the assumption that the population network has been evolving for an indefinite (i.e., potentially infinite) amount of time into the past. If the process has been changing according to the same dynamics for such a period of time (i.e., homogeneous) and if the transition probability P is sufficiently well-behaved so that it has a unique stationary distribution π (i.e., ergodic), then the initial state of $(Y(t))_{t=0,1,\ldots,T}$ is distributed according to π. Thus, without a compelling reason to the contrary, we model the initial distribution of $\mathbf{Y}$ by the stationary distribution π. Under this assumption, the marginal distribution of every $Y(t)$, for $t=0,1,\ldots,T$, is π, and the marginal distribution of each observation $Y(t)$ is unaffected by the arbitrary time at which we started observing $\mathbf{Y}$.

I expect that both the rewiring and Lévy process models presented in Sections 11.3 and 11.4 below possess a unique stationary distribution under non-restrictive conditions on their transition behavior, but so far their convergence behavior has not been explored in detail. Readers are referred to any graduate text on stochastic processes for a review of basic Markov chain theory.

Research Problem 11.1 *Study the convergence to stationary distribution for the Lévy processes and exchangeable rewiring processes presented below. For both classes give necessary and sufficient conditions under which a unique stationary distribution exists for the process evolving on* $\{0,1\}^{\mathbb{N}\times\mathbb{N}}$. *If possible, give upper and lower bounds on the rate of convergence to the stationary distribution, and find extra* *conditions under which these processes exhibit the cutoff phenomenon. See [56] for a discussion of the cutoff phenomenon for a related class of partition-valued Markov chains.*

11.2.1.2 Is the Markov property a good assumption?

When faced with a specific application, the appropriateness of the Markov assumption warrants careful consideration. In Section 11.1, Markovian dynamics seem reasonable for describing many of the interactions that occur. For example, an active user who 'likes' another user's content at time t may be inclined to 'like' that same user's content again at time $t+1$, especially if the content at times t and $t+1$ are related to one another. But as we have already noted above, the time scale on which interactions occur is likely to vary from user to user. Interaction patterns for a user who monitors social media several times per day are likely to differ from those of someone who looks at social media only a few times a week or a few times a month. A complete theory for modeling such networks is far beyond the reach of current capabilities. But even without such a theory, practical methodologies which account for such inhomogeneities in dynamic networks could be worthwhile for certain applications.

Research Problem 11.2 *Extend any of the models presented below to account for inhomogeneity in how individuals interact over time. An ideal extension should be tailored to a specific application domain, but this is not a strict requirement.*

Research Problem 11.3 *In light of the rewiring processes discussed in Section 11.3, bear in mind that a dynamic network model for which vertices exhibit inhomogeneous behavior cannot be exchangeable in the sense defined below. Explore the possibility of bringing relative exchangeability (Chapter 8) and/or relational exchangeability (Chapters 9–10) to bear on this problem.*

11.2.1.3 Temporal Exponential Random Graph Model (TERGM)

The *temporal exponential random graph model* (TERGM) [63, 87, 86, 109, 140] has been the most widely studied statistical model for dynamic networks to date. Let $\{0,1\}^{n\times n}$ be the state space for binary relational data of size n, let Θ be a parameter space, and define a joint sufficient statistic $T : \{0,1\}^{n\times n} \times \{0,1\}^{n\times n} \to \mathbb{R}^d$, where $d \geq 1$ is the length of the sufficient statistic vector $T = (T_1,\dots,T_d)$. The transition probabilities of the TERGM are given by

$$\Pr(Y(t+1) = \mathbf{y}' \mid Y(t) = \mathbf{y}; \theta, T) \propto \exp\{\eta(\theta)\cdot T(\mathbf{y},\mathbf{y}')\}, \quad \mathbf{y},\mathbf{y}' \in \{0,1\}^{n\times n}, \tag{11.3}$$

where $\eta(\theta)$ is the natural parameter for the exponential family and $\eta(\theta)\cdot T(\mathbf{y},\mathbf{y}') = \sum_{i=1}^{d} \eta_i(\theta)T_i(\mathbf{y},\mathbf{y}')$. Note that the transition probabilities for the TERGM in (11.3) incorporate temporal (i.e., Markovian) dependence into the ERGM defined in Section 2.3, with the main difference between the ERGM distribution (2.8) and the TERGM (11.3) being the joint dependence of the sufficient statistic T on both $\mathbf{y}$ and $\mathbf{y}'$ in the latter.

As in Chapter 2, the TERGM may be appropriate for modeling the network dynamics of a fully observed population, as in a small community of friends, a high school, or a company. But, as we also observed in Chapter 2, exponential random graph models struggle with sampled networks. In the Twitter scenario of Section 11.1, for example, the model should also account for the fact that the observed dynamics are sampled. Sampling issues in dynamic networks are among the most technically challenging topics covered in this book. Understanding the interplay between selection sampling and the Markov assumption reflects a long line of technical work in mathematical probability [43, 44, 48, 49, 50, 57], whose practical implications remain poorly understood and are among of the primary topics discussed in this chapter.

11.2.2 Projectivity and sampling

The population process in the scenario of Section 11.1 evolves $\{0,1\}^{N\times N}$, where N is the number of all Twitter users worldwide. With N on the order of hundreds of millions, it is computationally infeasible to analyze the complete evolution of Twitter interactions, even over a short period of time. So while we are interested in learning the dynamics of the population network, practical limitations require that such inferences be based on the dynamics observed for a sample of $n \ll N$ vertices at a sample of times.

Since the sampled network is often much smaller than the population, and most Twitter users interact with only a negligible fraction of the population, it is likely that the sampled vertices are not representative of the population as a whole and that the observed dynamics depend on interactions between unsampled vertices in a way that is hard to incorporate into a tractable statistical model. Gaining a better understanding of how dynamic networks are sampled, e.g., by extending the sampling schemes of Chapter 3 to the dynamic network case, is an important problem for future study.

Research Problem 11.4 *For a dynamic network* $\mathbf{Y} = (Y(t))_{t=0,1,\ldots,T}$ *on a population of N vertices, consider observing* $\mathbf{Y}^* = (\Sigma Y(t))_{t=0,1,\ldots,T}$*, where* Σ *is a generic (possibly random) network sampling operation as in Section 3.9.*

- *What choices for* Σ *make sense in practical applications?*
- *How is the relationship between observed network dynamics and population level dynamics affected by different distributions for* Σ*?*

To formulate a precise sampling context for dynamic networks, we regard $\mathbf{Y} = (Y(s))_{s=0,1,\ldots,T}$ as a map $Y : \{0,1,\ldots,T\} \to \{0,1\}^{N\times N}$, i.e., $Y(s) \in \{0,1\}^{N\times N}$ is the state of the network at time s. We obtain a sample of size (T',n), for $T' \leq T$ and $n \leq N$, from $\mathbf{Y}$ by specifying a pair (γ,ψ) such that $\gamma : \{0,1,\ldots,T'\} \to \{0,1,\ldots,T\}$ and $\psi : [n] \to [N]$ are injections and γ is order-preserving (i.e., $s \leq s'$ implies $\gamma(s) \leq \gamma(s')$). We then define $\mathbf{S}^{\gamma,\psi}_{(T',n),(T,N)}$ as the sampling operation that acts on $\mathbf{Y}$ by

$$\mathbf{S}^{\gamma,\psi}_{(T',n),(T,N)}\mathbf{Y} = (\mathbf{S}^{\psi}_{n,N}Y(\gamma(s)))_{s=0,1,\ldots,T'}, \tag{11.4}$$

where $\mathbf{S}^{\psi}_{n,N} : \{0,1\}^{N\times N} \to \{0,1\}^{n\times n}$ is the ψ-selection map defined in (3.17). In words, $\mathbf{S}^{\gamma,\psi}_{(T',n),(T,N)}\mathbf{Y}$ is the dynamic network obtained by sampling vertices according to ψ and times according to γ. If instead time is indexed continuously in $\mathbf{Y}$, so that $\mathbf{Y} = (Y(s))_{s\in[0,T)}$, then we define a sampling mechanism (γ,ψ) by taking $\gamma\colon [0,T') \to [0,T)$ to be a Lèbesgue measure-preserving map as in Section 7.3.7 which also preserves the order of $[0,T')$, and define $\mathbf{S}^{\gamma,\psi}_{(T',n),(T,N)}$ just as in (11.4). As we have in earlier chapters, in particular Section 3.9, we can consider dynamic networks observed as in (11.4) for (γ,ψ) chosen randomly according to some joint distribution possibly depending on the population process $\mathbf{Y}$. Following suit with earlier chapters, I highlight this general sampling context as interesting and important for future research in network analysis, cf. the major open questions cited in Section 1.7.3, and proceed for the rest of this chapter under the assumption that the observed network sequence is obtained by selection sampling jointly across all time points. To be specific, for a population process $\mathbf{Y} = (Y(t))_{t=0,1,\ldots}$ on $\{0,1\}^{N\times N}$ indexed over an indefinite time horizon $\{0,1,\ldots\}$, we assume that $\mathbf{Y}_n = (\mathbf{S}_{n,N}Y(t))_{t=0,1,\ldots,T}$ is observed for some $n < N$.

Under the assumptions of Section 11.2.1, the population structure $\mathbf{Y} = (Y(t))_{t=0,1,\ldots,T}$ is modeled by a time homogeneous Markov chain as in (11.2). The observed network is obtained from $\mathbf{Y}$ by choosing vertices $1,\ldots,n$ via selection sampling. Since the observed network is modeled as a sample from a Markov chain, and not directly as a Markov chain itself, we must consider the possibility that the Markov property assumed for $\mathbf{Y}$ may not be preserved in the sampled process. For example, suppose that users i and k both retweet a post by j at time t, i.e., $Y_{ij}(t) = 1$ and $Y_{kj}(t) = 1$, and that neither i nor k retweets j at time $t+1$. Further suppose that user i is a follower of k and k is a follower of j, so that user i is exposed to the activity of j only through k. In this case, the change from $Y_{ij}(t) = 1$ to $Y_{ij}(t+1) = 0$ is directly correlated to the change from $Y_{kj}(t) = 1$ to $Y_{kj}(t+1) = 0$, but this information would be unavailable in a sample that only includes i and j but not k.

To summarize, since *a function of a Markov chain need not be a Markov chain*, see, e.g., [28], it is possible that the Markovian dynamics of the population network are not preserved under sampling. In practice, however, when the population dynamics are assumed to follow the Markov property, it is common to also assume that the dynamics of the sampled network are Markovian. Before identifying which dynamic network models satisfy this property and what other practical implications this assumption might have (Sections 11.3–11.4), I first show an example in which this property fails.

11.2.2.1 Example: A TERGM for triangle counts

To illustrate the above point in the case of TERGMs, let the joint sufficient statistic $T(\mathbf{y},\mathbf{y}')$ in (11.3) be given by

$$T(\mathbf{y},\mathbf{y}') = \log(1 + T_\Delta(\mathbf{y},\mathbf{y}')), \tag{11.5}$$

where $T_\Delta(\mathbf{y},\mathbf{y}')$ is the number of triangles that $\mathbf{y}$ and $\mathbf{y}'$ have in common, i.e.,

$$T_\Delta(\mathbf{y},\mathbf{y}') = \sum_{1\le i<j<k\le N} y_{ij}y_{ik}y_{jk}y'_{ij}y'_{ik}y'_{jk}, \quad \mathbf{y},\mathbf{y}' \in \{0,1\}^{N\times N}.$$

Assume that the natural parameter $\eta(\theta)$ is set to 1 so that the transition probabilities governing $\mathbf{Y}_N$, with $N = 4$, are of the form

$$P(\mathbf{y},\mathbf{y}') \propto \exp\{T(\mathbf{y},\mathbf{y}')\}, \quad \mathbf{y},\mathbf{y}' \in \{0,1\}^{4\times 4}. \tag{11.6}$$

Suppose further that $\mathbf{Y}_n = (Y_n(t))_{t=0,1,\ldots}$ is obtained by selecting the vertices labeled $1,2,3$ from $\{1,2,3,4\}$. What is the probability of the transition from $\mathbf{y}$ to $\mathbf{y}'$ in $\mathbf{Y}_n$ when

$$\mathbf{y} = \mathbf{y}' = \begin{pmatrix} 0 & 1 & 1 \\ 1 & 0 & 0 \\ 1 & 0 & 0 \end{pmatrix}?$$

First, note that if $\mathbf{Y}_n$ were to evolve according to the TERGM with sufficient statistic T in (11.5), then the transition probability from $\mathbf{y}$ to $\mathbf{y}'$ must be $1/8$ since there are 8 undirected graphs in $\{0,1\}^{3\times 3}$ and the current state $\mathbf{y}$ has 0 triangles to start with; and therefore $T_\Delta(\mathbf{y},\mathbf{y}') = 0$ for all $\mathbf{y}'$. But since $\mathbf{Y}_n$ was obtained from $\mathbf{Y}_N$ by selection, it is possible that the transition from $\mathbf{y}$ to $\mathbf{y}'$ in $\mathbf{Y}_n$ depends on the states in $\{0,1\}^{4\times 4}$ from which $\mathbf{y}$ and $\mathbf{y}'$ were sampled. Now, the observation $\mathbf{y}'$ would have been obtained by selection sampling as long as the population process $\mathbf{Y}_N$ transitioned into any of the following 8 extensions of $\mathbf{y}'$:

$$\begin{pmatrix} 0 & 1 & 1 & \mathbf{0} \\ 1 & 0 & 0 & \mathbf{0} \\ 1 & 0 & 0 & \mathbf{0} \\ \mathbf{0} & \mathbf{0} & \mathbf{0} & \mathbf{0} \end{pmatrix}, \begin{pmatrix} 0 & 1 & 1 & \mathbf{1} \\ 1 & 0 & 0 & \mathbf{0} \\ 1 & 0 & 0 & \mathbf{0} \\ \mathbf{1} & \mathbf{0} & \mathbf{0} & \mathbf{0} \end{pmatrix}, \begin{pmatrix} 0 & 1 & 1 & \mathbf{0} \\ 1 & 0 & 0 & \mathbf{1} \\ 1 & 0 & 0 & \mathbf{0} \\ \mathbf{0} & \mathbf{1} & \mathbf{0} & \mathbf{0} \end{pmatrix}, \begin{pmatrix} 0 & 1 & 1 & \mathbf{0} \\ 1 & 0 & 0 & \mathbf{0} \\ 1 & 0 & 0 & \mathbf{1} \\ \mathbf{0} & \mathbf{0} & \mathbf{1} & \mathbf{0} \end{pmatrix},$$
$$\begin{pmatrix} 0 & 1 & 1 & \mathbf{1} \\ 1 & 0 & 0 & \mathbf{1} \\ 1 & 0 & 0 & \mathbf{0} \\ \mathbf{1} & \mathbf{1} & \mathbf{0} & \mathbf{0} \end{pmatrix}, \begin{pmatrix} 0 & 1 & 1 & \mathbf{1} \\ 1 & 0 & 0 & \mathbf{0} \\ 1 & 0 & 0 & \mathbf{1} \\ \mathbf{1} & \mathbf{0} & \mathbf{1} & \mathbf{0} \end{pmatrix}, \begin{pmatrix} 0 & 1 & 1 & \mathbf{0} \\ 1 & 0 & 0 & \mathbf{1} \\ 1 & 0 & 0 & \mathbf{1} \\ \mathbf{0} & \mathbf{1} & \mathbf{1} & \mathbf{0} \end{pmatrix}, \begin{pmatrix} 0 & 1 & 1 & \mathbf{1} \\ 1 & 0 & 0 & \mathbf{1} \\ 1 & 0 & 0 & \mathbf{1} \\ \mathbf{1} & \mathbf{1} & \mathbf{1} & \mathbf{0} \end{pmatrix}.$$

Notice that the upper 3×3 submatrix of these 8 arrays all coincide with $\mathbf{y}'$ and, furthermore, that these 8 matrices list all of the possible states into which $\mathbf{Y}_N$ could have transitioned at time $t+1$ to produce the event '$\mathbf{S}_{n,N}\mathbf{Y}_N(t+1) = \mathbf{y}'$'.

We compute the induced transition probability from $\mathbf{y}$ to $\mathbf{y}'$ in $\mathbf{Y}_n$ by aggregating the probabilities that $\mathbf{Y}_N$ transitioned into any one of the above 8 states. Since $\mathbf{y}$ is fixed, the dynamics of $\mathbf{Y}_n$ are Markovian only if the transition probability of $\mathbf{Y}_N$ into the above 8 states does not depend on the choice of representative for $\mathbf{y}$ from which to generate the population level transitions. (In other words, the transition $\mathbf{y} \mapsto \mathbf{y}'$ in $\mathbf{Y}_n$ results from a Markovian transition $\mathbf{Y}_N(t) \mapsto \mathbf{Y}_N(t+1)$ at the population level followed by selection:

$$\begin{aligned} \mathbf{S}_{n,N}\mathbf{Y}_N(t) &\mapsto \mathbf{S}_{n,N}\mathbf{Y}_N(t+1) \\ \mathbf{y} &\mapsto \mathbf{y}'. \end{aligned}$$

The Markov property holds for $\mathbf{Y}_n = (\mathbf{S}_{n,N}\mathbf{Y}_N(t))_{t=0,1,\ldots}$ only if the conditional distribution of $\mathbf{Y}_n(t+1)$, given $\mathbf{Y}_N(t)$, is a measurable function of $\mathbf{S}_{n,N}\mathbf{Y}_N(t) = \mathbf{y}$. This condition does not hold for all Markov chains, as this example shows.)

Suppose first that $\mathbf{y} = \mathbf{S}_{n,N}\mathbf{y}^*$ for

$$\mathbf{y}^* = \begin{pmatrix} 0 & 1 & 1 & \mathbf{0} \\ 1 & 0 & 0 & \mathbf{0} \\ 1 & 0 & 0 & \mathbf{0} \\ \mathbf{0} & \mathbf{0} & \mathbf{0} & \mathbf{0} \end{pmatrix}.$$

Since $\mathbf{y}^*$ has no triangles, it has no triangles in common with any of the 8 possible states listed above. It follows that all transitions are uniformly distributed and that the induced transition from $\mathbf{y}$ to $\mathbf{y}'$ has probability $1/8$, just as it does if the transitions on $\{0,1\}^{3\times 3}$ are modeled as in (11.5). If, on the other hand, $\mathbf{y} = \mathbf{S}_{n,N}\mathbf{y}^*$ for

$$\mathbf{y}^* = \begin{pmatrix} 0 & 1 & 1 & \mathbf{1} \\ 1 & 0 & 0 & \mathbf{1} \\ 1 & 0 & 0 & \mathbf{1} \\ \mathbf{1} & \mathbf{1} & \mathbf{1} & \mathbf{0} \end{pmatrix}$$

then the transition probabilities are no longer uniform. The induced transition probability from $\mathbf{y}$ to $\mathbf{y}'$ is

$$\frac{12}{8+10+8+10+10+12+10+12} = \frac{12}{80} \neq \frac{1}{8}.$$

Since the sampled state $\mathbf{y}$ does not maintain all of the information about the network from which it was sampled, the transition behavior in the sampled process $(\mathbf{S}_{n,N}\mathbf{Y}_N(t))_{t=0,1,\ldots}$ is not Markovian. In particular, by the Markov assumption, the conditional probability of $\mathbf{Y}_N(t+1)$ depends only on $\mathbf{Y}_N(t)$, but here the conditional probability of $\mathbf{Y}_N(t+1)|_{[3]}$ is not measurable with respect to $\mathbf{Y}_N(t)|_{[3]} = \mathbf{y}$. This example illustrates a failure of the *projective Markov property* [43, 44, 48].

11.2.2.2 Projective Markov property

A Markov chain $\mathbf{Y} = (Y(t))_{t=0,1,\ldots}$ on $\{0,1\}^{N\times N}$, for $N = 1,2,\ldots,\infty$,[2] has the *projective Markov property* if

$$\mathbf{Y}_n = (\mathbf{S}_{n,N}Y(t))_{t=0,1,\ldots}\text{is a Markov chain for every } n = 1,\ldots,N. \tag{11.7}$$

Equivalently, $\mathbf{Y}$ has the projective Markov property if its transition probabilities satisfy

$$\Pr(\mathbf{S}_{n,N}Y(t+1) = \mathbf{y} \mid Y(t) = \mathbf{y}') = \Pr(\mathbf{S}_{n,N}Y(t+1) = \mathbf{y} \mid Y(t) = \mathbf{y}'')$$
$$\text{for all } \mathbf{y}',\mathbf{y}'' \text{ with } \mathbf{S}_{n,N}\mathbf{y}' = \mathbf{S}_{n,N}\mathbf{y}''. \tag{11.8}$$

Exercise 11.1 *Show that conditions* (11.7) *and* (11.8) *are equivalent.*

[2] When $N = \infty$ the state space is taken to be $\{0,1\}^{\mathbb{N}\times\mathbb{N}}$.

The projective Markov property can be extended by extrapolating from the discussion of Chapters 3 and 5 to define the notion of *Markovian coherence with respect to subsampling*. For each $n \geq 1$, let $\mathscr{M}_n$ be a family of Markovian transition probabilities on $\{0,1\}^{n\times n}$ and let $\{\Sigma_{m,n}\}_{n\geq m\geq 1}$ be a family of (possibly random) sampling operations. Then the model $(\{\mathscr{M}_n\}_{n\geq 1}, \{\Sigma_{m,n}\}_{n\geq m\geq 1})$ is *coherent* if every $\mathbf{Y}_n = (Y(t))_{t=0,1,...}$ with transition probabilities given by some $P_n \in \mathscr{M}_n$ projects to a Markov chain $(\Sigma_{m,n}\mathbf{Y}_n(t))_{t=0,1,...}$ with transition probability $P_m \in \mathscr{M}_m$, and conversely every $P_m \in \mathscr{M}_m$ is the transition probability of $(\Sigma_{m,n}Y_n(t))_{t=0,1,...}$ sampled from a Markov chain $\mathbf{Y}_n = (Y_n(t))_{t=0,1,...}$ governed by some transition probability $P_n \in \mathscr{M}_n$. At present there is no mathematical machinery available for studying systems of Markov chains in a sampling context other than selection, and so I leave this general case as an open area of study.

11.3 Rewiring chains and Markovian graphons

In [44] I introduced rewiring processes as the class of Markov chains which evolve by successive application of randomly chosen 'rewiring maps', defined as follows. For $n \geq 1$ and $W \in (\{0,1\} \times \{0,1\})^{n\times n}$, write the ij entry of W as $W(i,j) = (W_0(i,j), W_1(i,j))$ so that both $W_0(i,j)$ and $W_1(i,j)$ are elements of $\{0,1\}$ and $W_0 = (W_0(i,j))_{1\leq i,j\leq n}$ and $W_1 = (W_1(i,j))_{1\leq i,j\leq n}$ can each be regarded as $\{0,1\}$-valued arrays in their own right. The *rewiring map* determined by W is a function

$$\begin{aligned} W :& \{0,1\}^{n\times n} \to \{0,1\}^{n\times n} \\ & \mathbf{y} \mapsto W(\mathbf{y}) = \mathbf{y}', \end{aligned}$$

defined by

$$y'_{ij} = \begin{cases} W_1(i,j), & y_{ij} = 1, \\ W_0(i,j), & y_{ij} = 0. \end{cases} \tag{11.9}$$

In words, W acts on $\mathbf{y}$ by replacing each y_{ij} by $W_1(i,j)$ if there is an edge between i and j in $\mathbf{y}$ and by $W_0(i,j)$ if there is not an edge between i and j in $\mathbf{y}$. In [44], I called $\mathbf{y} \mapsto W(\mathbf{y})$ a *rewiring* operation and W a *rewiring map* because it acts on $\{0,1\}^{n\times n}$ by 'rewiring' each entry of the network represented by $\mathbf{y}$ according to the configuration of 0s and 1s in W. For any $n = 1,2,\ldots$, let $\mathscr{W}_n$ be the space of *rewiring maps*, each of which corresponds to an $n \times n$ array taking values in $\{0,1\} \times \{0,1\}$.

For a concrete example, consider the operation $\mathbf{y} \mapsto W(\mathbf{y})$ given by

$$
\begin{array}{ccc}
\mathbf{y} & & W \\
\begin{pmatrix} 0 & 1 & 1 & 0 & 1 \\ 1 & 0 & 0 & 0 & 1 \\ 1 & 0 & 0 & 1 & 0 \\ 0 & 0 & 1 & 0 & 0 \\ 1 & 1 & 0 & 0 & 0 \end{pmatrix} & &
\begin{pmatrix}
(\underline{\mathbf{0}},0) & (1,\underline{\mathbf{0}}) & (0,\underline{\mathbf{1}}) & (\underline{\mathbf{0}},0) & (0,\underline{\mathbf{1}}) \\
(1,\underline{\mathbf{0}}) & (\underline{\mathbf{0}},0) & (\underline{\mathbf{1}},0) & (\underline{\mathbf{1}},1) & (1,\underline{\mathbf{0}}) \\
(0,\underline{\mathbf{1}}) & (\underline{\mathbf{1}},0) & (\underline{\mathbf{0}},0) & (0,\underline{\mathbf{1}}) & (\underline{\mathbf{0}},0) \\
(\underline{\mathbf{0}},0) & (\underline{\mathbf{1}},1) & (0,\underline{\mathbf{1}}) & (\underline{\mathbf{0}},0) & (\underline{\mathbf{1}},0) \\
(0,\underline{\mathbf{1}}) & (1,\underline{\mathbf{0}}) & (\underline{\mathbf{0}},0) & (\underline{\mathbf{1}},0) & (\underline{\mathbf{0}},0)
\end{pmatrix}
\end{array}
$$

$$
\mapsto \quad \begin{array}{c} \mathbf{y}' = W(\mathbf{y}) \\ \begin{pmatrix} 0 & 0 & 1 & 0 & 1 \\ 0 & 0 & 1 & 1 & 0 \\ 1 & 1 & 0 & 1 & 0 \\ 0 & 1 & 1 & 0 & 1 \\ 1 & 0 & 0 & 1 & 0 \end{pmatrix} \end{array}. \tag{11.10}
$$

In this demonstration, the elements of W written in **bold** are those which have been copied from W into the image $\mathbf{y}' = W(\mathbf{y})$ in accordance with (11.9). For example, since $y_{12} = 1$, the right-hand entry of $W_{12} = (1,0)$ is chosen as the corresponding entry in $W(\mathbf{y})$; since $y_{13} = 1$, the right-hand entry of $W_{13} = (0,1)$ is chosen as the corresponding entry in $W(\mathbf{y})$; since $y_{14} = 0$, the left-hand entry of $W_{14} = (0,0)$ is chosen as the corresponding entry in $W(\mathbf{y})$; and so on.

From the rewiring operation in (11.9), we construct a time homogeneous projective Markov chain on $\{0,1\}^{N\times N}$ as follows. Define any probability distribution ω on $\mathscr{W}_N$ and, for any initial state $\mathbf{y} \in \{0,1\}^{N\times N}$, set $Y(0) = \mathbf{y}$ and put

$$
Y(t+1) = W_{t+1}(Y(t)) = (W_{t+1} \circ \cdots \circ W_1)(\mathbf{y}), \quad t = 0,1,\ldots, \tag{11.11}
$$

for $W_1, W_2, \ldots$ i.i.d. from ω and $W(\mathbf{y})$ as defined in (11.9). For two rewiring maps W, W', the operation $W \circ W'$ denotes the usual composition of functions, so that $(W \circ W')(\mathbf{y}) = W(W'(\mathbf{y}))$. In general, we have

$$
(W_{t+1} \circ \cdots \circ W_1)(\mathbf{y}) = W_{t+1}(W_t(\cdots(W_1(\mathbf{y})))),
$$

as in (11.11). We call the process $\mathbf{Y}$ constructed in (11.11) a *rewiring chain directed by* ω.

For $W \in \mathscr{W}_N$, we define the restriction $W|_{[n]} \in \mathscr{W}_n$ in the usual way by $W|_{[n]} = (W(i,j))_{1\le i,j\le n}$, so that any probability distribution ω on $\mathscr{W}_N$ induces a distribution ω_n on $\mathscr{W}_n$ by

$$
\omega_n(\mathbf{w}) = \omega(\{\mathbf{w}^* \in \mathscr{W}_N : \mathbf{w}^*|_{[n]} = \mathbf{w}\}), \quad \mathbf{w} \in \mathscr{W}_n. \tag{11.12}
$$

The transition behavior of the chain constructed in (11.11) is determined by an i.i.d. sequence of random rewiring maps $W_1, W_2, \ldots$, and the action of each W on $\{0,1\}^{N\times N}$ is such that the distribution of $Y(m+1)|_{[n]}$ given W_{m+1} and $Y(m)$ depends only on the restrictions $W_{m+1}|_{[n]}$ and $Y(m)|_{[n]}$. Rewiring chains thus satisfy the projective Markov property (11.8) by default.

Theorem 11.1 (Crane [44]) *The restriction of a rewiring chain $\mathbf{Y}_N$ directed by ω to $\{0,1\}^{n\times n}$, i.e., $\mathbf{Y}_n = (\mathbf{S}_{n,N}Y(t))_{t=0,1,\dots}$, is a rewiring chain constructed as in* (11.11) *with $W_1', W_2', \dots$ i.i.d. from ω_n defined in* (11.12).

11.3.1 Exchangeable rewiring processes (Markovian graphons)

In keeping with the general theme of exchangeability observed through the Aldous–Hoover theorem in Chapter 6, the Ackerman–Crane–Towsner theorem in Chapter 8, and the Crane–Dempsey theorem in Chapters 9 and 10, we here see that any projective Markov chain on $\{0,1\}^{\mathbb{N}\times\mathbb{N}}$ that satisfies an additional exchangeability assumption must also be a rewiring chain. In fact, under the additional exchangeability assumption in (11.13) below, we observe a connection between rewiring chains and a Markovian generalization of the graphon models from Chapter 6.

We call a Markov chain $\mathbf{Y}_n = (Y(t))_{t=0,1,\dots}$ on $\{0,1\}^{n\times n}$ *exchangeable* if its transition probabilities satisfy

$$\Pr(Y(t+1) = \mathbf{y}'^{\sigma} \mid Y(t) = \mathbf{y}^{\sigma}) = \Pr(Y(t+1) = \mathbf{y}' \mid Y(t) = \mathbf{y}) \qquad (11.13)$$

for all $\mathbf{y}, \mathbf{y}' \in \{0,1\}^{n\times n}$ and all permutations $\sigma : [n] \to [n]$. In words, $\mathbf{Y}_n$ is exchangeable if the probability of a transition between two states depends only on the relative structure of the states, and not on how the structure manifests itself through the vertex labels. Notice that if $\mathbf{Y}_n = (Y(t))_{t=0,1,\dots}$ has an exchangeable initial state (i.e., $Y(0)^{\sigma} =_{\mathscr{D}} Y(0)$ for all permutations $\sigma : [n] \to [n]$) and exchangeable transition probabilities, as in (11.13), then the process is exchangeable jointly at all times, in the sense that $\mathbf{Y}_n^{\sigma} = (Y(t)^{\sigma})_{t=0,1,\dots} =_{\mathscr{D}} \mathbf{Y}_n$ for all permutations $\sigma : [n] \to [n]$.

Theorem 11.2 below says that every exchangeable, projective Markov chain on $\{0,1\}^{\mathbb{N}\times\mathbb{N}}$ corresponds to a *rewiring chain* constructed as in (11.11) for some exchangeable distribution ω on $(\{0,1\}\times\{0,1\})^{\mathbb{N}\times\mathbb{N}}$, where we call ω *exchangeable* if $W \sim \omega$ satisfies

$$W^{\sigma} = (W(\sigma(i), \sigma(j)))_{i,j\geq 1} =_{\mathscr{D}} W$$

for all permutations $\sigma : \mathbb{N} \to \mathbb{N}$. Note that this characterization only holds for chains on countable arrays, just as for the characterizations of vertex exchangeable, relatively exchangeable, and relationally exchangeable structures in Chapters 6–10.

Theorem 11.2 (Crane [44, 48]) *Let $\mathbf{Y} = (Y(t))_{t=0,1,\dots}$ be an exchangeable, projective Markov chain on $\{0,1\}^{\mathbb{N}\times\mathbb{N}}$ with initial state $\mathbf{y} \in \{0,1\}^{\mathbb{N}\times\mathbb{N}}$. Then there exists an exchangeable probability distribution ω on $\mathscr{W}_{\mathbb{N}}$ such that $\mathbf{Y} =_{\mathscr{D}} \mathbf{Y}^* = (Y^*(t))_{t=0,1,\dots}$, for $\mathbf{Y}^*$ generated by putting $Y^*(0) = \mathbf{y}$ and*

$$Y^*(t) = W_t(Y^*(t-1)) = (W_t \circ \cdots \circ W_1)(\mathbf{y}), \quad t = 1, 2, \dots,$$

for $W_1, W_2, \dots$ i.i.d. from ω.

The correspondence between $\mathscr{W}_{\mathbb{N}}$ and infinite $\{0,1\}\times\{0,1\}$-valued arrays allows us to restate Theorem 11.2 as a temporal extension of the graphon models from Section 6.4.1. Recall that any $\mathbf{Y} = (Y_{ij})_{1\leq i,j\leq N}$ distributed according to a graphon model with parameter $\phi : [0,1] \times [0,1] \to [0,1]$ can be generated by taking $U_1, \dots, U_N$ i.i.d. Uniform$[0,1]$ and putting

$$\Pr(Y_{ij} = 1 \mid U_1, \dots, U_N) = \phi(U_i, U_j) \quad \text{and} \quad \Pr(Y_{ij} = 0 \mid U_1, \dots, U_N) = 1 - \phi(U_i, U_j)$$

conditionally independently for all $1 \leq i, j \leq N$. In the present setting of dynamic networks, we define a *Markovian graphon* as a function $\phi : [0,1] \times [0,1] \to [0,1] \times [0,1]$ so that $\phi(u,v) = (\phi_0(u,v), \phi_1(u,v))$ determines a transition probability matrix $\{0,1\} \to \{0,1\}$ given by

$$\begin{array}{c} \\ 0 \\ 1 \end{array} \begin{array}{c} \begin{array}{cc} \quad 0 \quad\quad\quad & \quad 1 \end{array} \\ \begin{pmatrix} 1 - \phi_0(u,v) & \phi_0(u,v) \\ 1 - \phi_1(u,v) & \phi_1(u,v) \end{pmatrix} \end{array}. \tag{11.14}$$

To be specific, (11.14) determines the transition probabilities of a Markov chain $\mathbf{Y} = (Y(t))_{t=0,1,\dots}$ with transition probabilities

$$\begin{aligned} &\Pr(Y_{ij}(t+1) = 1 \mid Y(t), Y_{ij}(t) = y, U_1^t, \dots, U_N^t) = \phi_y(U_i^t, U_j^t) \quad \text{and} \\ &\Pr(Y_{ij}(t+1) = 0 \mid Y(t), Y_{ij}(t) = y, U_1^t, \dots, U_N^t) = 1 - \phi_y(U_i^t, U_j^t), \end{aligned}$$

conditionally independently for all $1 \leq i, j \leq N$ and all $t = 0, 1, \dots,$ for $(U_i^t)_{i \geq 1; t=1,2,\dots}$ i.i.d. Uniform$[0,1]$.

The representation in Theorem 11.2 and the connection to graphons just observed have several consequences for the possible behaviors of exchangeable, projective Markov chains on $\{0,1\}^{\mathbb{N} \times \mathbb{N}}$. Since rewiring processes allow for non-exchangeable initial states $Y(0)$, it is possible for the state of the chain to be sparse at any given time; however, the exchangeable transitions are such that the sparsity will become more and more homogeneous as time goes on. Consequently, the process converges either to a dense or empty state. The effects of exchangeability are even more pronounced for Markov processes indexed by continuous time. I discuss these briefly in Section 11.5. The reader is referred to [44, 48] for many more details about discrete and continuous time rewiring chains.

11.4 Graph-valued Lévy processes

By Theorem 11.2 every exchangeable, time homogeneous, projective Markov chain on $\{0,1\}^{\mathbb{N} \times \mathbb{N}}$ can be constructed from an i.i.d. sequence of randomly generated rewiring maps as in (11.11). The explicit connection to graphon models discussed after the statement of the theorem highlights potential limitations of such processes as models for dynamic network data, cf. Section 6.5. Since the exchangeability condition in (11.13) and/or infinite population size may not be suitable for any given application, it behooves us to explore models that relax one or both of these assumptions. Graph-valued Lévy processes comprise one such model class.

As in Section 11.1, we assume the population process $\mathbf{Y} = (Y(t))_{t=0,1,\dots}$ is a sequence in $\{0,1\}^{N \times N}$. Whereas the rewiring chains of Section 11.3 are defined by repeated composition of rewiring maps, as in (11.11), graph-valued Lévy processes are defined by how their *increments* behave. Informally, the *increment* between $\mathbf{y}$ and $\mathbf{y}'$ in $\{0,1\}^{N\times N}$ is the 'difference' or 'change' necessary to convert $\mathbf{y}$ into $\mathbf{y}'$. Formally, the *increment between* $\mathbf{y}$ *and* $\mathbf{y}'$ is defined as the symmetric difference between the edge sets of $\mathbf{y}$ and $\mathbf{y}'$, as expressed by the array $\mathbf{y} \triangle \mathbf{y}' = (\Delta_{ij})_{1\leq i,j\leq N}$ with ij-entry given by $\Delta_{ij} = |y_{ij} - y'_{ij}|$ for each $1 \leq i,j \leq N$. Thus, $\mathbf{y} \triangle \mathbf{y}'$ is a graph whose edges record whether or not there is a difference between the ij-entries of $\mathbf{y}$ and $\mathbf{y}'$. For example, the increment between

$$\mathbf{y} = \begin{pmatrix} 0 & 0 & 0 & 0 & 0 & 0 & 1 \\ 0 & 0 & 1 & 0 & 0 & 1 & 0 \\ 0 & 1 & 0 & 1 & 1 & 1 & 0 \\ 0 & 0 & 1 & 0 & 0 & 1 & 0 \\ 0 & 0 & 1 & 0 & 0 & 0 & 0 \\ 0 & 1 & 1 & 1 & 0 & 0 & 1 \\ 1 & 0 & 0 & 0 & 0 & 1 & 0 \end{pmatrix}$$

and

$$\mathbf{y}' = \begin{pmatrix} 0 & 1 & 0 & 0 & 0 & 0 & 0 \\ 1 & 0 & 1 & 0 & 0 & 0 & 0 \\ 0 & 1 & 0 & 0 & 0 & 1 & 0 \\ 0 & 0 & 0 & 0 & 1 & 1 & 0 \\ 0 & 0 & 0 & 1 & 0 & 1 & 0 \\ 0 & 0 & 1 & 1 & 1 & 0 & 0 \\ 0 & 0 & 0 & 0 & 0 & 0 & 0 \end{pmatrix}$$

is

$$\Delta = \mathbf{y} \triangle \mathbf{y} = \begin{pmatrix} 0 & 1 & 0 & 0 & 0 & 0 & 1 \\ 1 & 0 & 0 & 0 & 0 & 1 & 0 \\ 0 & 0 & 0 & 1 & 1 & 0 & 0 \\ 0 & 0 & 1 & 0 & 1 & 0 & 0 \\ 0 & 0 & 1 & 1 & 0 & 1 & 0 \\ 0 & 1 & 0 & 0 & 1 & 0 & 1 \\ 1 & 0 & 0 & 0 & 0 & 1 & 0 \end{pmatrix}. \tag{11.15}$$

See Figure 11.2 for a visual illustration of the increment $\mathbf{y} \triangle \mathbf{y}'$ calculated in (11.15).

Definition 11.1 (Graph-valued Lévy process [50]) *A collection* $\mathbf{Y} = (Y(t))_{t=0,1,\ldots}$ *of random arrays indexed by* $t = 0,1,\ldots$ *is a* graph-valued Lévy process *if it has*

- *initial state* $Y(0) = \mathbf{0}_N$*, i.e.,* $Y_{ij}(0) = 0$ *for all* $1 \leq i,j \leq N$*,*
- *stationary increments, i.e.,*

$$Y(t+s) \triangle Y(t) =_{\mathscr{D}} Y(s) \text{ for all } s,t = 0,1,\ldots, \text{ and} \tag{11.16}$$

- *independent increments, i.e.,*

$$Y(t_1) \triangle Y(t_0), \ldots, Y(t_k) \triangle Y(t_{k-1}) \text{ are independent for all } 0 \leq t_1 \leq \cdots \leq t_k < \infty. \tag{11.17}$$

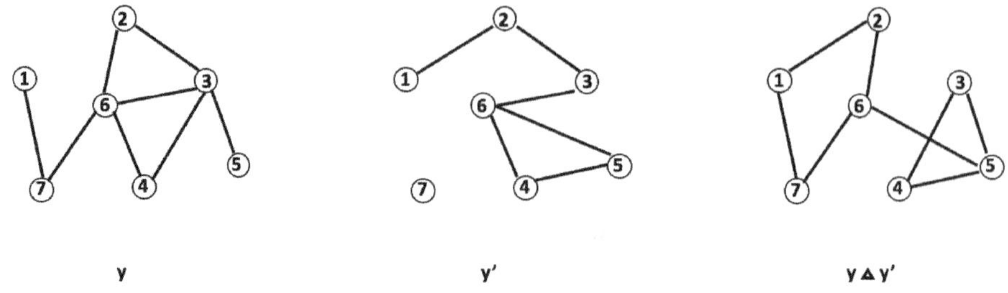

Figure 11.2 *Visual illustration of increment operation shown in* (11.15).

By [50, Theorem 4.5], any discrete-time process $\mathbf{Y}^* = (Y^*(t))_{t=0,1,...}$ satisfying Definition 11.1 can be constructed from a probability distribution μ on $\{0,1\}^{N\times N}$ by taking $Z_1, Z_2, \ldots$ i.i.d. from μ and putting

$$\begin{aligned} Y^*(0) &= \mathbf{0}_N \quad \text{and} \\ Y^*(t+1) &= Y^*(t) \,\triangle\, Z_{t+1} \quad \text{for } t = 0,1,\ldots. \end{aligned} \tag{11.18}$$

Alternatively, $\mathbf{Y}^*$ can be constructed as a rewiring process (11.11) with $W_1, W_2, \ldots$ defined from $Z_1, Z_2, \ldots$ by

$$W_t(i,j) = (Z_t(i,j), 1 - Z_t(i,j)), \quad 1 \le i,j \le N.$$

Since Lévy processes are a special case of rewiring chains from Section 11.3, they automatically satisfy the projective Markov property.

Theorem 11.3 (Crane [50]) *Let* $\mathbf{Y} = (Y(t))_{t=0,1,...}$ *be a graph-valued Lévy process on* $\{0,1\}^{N\times N}$. *Then there exists a unique probability distribution* μ *on* $\{0,1\}^{N\times N}$ *such that* $\mathbf{Y} =_{\mathscr{D}} \mathbf{Y}^*$, *for* $\mathbf{Y}^*$ *constructed as in* (11.18).

The requirement on the initial state ($Y(0) = \mathbf{0}_N$) in Definition 11.1 is merely a convention. Since the behavior of the process is determined by its increments, we can consider Lévy processes with arbitrary initial states $Y(0) = \mathbf{y}$ by defining $\mathbf{Y}^{\mathbf{y}} = (Y^{\mathbf{y}}(t))_{t=0,1,...}$ by

$$Y^{\mathbf{y}}(t) = Y(t) \,\triangle\, \mathbf{y}, \quad t = 0,1,\ldots,$$

for $\mathbf{Y} = (Y(t))_{t=0,1,...}$ a Lévy process with initial state $Y(0) = \mathbf{0}_N$. See [50] for further discussion of graph-valued Lévy processes.

11.4.1 Inference from graph-valued Lévy processes

On the one hand, Lévy processes are a special kind of rewiring chain, and thus can only model local dynamics. On the other hand, the description of Lévy processes in terms of a distribution on $\{0,1\}$-valued arrays, instead of $\{0,1\} \times \{0,1\}$-valued arrays as in Theorem 11.2, makes them more amenable to statistical inference. For instance, suppose $\mathbf{Y}$ is observed at times $t = 0, 1, \ldots, T$. If $\mathbf{Y}$ is modeled as a graph-valued Lévy process, then its increment distribution μ can be estimated by the empirical distribution

$$\hat{\mu}_T(\mathbf{x}) = \frac{1}{T}\sum_{t=1}^{T} \mathbf{1}(\mathbf{y}_t \triangle \mathbf{y}_{t-1} = \mathbf{x}), \quad \mathbf{x} \in \{0,1\}^{N\times N}, \tag{11.19}$$

where $(Y(t))_{t=0,1,\ldots,T} = (\mathbf{y}_t)_{t=0,1,\ldots,T}$ is the observed network data and

$$\mathbf{1}(a = b) = \begin{cases} 1, & \text{if } a = b \text{ is true}, \\ 0, & \text{otherwise}, \end{cases}$$

is the indicator function. The empirical distribution (11.19) counts the number of times that each array $\mathbf{x}$ occurs as an increment of the observed process. Since these increments are i.i.d., the strong law of large numbers immediately implies that $\hat{\mu}_T \to \mu$ with probability 1 as $T \to \infty$, under the assumption that $\mathbf{Y}$ behaves according to some graph-valued Lévy process with increment distribution μ. In practice, however, since $\mathbf{Y}$ evolves on a space with $O(2^{N^2})$ states, this approximation is likely to be reliable only if the process is observed at a large set of times.

The simple description of Lévy processes in terms of independent increments also gives a straightforward way to test for exchangeability. In addition to (11.19), one can compute the exchangeable empirical measure by averaging over equivalence classes:

$$\hat{\mu}_T^{\text{ex}}(\mathbf{x}) = \frac{1}{|\{\mathbf{x}' \in \{0,1\}^{N\times N} : \mathbf{x}' \cong \mathbf{x}\}|} \sum_{\mathbf{x}' \in \{0,1\}^{N\times N} : \mathbf{x}' \cong \mathbf{x}} \hat{\mu}_T(\mathbf{x}'), \quad \mathbf{x} \in \{0,1\}^{N\times N}, \tag{11.20}$$

where $\mathbf{x}' \cong \mathbf{x}$ indicates that $\mathbf{x}'$ and $\mathbf{x}$ are equivalent up to relabeling, i.e., there exists a permutation $\sigma : [N] \to [N]$ such that $\mathbf{x}' = \mathbf{x}^{\sigma}$.

Much more generally, since the dynamics of a Lévy process are determined by a probability distribution on $\{0,1\}^{N\times N}$, these models can be fit using any existing method for relational or network data. For example, covariates could possibly be incorporated into dynamic network modeling by appealing to the latent space models of Section 8.4. I leave these questions to future work.

Research Problem 11.5 *Can the crude estimates in* (11.19) *and* (11.20) *be enhanced so as to not be so 'discrete'. In particular, notice that the estimate in* (11.19) *assigns a value to each of the* $O(2^{N^2})$ *different elements of* $\{0,1\}^{N\times N}$ *and* (11.20) *involves an average over the equivalence class* $\{\mathbf{x}' : \mathbf{x}' \cong \mathbf{x}\}$. *Both of these calculations are likely to be inefficient for practical purposes. Is there a natural (non-uniform) topology on* $\{0,1\}^{N\times N}$ *under which these estimates can be 'smoothed out' in a manner similar to kernel density estimation?*

11.5 Continuous time processes

The scenario of Section 11.1 and the models discussed in Sections 11.3 and 11.4 are tailored to discrete time network dynamics. In Section 11.1, for example, social media interactions are observed at evenly spaced time intervals of one day. For dynamic network data observed at unevenly spaced times, it may be better to model the evolution of the underlying interactions by a continuous time process. In this case, we assume that the population process is indexed by all times $t \in [0,\infty)$ but that the data $\mathbf{Y}[S] = (Y(s))_{s\in S}$ is observed only at a finite set of times $S \subset [0,\infty)$.[3] As a general principle, the mechanism used in obtaining the observed set of times S may depend on the process $\mathbf{Y}$ and should be accounted for when modeling the data, but for the sake of our discussion here we assume that S is chosen independently of $\mathbf{Y}$. See Section 11.2.2 for prior discussion on general sampling contexts for dynamic networks.

11.5.1 Poissonian construction

For $n \geq 1$, let $\mathscr{W}_n$ be the space of rewiring maps acting on $\{0,1\}^{n\times n}$. To construct a continuous time rewiring process $\mathbf{Y} = (Y(t))_{t\geq 0}$ on $\{0,1\}^{n\times n}$, we let ω be a finite measure on $\mathscr{W}_n$ and write dt to denote Lebesgue measure on $[0,\infty)$.[4] Let $\mathbf{W} = \{(t,W_t)\} \subseteq [0,\infty) \times \omega$ be a Poisson point process with intensity measure $dt \otimes \omega$, where dt denotes Lebesgue measure on $[0,\infty)$ and $dt \otimes \omega$ denotes the product measure of dt and ω. Given $\mathbf{W}$, construct $\mathbf{Y}$ by fixing any initial state $Y(0)$ and defining $Y(t)$ for each $t > 0$ by

- $Y(t) = W_t(Y(t-))$ if t is an atom time of $\mathbf{W}$, i.e., if $(t,W_t) \in \mathbf{W}$ for some $W_t \in \mathscr{W}_n$, where $Y(t-) = \lim_{s\uparrow t} Y(s)$ is the state of the process in the instant immediately preceding time t, and
- $Y(t) = Y(t-)$ otherwise.

By this description, the atoms of $\mathbf{W}$ determine the jumps of $\mathbf{Y}$. Whenever an atom (t,W_t) occurs in $\mathbf{W}$, the process $\mathbf{Y}$ makes a transition according to the same procedure as in (11.11). Otherwise, $\mathbf{Y}$ is constant between the atom times of $\mathbf{W}$. This generic construction in terms of the Poisson process $\mathbf{W}$ makes rewiring processes amenable to simulation on a computer.

Just as in Section 11.3, the above construction characterizes all continuous time projective Markov processes $\mathbf{Y}$ which are exchangeable and evolve on $\{0,1\}^{\mathbb{N}\times\mathbb{N}}$.

Theorem 11.4 (Crane [48]) *Let* $\mathbf{Y}$ *be an exchangeable, projective Markov process on* $\{0,1\}^{\mathbb{N}\times\mathbb{N}}$. *Then there exists an exchangeable measure* ω *on* $\mathscr{W}_{\mathbb{N}}$ *such that*

$$\omega(\{id_{\mathbb{N}}\}) = 0 \quad \text{and} \tag{11.21}$$

$$\omega(\{\mathbf{w} \in \mathscr{W}_{\mathbb{N}} : \mathbf{w}|_{[n]} = \mathbf{w}'\}) < \infty \quad \text{for all } \mathbf{w}' \in \{0,1\}^{n\times n} \text{ and all } n \geq 1, \tag{11.22}$$

for which $\mathbf{Y} =_{\mathscr{D}} \mathbf{Y}^*$, *for* $\mathbf{Y}^*$ *constructed from a Poisson point process* $\mathbf{W}$ *with intensity* $dt \otimes \omega$ *as given above.*

---

[3] $\mathbf{Y}[S] = (Y(s))_{s\in S}$ indicates the process observed at the subset of times $S \subset [0,\infty)$ and should not be confused with $\mathbf{Y}|_{[n]}$, which indicates the process restricted to vertices $[n]$ over all times.

[4] In continuous time, ω need only be a finite, positive measure on $\mathscr{W}_n$. When extending this construction to rewiring processes on $\{0,1\}^{\mathbb{N}\times\mathbb{N}}$, ω must satisfy an additional σ-finiteness constraint, as given in (11.22).

Note that in (11.21), $\mathrm{id}_{\mathbb{N}}$ is the identity map $\{0,1\}^{\mathbb{N}\times\mathbb{N}} \to \{0,1\}^{\mathbb{N}\times\mathbb{N}}$, which corresponds to the $\{0,1\} \times \{0,1\}$-valued array with all entries equal to $(0,1)$. Conditions (11.21) and (11.22) are necessary to ensure that $\mathbf{Y}$ is well behaved:

- Condition (11.21) forbids trivial atoms in $\mathbf{W}$ (i.e., atoms corresponding to the identity rewiring map $\mathrm{id}_{\mathbb{N}}$), because the action of the identity does not affect the state of the process.
- Condition (11.22) makes sure that each finite sample process $\mathbf{Y}|_{[n]}$ jumps at most finitely often in bounded time intervals. Since $\{0,1\}^{n\times n}$ is a finite state space for every $n \geq 1$ and a Markov chain on a finite state space remains in each state it visits for a strictly positive amount of time with probability 1, (11.22) is required to avoid pathological behaviors.

Graph-valued Lévy processes on $\{0,1\}^{\mathbb{N}\times\mathbb{N}}$ can be described in much the same way as in Theorem 11.4 with the exception that the Poisson point process $\mathbf{W}$ on $[0,\infty) \times \mathscr{W}_{\mathbb{N}}$ is replaced by a Poisson point process on $[0,\infty) \times \{0,1\}^{\mathbb{N}\times\mathbb{N}}$. Several other properties of rewiring and Lévy processes have been proven rigorously in [44, 48, 49, 50].

11.6 Further reading

In this chapter I have focused primarily on two classes of processes (rewiring processes and graph-valued Lévy processes) which are meant to provide the starting point for a coherent foundation of dynamic network analysis. The statistician seeking a well-established theory of dynamic network modeling is likely to have found the above discussion lacking in the necessary technical detail. To date there has been remarkably little effort in establishing a statistical theory for dynamic network analysis, and I cannot claim to have made up for such lacking in this short chapter. As far as I know, my own study of dynamic networks [44, 48, 49, 50] is among the first mathematical treatments of dynamic network modeling in the presence of sampling, and it may be the only such treatment at the time of this writing. For the most part, these ideas are little known in the statistical literature, due to both their technical nature and the relative lack of interest in statistical modeling of dynamic networks until recently.

By analogy to the discussion of consistency under subsampling for non-dynamic networks (Chapter 3), I have focused exclusively here on models that exhibit the so-called *projective Markov property*, by which the Markov property is preserved under selection sampling [44, 48]. Many of the same questions about the validity of selection sampling (see Chapter 3) can also be leveled against the projective Markov property. But given the underdevelopment of statistical analysis for dynamic networks, it is important to start somewhere. Almost any statistical or probabilistic question one could ask for these processes remains open and worthy of study.

Several open problems are stated throughout this chapter. Problem 11.4 poses an open-ended challenge to dynamic network modeling which has not yet been addressed and which provides a good starting point for any readers interested in developing the theory of dynamic network models further.

Despite recent interest in temporal networks among applied mathematicians, physicists, and epidemiologists, e.g., [91, 119], most statistical efforts in analyzing dynamic network data have been confined to applications of the temporal exponential random graph model (see Section 11.2.1.3) and other *ad hoc* approaches. Kolaczyk and Csárdi [108, Chapter 10] mention dynamic network modeling in their final chapter, but do not go into detail. See also [65, 66, 156, 159] and references therein for other recent work on dynamic network analysis.

11.7 Solutions to exercises

11.7.1 Exercise 11.1

To see why (11.7) and (11.8) are equivalent, suppose first that (11.8) holds. Then for any $n = 1, \ldots, N$, define a transition probability $P_n(\cdot,\cdot)$ on $\{0,1\}^{n\times n}$ by

$$P_n(\mathbf{y},\mathbf{y}') = \Pr(Y(t+1) \in \{\mathbf{y}'' \in \{0,1\}^{N\times N} : \mathbf{y}''|_{[n]} = \mathbf{y}'\} \mid Y(t) = \mathbf{y}^*), \qquad (11.23)$$

for any $\mathbf{y}^* \in \{0,1\}^{N\times N}$ such that $\mathbf{y}^*|_{[n]} = \mathbf{y}$, for each $\mathbf{y},\mathbf{y}' \in \{0,1\}^{n\times n}$. But since P_n is a transition probability governing $\mathbf{Y}_n$, we have proven (11.7). Conversely, if (11.7) holds then so must (11.8) by the necessary condition of Burke and Rosenblatt [28] for determining whether a function of a Markov chain is a Markov chain.

References

[1] J. Abello, A. Buchsbaum, and J. Westbrook. A functional approach to external graph algorithms. *Proceedings of the 6th European Symposium on Algorithms*, pages 332–343, 1998.

[2] D. Achlioptas, A. Clauset, D. Kempe, and C. Moore. On the Bias of Traceroute Sampling or, Power-law Degree Distributions in Regular Graphs. *STOC '05*, 2005.

[3] N. Ackerman. Representations of Aut($\mathscr{M}$)-Invariant Measures: Part 1. *Accessed at arXiv:1509.0617*, 2015.

[4] N. Ahmed, J. Neville, and R. Kompella. Network Sampling: From Static to Streaming Graphs. *ACM Transactions on Knowledge Discovery from Data*, 8, 2014.

[5] W. Aiello, F. Chung, and L. Lu. A random graph model for massive graphs. In *Proceedings of the Thirty-Second Annual ACM Symposium on Theory of Computing*, pages 171–180, New York, 2000. ACM Press.

[6] E.M. Airoldi, D.M. Blei, S.E. Fienberg, and E.P. Xing. Mixed Membership Stochastic Blockmodels. *Journal of Machine Learning Research*, 9:1981–2014, 2008.

[7] R. Albert and A.-L. Barabási. Statistical mechanics of complex networks. *Rev. Modern Phys.*, 74(1):47–97, 2002.

[8] D.J. Aldous. Representations for partially exchangeable arrays of random variables. *J. Multivariate Anal.*, 11(4):581–598, 1981.

[9] D.J. Aldous. Exchangeability and related topics. In *École d'été de probabilités de Saint-Flour, XIII—1983*, volume 1117 of *Lecture Notes in Math.*, pages 1–198. Springer, Berlin, 1985.

[10] A.M. Antonopoulos. *Mastering Bitcoin: Programming the Open Blockchain, 2nd Edition*. O'Reilly Media, 2017.

[11] A. Athreya, D.E. Fishkind, K. Levin, V. Lyzinski, Y. Park, Y. Qin, D.L. Sussman, M. Tang, J.T. Vogelstein, and C.E. Priebe. Statistical inference on random dot product graphs: a survey. *Journal of Machine Learning Research*, in press, 2017.

[12] T. Austin. On exchangeable random variables and the statistics of large graphs and hypergraphs. *Probability Surveys*, 5:80–145, 2008.

[13] A.-L. Barabási. *Linked: How Everything is Connected to Everything Else and*

What It Means for Business, Science, and Everyday Life. Plume, 2003.

[14] A.-L. Barabási and R. Albert. Emergence of scaling in random networks. *Science*, 286(5439):509–512, 1999.

[15] S. Basu, A. Shojaie, and G. Michailidis. Network Granger causality with inherent grouping structure. *The Journal of Machine Learning Research*, 16(1):417–453.

[16] J. Bertoin. *Random Fragmentation and Coagulation Processes*, volume 102 of *Cambridge Studies in Advanced Mathematics*. Cambridge University Press, Cambridge, 2006.

[17] S. Bhamidi, G. Bresler, and A. Sly. Mixing time of exponential random graphs. *IEEE 49th Annual IEEE Symposium on Foundations of Computer Science (FOCS)*, pages 803–812, 2008.

[18] S. Bhamidi, J.M. Steele, and T. Zaman. Twitter event networks and the superstar model. *Annals of Applied Probability*, 25(5):2462–2502, 2015.

[19] P. Bickel and A. Chen. A nonparametric view of network models and Newman–Girvan and other modularities. *Proceedings of the National Academy of Sciences of the United States of America*, 106(50):21068–21073, 2009.

[20] B. Bloem-Reddy and P. Orbanz. Random walk models of network formation and sequential Monte Carlo methods for graphs. *Accessed at arXiv:1612.06404*, 2016.

[21] B. Bollobás. *Random Graphs, 2nd Edition*, volume 73 of *Cambridge Series in Mathematics*. Cambridge University Press, 2001.

[22] B. Bollobás, C. Borgs, J. Chayes, and O. Riordan. Directed scale-free graphs. In *In Proceedings of the Fourteenth Annual ACM-SIAM Symposium on Discrete Algorithms (Baltimore)*, pages 132–139. New York, 2003.

[23] C. Borgs, J.T. Chayes, H. Cohn, and N. Holden. Sparse exchangeable graphs and their limits via graphon processes. *Accessed at arXiv:1601.07134*, 2016.

[24] C. Borgs, J.T. Chayes, H. Cohn, and V. Veitch. Sampling perspectives on sparse exchangeable graphs. *Accessed at arXiv:1708.03237*, 2017.

[25] C. Borgs, J.T. Chayes, H. Cohn, and Y. Zhao. An L^p theory of sparse graph convergence I: Limits, sparse random graph models, and power law distributions. *Accessed at arXiv:1401.2906*, 2014.

[26] G.E.P. Box and N.R. Draper. *Empirical Model Building and Response Surfaces*. John Wiley & Sons, New York, NY, 1987.

[27] A.D. Broido and A. Clauset. Scale-free networks are rare. *Accessed at https://arxiv.org/pdf/1801.03400.pdf on February 16, 2018*, 2018.

[28] C. J. Burke and M. Rosenblatt. A Markovian function of a Markov chain. *Ann. Math. Statist.*, 29:1112–1122, 1958.

[29] V. Buterin. Ethereum “white paper”. *Accessed at https://github.com/ethereum/wiki/wiki/White-Paper on February 13, 2018.*

[30] D. Cai, T. Campbell, and T. Broderick. Edge-exchangeable graphs and sparsity. In D. D. Lee, U. V. Luxburg, I. Guyon, and R. Garnett, editors, *Advances in Neural Information Processing Systems 29*, pages 4242–4250. Curran Associates, Inc., 2016 (appeared online November 20, 2016).

[31] F. Caron and E.B. Fox. Sparse graphs using exchangeable random measures. *Accessed at arXiv:1401.1137*, 2014.

[32] F. Caron and E.B. Fox. Sparse graphs using exchangeable random measures. *Journal of the Royal Statistical Society, Series B*, 79(5), 2017.

[33] F. Caron and J. Rousseau. On sparsity and power-law properties of graphs based on exchangeable point processes. *Accessed at arXiv:1708.03120*, 2017.

[34] S. Chatterjee and P. Diaconis. Estimating and understanding exponential random graph models. *Annals of Statistics*, 41(5):2428–2461, 2013.

[35] D.S. Choi, P.J. Wolfe, and E.M. Airoldi. Stochastic blockmodels with a growing number of classes. *Biometrika*, 99(2):273–284, 2012.

[36] F. Chung and L. Lu. *Complex Graphs and Networks*, volume 107 of *CBMS Regional Conference Series in Mathematics*. Published for the Conference Board of the Mathematical Sciences, Washington, DC, 2006.

[37] R.A. Clarke and R.K. Knake. *Cyber War: The Next Threat to National Security and What to Do About It.* HarperCollins, New York, 2010.

[38] A. Clauset, C. Moore, and M.E.J. Newman. Hierarchical structure and the prediction of missing links in networks. *Nature*, 453:98–101, 2008.

[39] A. Clauset, C.R. Shalizi, and M.E.J. Newman. Power-law distributions in empirical data. *SIAM Review*, 51(4):661–703, 2009.

[40] RChain Cooperative. *Accessed at https://medium.com/rchain-cooperative on February 13, 2018.*

[41] O.T. Courtney and G. Bianconi. Dense Power-law Networks and Simplicial Complexes. *Accessed at arXiv:1802.01465*, 2018.

[42] D.R. Cox and D.V. Hinkley. *Theoretical Statistics.* Chapman & Hall, London, 1974.

[43] H. Crane. The cut-and-paste process. *Annals of Probability*, 42(5):1952–1979, 2014.

[44] H. Crane. Time-varying network models. *Bernoulli*, 21(3):1670–1696, 2014.

[45] H. Crane. Rejoinder: The ubiquitous Ewens sampling formula. *Statistical Science*, 31(1):37–39, 2016.

[46] H. Crane. The ubiquitous Ewens sampling formula (with discussion and a rejoinder by the author). *Statistical Science*, 31(1):1–39, 2016.

[47] H. Crane. Comment on F. Caron and E.B. Fox. Sparse graphs using exchangeable random measures. *Journal of the Royal Statistical Society, Series B*, 79(5), 2017.

[48] H. Crane. Exchangeable graph-valued Feller processes. *Probability Theory and Related Fields*, 168(3–4):849–899, 2017.

[49] H. Crane. Randomly evolving graphs and their graph limits. *Annals of Applied Probability*, 26(2):691–721, 2017.

[50] H. Crane. Combinatorial Lévy processes. *Annals of Applied Probability*, 28(1):285–339, 2018.

[51] H. Crane and W. Dempsey. Community detection for interaction networks. *Accessed at arXiv:1509.09254*, 2015.

[52] H. Crane and W. Dempsey. A framework for statistical network modeling. *Accessed at arXiv:1509.08185*, 2015.

[53] H. Crane and W. Dempsey. Relational exchangeability. Accessed at arXiv:1607.06762, 2016.

[54] H. Crane and W. Dempsey. Edge exchangeable models for interaction networks. *Journal of the American Statistical Association*, in press, 2017.

[55] H. Crane and W. Dempsey. A framework for statistical network modeling. *First version, Accessed at arXiv:1509.08185v1*, September 28, 2015.

[56] H. Crane and S.P. Lalley. Convergence rates of Markov chains on spaces of partitions. *Electronic Journal of Probability*, 18(paper no. 61):1–23, 2013.

[57] H. Crane and H. Towsner. The structure of combinatorial Markov processes. *Accessed at arXiv:1603.05954*, 2016.

[58] H. Crane and H. Towsner. Relative exchangeability with equivalence relations. *Archive of Mathematical Logic*, in press, 2017.

[59] H. Crane and H. Towsner. Relatively exchangeable structures. *Journal of Symbolic Logic*, in press, 2017.

[60] B. de Finetti. La prévision: ses lois logiques, ses sources subjectives. *Annales de l'Institut Henri Poincaré*, 7:1–68.

[61] P. Diaconis and D. Freedman. On the statistics of vision: The Julesz conjecture. *J. Math. Psychol.*, pages 112–138, 1981.

[62] P. Diaconis and S. Janson. Graph limits and exchangeable random graphs. *Rend. Mat. Appl. (7)*, 28(1):33–61, 2008.

[63] P. Doreian and F. N. Stokman eds. *Evolution of Social Networks*. Routledge, Mahway, NJ, 1997.

[64] S. N. Dorogovtsev and J. F. F. Mendes. *Evolution of networks: From biological nets to the Internet and WWW*. Oxford University Press, Oxford, 2003.

[65] D. Durante and D.B. Dunson. Nonparametric Bayes dynamic modelling of relational data. *Biometrika*, 101(4):883–898, 2014.

[66] D. Durante, N. Mukherjee, and R.C. Steorts. Bayesian Learning of Dynamic Multilayer Networks.

[67] R. Durrett. *Random Graph Dynamics*. Cambridge Series in Statistical and Probabilistic Mathematics. Cambridge University Press, Cambridge, 2007.

[68] P. Erdős and A. Rényi. On random graphs. I. *Publ. Math. Debrecen*, 6:290–297, 1959.

[69] W. J. Ewens. The sampling theory of selectively neutral alleles. *Theoret. Population Biology*, 3:87–112, 1972.

[70] M. Faloutsos, P. Faloutsos, and C. Faloutsos. On power-law relationships of the Internet topology. *ACM Comp. Comm. Review*, 29, 1999.

[71] W. Feller. *An Introduction to Probability Theory and Its Applications.* John Wiley, 1957.

[72] S. Feng. *The Poisson-Dirichlet Distribution and Related Topics.* Probability and its Applications. Springer-Verlag, Berlin, 2010.

[73] T. Ferguson. A Bayesian Analysis of Some Nonparametric Problems. *Annals of Statistics*, 1(2):209–230, 1973.

[74] S.E. Fienberg. A Brief History of Statistical Models for Network Analysis and Open Challenges. *Journal of Computational and Graphical Statistics*, 21(4):825–839, 2012.

[75] O. Frank. Network sampling and model fitting. In *Models and Methods in Social Network Analysis*, pages 31–56. Cambridge University Press, New York, 2005.

[76] O. Frank. Estimation and sampling in social network analysis. In *Encyclopedia of Complexity and Systems Science*, pages 8213–8231. Springer, New York, 2009.

[77] O. Frank. Survey sampling in networks. In *Handbook of Social Network Analysis*. Sage, London, 2011.

[78] O. Frank and D. Strauss. Markov graphs. *Journal of the American Statistical Association*, 81(395):832–842, 1986.

[79] C. Gao, Lu. Y., and H.H. Zhou. Rate optimal graphon estimation. *Annals of Statistics*, 43(6):2624–2652, 2015.

[80] F. Gao and A. van der Vaart. On the asymptotic normality of estimating the affine preferential attachment network models with random initial degrees. *Stochastic Processes and Their Applications*, 2017.

[81] E.N. Gilbert. Random graphs. *Annals of Mathematical Statistics*, 30(4):1141–1144, 1959.

[82] C.E. Ginestet, P. Balachandran, S. Rosenberg, and E.D. Kolaczyk. Hypothesis testing for network data in functional neuroimaging. *Annals of Applied Statistics*, 11(2):725–750, 2017.

[83] M. Girvan and M.E.J. Newman. Community structure in social and biological networks. *Proceedings of the National Academy of Sciences*, 99(12):7821–7826, 2002.

[84] A. Goldenberg, A.X. Zheng, S.E. Fienberg, and E.M. Airoldi. A survey of statistical network models. *Foundations and Trends in Machine Learning*, 2(2):1–117, 2009.

[85] M.S. Handcock and K.J. Gile. Modeling social networks from sampled data. *Ann. Appl. Stat.*, 4(1):5–25, 2010.

[86] S. Hanneke, W. Fu, and E.P. Xing. Discrete temporal models of social networks. *Electronic Journal of Statistics*, 4:585–605, 2010.

[87] S. Hanneke and E.P. Xing. Discrete temporal models of social networks. In In E. Airoldi, David M. Blei, S.E. Fienberg, A. Goldenberg, E.P. Xing, and A.X. Zheng, eds., *Statistical Network Analysis: Models, Issues, and New Directions: ICML 2006 Workshop on Statistical Network Analysis*, volume 4503 of *Lecture Notes in Computer Science*, pages 115–125. Springer, 2007.

[88] P.D. Hoff, A.E. Raftery, and M.S. Handcock. Latent space approaches to social network analysis. *J. Amer. Statist. Assoc.*, 97(460):1090–1098, 2002.

[89] P.W. Holland, K.B. Laskey, and S. Leinhardt. Stochastic blockmodels: First steps. *Social Networks*, 5(2):109–137, 1983.

[90] P.W. Holland and S. Leinhardt. An exponential family of probability distributions for directed graphs. *Journal of the American Statistical Association*, pages 33–65, 1981.

[91] P. Holme and J. Saramäki (eds.). *Temporal Networks*. Understanding Complex Systems. Springer, 2013.

[92] D.N. Hoover. Relations on Probability Spaces and Arrays of Random Variables. Preprint, Institute for Advanced Studies, 1979.

[93] D.R. Hunter, S.M. Goodreau, and M.S. Handcock. Goodness of Fit of Social Network Models. *Journal of the American Statistical Association*, 2008.

[94] H. Ishwaran and L.F. James. Gibbs sampling methods for stick-breaking priors. *Journal of the American Statistical Association*, 96:161–173, 2001.

[95] S. Janson. On edge exchangeable random graphs. *Accessed at arXiv:1702.06396*, 2017.

[96] P. Ji and J. Jin. Coauthorship and citation networks for statisticians (with discussion and rejoinder by the authors). *Annals of Applied Statistics*, 10(4):1779–1812, 2016.

[97] O. Kallenberg. Exchangeable random measures in the plane. *Journal of Theoretical Probability*, 3(1):81–136, 1990.

[98] O. Kallenberg. *Probabilistic Symmetries and Invariance Principles*. Probability and Its Applications. Springer, 2005.

[99] B. Karrer and M.E.J. Newman. Stochastic blockmodels and community structure in networks. *Physical Review E*, 83:016107, 2011.

[100] N. Katenka, E. Levina, and G. Michailidis. Local vote decision fusion for target detection in wireless sensor networks. *IEEE Transactions on Signal Processing*, 56(1):329–338.

[101] N. Katenka, E. Levina, and G. Michailidis. Detection, Localization, and Tracking of a Single and Multiple Targets with Wireless Sensor Networks.

Computational Network Theory: Theoretical Foundations and Applications, 5, 2015.

[102] M. Khabbazian, B. Hanlon, Z. Russek, and K. Rohe. Novel Sampling Design for Respondent-driven Sampling. *Accessed at arXiv:1606:00387*, 2016.

[103] M. Kivelä, A. Arenas, M. Barthelemy, J.P. Gleeson, Y. Moreno, and M.A. Porter. Multilayer Networks. *Journal of Complex Networks*, 2(3):203–271, 2014.

[104] B. Klimt and Y. Yang. Introducing the Enron corpus. *CEAS*, 2004.

[105] J.M. Klusowski and Y. Wu. Estimating the number of connected components in a graph via subgraph sampling. *Accessed at https://arxiv.org/pdf/1801.04339.pdf on February 16, 2018*, 2018.

[106] E.D. Kolaczyk. *Statistical Analysis of Network Data*. Springer Series in Statistics. Springer, New York, 2009. Methods and models.

[107] E.D. Kolaczyk. *Topics at the Frontier of Statistics and Network Analysis (Re)Visiting the Foundations*. SemStat Elements. Cambridge, 2017.

[108] E.D. Kolaczyk and G. Csárdi. *Statistical Analysis of Network Data with R*. Use R! Springer, 2014.

[109] P.N. Krivitsky and M.S. Handcock. A Separable Model for Dynamic Networks. *Journal of the Royal Statistical Society Series B*, 76(1):29–46, 2014.

[110] P.N. Krivitsky and E.D. Kolaczyk. On the question of effective sample size in network modeling: An asymptotic inquiry. *Statistical Science*, 30(2):184–198, 2014.

[111] R. Kumar, P. Raghavan, S. Rajagopalan, D. Sivakumar, A. Tomkins, and E. Upfal. Stochastic models for the Web graph. *Proceedings of the 41st Annual Symposium on Foundations of Computer Science*, pages 57–65, 2000.

[112] S. H. Lee, P. Kim, and H. Jeong. Statistical properties of sampled networks. *Physical Review E*, 73:016102, 2006.

[113] L. Li, D. Alderson, J.C. Doyle, and W. Willinger. Towards a theory of scale-free graphs: Definition, properties, and implications. *Internet Math.*, 2(4):431–523, 2005.

[114] X. Li and K. Rohe. Central limit theorems for network driven sampling. *Accessed at arXiv:1509.04704*, 2015.

[115] L. Lovász. *Large Networks and Graph Limits*, volume 60 of *AMS Colloquium Publications*. American Mathematical Society, Providence, RI, 2012.

[116] L. Lovász and B. Szegedy. Limits of dense graph sequences. *Journal of Combinatorial Theory, Series B*, 96:933–957, 2006.

[117] S. Mankad and G. Michailidis. Analysis of multiview legislative networks with structured matrix factorization: Does Twitter influence translate to the real world? *Annals of Applied Statistics*, 9(4):1950–1972, 2015.

[118] R. Martin and C. Liu. *Inferential Models: Reasoning with Uncertainty*. Chapman & Hall, 2016.

[119] N. Masuda and P. Holme (eds.). *Temporal Network Epidemiology*. Springer Nature, Singapore, 2017.

[120] P. McCullagh. What is a statistical model? *Ann. Statist.*, 30(5):1225–1310, 2002. With comments and a rejoinder by the author.

[121] J.L. Moreno. *Who Shall Survive? A New Approach to the Problem of Human Interrelations*. Beacon House, 1934.

[122] S. Nakamoto. Bitcoin: A Peer-to-Peer Electronic Cash System. *Accessed at https://bitcoin.org/bitcoin.pdf on February 13, 2018.*

[123] M. E. J. Newman. The structure and function of complex networks. *SIAM Rev.*, 45(2):167–256 (electronic), 2003.

[124] Y.C. Ng and R. Silva. A Dynamic Edge Exchangeable Model for Sparse Temporal Networks. *Accessed at arXiv:1710:04008*, 2017.

[125] K. Okike, K.T. Hug, M.S. Kocher, and S.S. Leopold. Single-blind vs Double-blind Peer Review in the Setting of Author Prestige. *Journal of the American Medical Association*, 316(12):1315, 2016.

[126] P. Orbanz. Comment on F. Caron and E.B. Fox. Sparse graphs using exchangeable random measures. *Journal of the Royal Statistical Society, Series B*, 79(5), 2017.

[127] P. Orbanz. Subsampling large graphs and invariance in networks. *arXiv:1710.04217*, 2017.

[128] P. Orbanz and D.M. Roy. Bayesian Models of Graphs, Arrays and Other Exchangeable Random Structures. *IEEE Transactions on Pattern Analysis and Machine Intelligence*, 37(2):437–461, 2015.

[129] K. Palla, F. Caron, and Y.W. Teh. Bayesian nonparametrics for sparse dynamic networks. *arXiv:1607.01624*, 2016.

[130] M. Perman, J. Pitman, and M. Yor. Size-biased sampling of poisson point processes and excursions. *Probab. Th. Relat. Fields*, 92:21–39, 1992.

[131] P.O. Perry and P.J. Wolfe. Point process modelling for directed interaction networks. *Journal of the Royal Statistical Society, Series B*, 75:821–849, 2013.

[132] J. Pitman. *Combinatorial Stochastic Processes*, volume 1875 of *Lecture Notes in Mathematics*. Springer-Verlag, Berlin, 2006. Lectures from the 32nd Summer School on Probability Theory held in Saint-Flour, July 7–24, 2002, With a foreword by Jean Picard.

[133] S.I. Resnick. *Heavy Tail Phenomena: Probabilistic and Statistical Modeling*. Springer Series in Operations Research and Financial Engineering. Springer-Verlag, 2007.

[134] R. A. Rossi and N. K. Ahmed. The network data repository with interactive graph analytics and visualization. In *Proceedings of the Twenty-Ninth AAAI Conference on Artificial Intelligence*, 2015.

[135] M. Sageman. *Understanding Terror Networks*. University of Pennsylvania Press, 2004.

[136] M. Schweinberger, P.N. Krivitsky, and C.T. Butts. Foundations of Finite-, Super-, and Infinite-Population Random Graph Inference. *Accessed at arXiv:1707.04800*, 2017.

[137] J. Sethuraman. A constructive definition of Dirichlet priors. *Statistica Sinica*, 4:639–650, 1994.

[138] C.R. Shalizi and A. Rinaldo. Consistency under subsampling of exponential random graph models. *Annals of Statistics*, 41:508–535, 2013.

[139] H.A. Simon. On a class of skew distribution functions. *Biometrika*, 42:425–440, 1955.

[140] T.A.B. Snijders. Stochastic Actor-Oriented Models for Network Dynamics. *Annual Review of Statistics and Its Application*, 4:343–363, 2017.

[141] T.A.B. Snijders and K. Nowicki. Estimation and prediction for stochastic block models for graphs with latent block structure. *Journal of Classification*, 14:75–100, 1997.

[142] D.L. Sussman and E.M. Airoldi. Elements of estimation theory for causal effects in the presence of network interference. *Accessed at arXiv:1702.03578*, 2017.

[143] S.K. Thompson and O. Frank. Model-based estimation with link-tracing sampling designs. *Survey Methodology*, 26:87–98.

[144] S.K. Thompson and G.A.F. Seber. *Adaptive Sampling*. Wiley, New York, 1996.

[145] A. Todeschini, X. Miscouridou, and F. Caron. Exchangeable Random Measures for Sparse and Modular Graphs with Overlapping Communities. *Accessed at arXiv:1602.0211*, 2016.

[146] R. van der Hofstad. *Random walks and complex networks*. Lecture notes. 2012.

[147] V. Veitch and D. Roy. The Class of Random Graphs Arising from Exchangeable Random Measures. *Accessed at arXiv:1512.03099*, 2015.

[148] V. Veitch and D.M. Roy. Sampling and Estimation for (Sparse) Exchangeable Graphs. *Accessed at arXiv:1611.00843*, 2016.

[149] P. Wan, T. Wang, R.A. Davis, and S.I. Resnick. Fitting the linear preferential attachment model. *Electronic Journal of Statistics*, 11(2):3738–3780, 2017.

[150] S.S. Wasserman and P.E. Pattison. Logit models and logistic regression for social networks: I. An introduction to Markov graphs and p^*. *Psychometrika*, 61(3):401–425, 1996.

[151] D. Watts. *Six Degrees: The Science of a Connected Age*. W.W. Norton, 2004.

[152] D. Watts and S. Strogatz. Collective dynamics of 'small-world' networks. *Nature*, 393:440–442, 1998.

[153] S.A. Williamson. Nonparametric Network Models for Link Prediction. *Journal of Machine Learning Research*, 17:1–21, 2016.

[154] W. Willinger, D. Alderson, and J.C. Doyle. Mathematics and the Internet: A source of enormous confusion and great potential. *Notices Amer. Math. Soc.*, 56(5):586–599, 2009.

[155] P.J. Wolfe and S.C. Olhede. Nonparametric graphon estimation. *Available at arXiv:1309.5936*, 2014.

[156] E.P. Xing, W. Fu, and L. Song. A state-space mixed membership blockmodel for dynamic network tomography. *Annals of Applied Statistics*, 4(2):535–566, 2010.

[157] M. Xu, V. Jog, and P.-L. Loh. Optimal rates for community estimation on the weighted stochastic block model. *Accessed at http://www-stat.wharton.upenn.edu/~ minx/docs/wsbm.pdf on February 22, 2018*, 2018.

[158] J. Yang, C. Han, and E. Airoldi. Nonparametric estimation and testing of exchangeable graph models. *Proceedings of the Seventeenth International Conference on Artificial Intelligence and Statistics, PMLR*, 33:1060–1067, 2014.

[159] T. Yang, Y. Chi, S. Zhu, Y. Gong, and R. Jin. Detecting communities and their evolutions in dynamic social networks–a Bayesian approach. *Machine Learning*, 82(2):157–189, 2011.

[160] S. Young and E. Scheinerman. Random dot product graph models for social networks. *Proceedings of the 5th International Conference on Algorithms and Models for the Web-Graph*, pages 138–149, 2007.

[161] W.W. Zachary. An information flow model for conflict and fission in small groups. *Journal of Anthropological Research*, 33:452–473, 1977.

[162] Y. Zhang, E.D. Kolaczyk, and B.D. Spencer. Estimating network degree distributions under sampling: an inverse problem, with applications to monitoring social media networks. *Annals of Applied Statistics*, 9(1):166–199, 2015.

Index

CPSIA information can be obtained
at www.ICGtesting.com
Printed in the USA
BVHW040526161218
535582BV00009B/38/P

9 781138 585997